Managing Quality in Biotechnology

A Complete Guide for Implementing Total Quality Management
in Bio-related Laboratories and Industry;
Designed for Students, Technicians and Managers

Val d'Or

Goldenthal Consulting Services
Serving Biopharma Since 1989

Dr. Allen E. Goldenthal
PhD, MBA, DVM, BSc

TQM, QMS, QA and GLP Preclinical Specialist
Certified ETRS Auditor, Medical Technologist

139 Estrada do Repouso, Suite 5B
Macau, Macau S.A.R China

Mobile: +853 623 75280 or +86 136 4141 3900
Email: biovet2@hotmail.com
Skype: 0064-889-8080
        allen.goldenthal

Managing Quality in Biotechnology

A Complete Guide for Implementing Total Quality Management
in Bio-related Laboratories and Industry;
Designed for Students, Technicians and Managers

Author	Allen E. Goldenthal
Publisher	Val d'Or
	Published and printed in Charleston, S.C.

ISBN 978-0-9942559-5-2

About the author

Dr. Allen E. Goldenthal was recently a Director and Senior Instructor of the Faculty of Health Sciences at the University of Macau but has now returned to consulting to the bio and pharmaceutical industries in a full time capacity.

He personally has over three decades of experience in the biotechnological industries having held former positions as either a Director or Chief Officer at companies such as Pasteur-Merieux Connaught, PA Biologicals, Invivotech, Estendart and H1K Biotechnology Shenzhen. Since his certification as an ETRS Quality Auditor almost 20 years ago, he has consulted in many countries around the world, lending his expertise to the vaccine industries in the United States, Canada, New Zealand, Australia, South Africa, Korea, Taiwan and the mainland China.

He was part of the team responsible for the first WHO pre-qualified vaccine (Japanese encephalitis) manufactured in China by CDIBP. His long time commitment to quality has been a driving force throughout the biopharmaceutical and biotechnological industry both within the commercial and academic spheres.

Apart from writing scientific reference books, Dr. Goldenthal is also an avid writer of historical novels and self-help guides.

Preface

Managing Quality in Biotechnology is designed purposely as a manual for introducing Total Quality Management (TQM) within the biotechnological industries with specific reference to the immunocellular therapy industry after my years of research in that particular industry. The units, will be primarily focused on cell therapy production when detailing technical examples but those sections which deal with the overall administration required in order to make the system work will generally not be industry specific. The overall structure of the text is designed to provide clear practical guidelines for the full range of activities required to make the facilities functional but should not be confused with Standard Operating Procedures (SOPs), which are designed to make the performance of tasks consistent. SOPs are dealt with as separate issues in the various chapters of this book. It should be noted that the development and issuance of SOPs are a requirement of this guidebook and neither should be considered as a replacement for the other but together they will provide the level of Total Quality Management desired.

Implementation of this TQM guidebook is required in order that the systems and personnel can operate to a GMP/GTP regulatory requirement and every section will include self-study multiple choice quizzes that relate to actual situations within a company. In order that a company can prepare well-structured implementation plans for TQM, it will be a requirement that the content of this guidebook be delivered under the auspices of a formal training structure to ensure that all personnel complete the TQM training. This guidebook works hand-in-hand with the regulations, a company's Quality Manual and Risk Management Plan. Together they form the framework for the company's management to function as a GMP facility. Whereas all those documents are part of a company's Quality Management System (QMS), it is the understanding and appreciation of how to apply QMS that is essentially the basis for Total Quality Management.

Having three decades of working experience in the biopharmaceutical industry, I have seen the approach to Quality Management change dramatically over that time period. Working with one of the world leading biological manufacturers in North America during the late 1980s and into the 1990s (now Sanofi-Aventis), I witnessed the dramatic change in focus towards Good Manufacturing Practices (GMP) which were initially introduced in the 1970s. It was not an easy transition and companies went through tremendous swings of successful introduction and failure through trial and error. Being in China since 2009, I have had the opportunity to experience all over again those trials and tribulations from the wilderness days of North America. However, the one thing that is common to the establishment of Quality Systems, whether then or now is "Common Sense". There is still this misconception that Quality is something you can buy; that it comes out of a package with ribbons and bows and only needs to be removed from the box and a company instantly has quality. Unfortunately, there are some quality services companies that advertise exactly that concept, asking for a one-time fee for which they can provide you with a quality system, complete with certification. Having a certificate is not the same as having quality. It cannot be bought, it cannot be borrowed and it certainly cannot be acquired through a simple monetary transaction. To have quality it has to be grown, it has to be instilled and personnel have to have the common sense to understand that safety and efficacy in production begins with them, not from a service outside the company. Unfortunately, there are still those that fail to understand this concept, assuming that as long as they throw enough money at the situation they will acquire their certification as a company practising QMS. They will probably get that certificate, but unless the training, comprehension and adoption of a quality production framework becomes engrafted into the spirit and soul of a company, facility or institution, they will not have sustainability, a long-term vision, or the ability to appreciate the true meaning of Total Quality Management.

Dr. Allen E. Goldenthal

Introduction

1. The importance of TQM at a company

There is a critical factor that a company can implement on its own in order to improve the quality of its products, thereby delivering therapies with a level of quality that meets customer/patient requirements. It is called Total Quality Management (TQM). Achievement of this level of implementation is essential in order to ensure business success and sustainability within a regulatory governed environment.

Facing the fierce competition of today's market for therapies and therapeutic products in China and elsewhere, the level of quality at a company actually needs to exceed what international customers/patients already expect, providing safety and efficacy at a competitive price. Achieving this level of quality involves everyone at the company, as well as our suppliers and hospital contract service providers. In order to achieve this level, it requires good management systems and practices throughout the organisation, from having a vision of the future of this company to maintaining a safe and healthy work environment in the laboratories and hospitals. It means having well-trained and motivated employees, standardised work procedures, and effective production control of the therapies. It also means ensuring the quality of incoming supplies, and operating an efficient after-treatment service. Above all, it requires the active participation of senior management if TQM is going to succeed. Every member of the company's staff can and must support the implementation of this level, hence the name Total Quality Management.

2. Key features of this guidebook

This guidebook is designed to enable a company, facility or institution to implement TQM within its own specific context and situation, using only the resources currently available. The requirement is that after personnel have read a specific section, they then undertake the task of compiling an action plan, in order to demonstrate their knowledge and understanding before progressing to the next section. By undertaking the writing of an action plan, they in turn will begin implementing the TQM programme within their own work area, by turning the action plan into a working reality.

The multiple choice questions at the end of each chapter serve to focus management, supervisors and laboratory employees on how the ideas in the text can be applied to their particular work area. The action plan they will develop, provides a framework for preparing well-structured and concrete plans by which they will implement their ideas.

The chapters are designed to be practical. The text, although often heavily detailed, is written in a language that is intended to be clear and easy to follow, which complements ease of translation into Chinese, and the learning activities are both concrete and practical for cell therapy laboratories.

The various chapters contain numerous examples, as well as some forms, tables and charts, which can be adapted to be used in almost any situation.

References for further reading are provided so that the student of TQM can gain further knowledge applicable to that chapter from sources that I deemed noteworthy.

3. TQM content

Managing Quality in Biotechnology covers the entire range of TQM activities in 20 chapters, presented in four focused sections for ease of training. These sections are as follows:

Section one: Governance and leadership in TQM

Initially, the Chief Executive Officer's overall responsibility in managing policy and ensuring the quality of the product as well as the overall performance of the company within a quality framework is examined. Direct control is performed at laboratory manager levels as they are responsible of not only ensuring that the appropriate systems and equipment are in place and functioning properly but that they have the right people in the right place at the right time.

Section two: The Quality departments

Different levels of quality management exist in a company but they must all come together under the framework of Quality Operations meetings which identify existing problems within the structure of the company as well as the processes. Problem solving is the primary function of these meetings, using statistical methods to determine the extent of any problems and the auditing process to ensure proper correction. Education and training is emphasised.

Section three: Working in the laboratory

Standardisation of methodologies is essential in order to have quality performance. This requires a high level of measurement control to ensure that the key checkpoints all meet acceptable levels. To provide a suitable laboratory environment in which to perform proper measurement control means having well developed cleaning, storage and disposal policies as well as an overall safety policy to protect the technicians.

Section four: Production, materials and development

The primary gauge of success within the biopharmaceutical/biotherapeutic industries is the overall acceptance and satisfaction by the customer/patient with the actual products produced. To be the best means having not only tight production control of the therapies but full control of the processes, the raw material suppliers, and providing a high level of patient aftercare which in turn determines the requirements for future products.

4. Chapter structure

Each chapter consists of:
- Several textual sections, each with specific focuses regarding the topic, and then the expectation that staff will undertake the writing of an action plan as the follow-up to the learning and any discussions.

- A multiple choice interactive test of 50 questions (Chapters 18 and 19 only have 30).
- Regulatory or preferred reading references are recommended.

Texts: Each chapter represents a different subtopic of the section's main theme. They vary in length and detail depending on the nature of the sub-topic: some are short and quite simple; others are long and heavily detailed.

Since this is a practical guidebook, the texts only become fully meaningful when students and/or company personnel actually discuss how to apply them with their teachers and/or supervisors. Therefore it is expected that the readers will take the time to sit down with their supervisors and managers and actually discuss the content of their lessons and their action plans before progressing to the next section.

It is anticipated that personnel will set their own learning pace and not feel pressured to advance too quickly, but it is still expected that they will implement the content derived from self-training in a reasonable time frame.

The need for discussion after learning a section serves two purposes:
- To encourage participants to reflect critically on what they have been doing at their company, particularly in the functional area as presented in the text but also to recognise how effective their work efforts are and whether or not there is a need for improvement. By discussing each section, the employee takes ownership of that knowledge and transforms it from words on paper into actual practice.
- Central to the discussion are what I have labelled as the **MADMEN** questions. These questions are expected to become the core issues for any fruitful discussions after reading each section in the guidebook. Until all questions can be answered properly, the student or trainee should refrain from moving on to the next section.

M: How can I use this discussion to make **MEANINGFUL** changes in the institution/facility?

A: How do I **APPLY** the changes in a complete and efficient manner?

D: What **DIFFICULTIES** do I anticipate in making the changes?

M: What can I do to **MEASURE** the results of any implementation?

E: Is there an alternative method to the one discussed that would be more **EFFECTIVE**?

N: What are the **NECESSARY** resources needed to make it happen?

In other words, for each trainee to develop a **meaningful application** of what they learn in order that any **difficulties** in quality can be **measured** and then **effectively** resolved as **necessary**. A full discussion of these questions leads employees to their own conclusions about how best to implement the Total Quality Management in their department and job.

The action plan can then be written based on the answers they arrive at from the discussion. Why have I instituted this approach?

- China can no longer be a nation of copiers but needs its scientists and technicians to be innovators. The same holds true for all developing countries. I am forcing the employees to focus their thinking, and as a result, they in turn will be clear about their understanding of the principles necessary in order to implement the TQM system in their laboratory.
- It also trains employees in how to crystallise their ideas, convert them in to a summarised

format that can then be presented to the decision-makers at a company as draft proposals with the intention of implementing this particular aspect of TQM in their company.

- It is a well known fact that the workforce in China, thinks as a collective, which is very different from the concept of the individual in western companies. By structuring the training in this manner, it encourages individualism to take place first and only after the presentation of individual thought is acceptance of any proposal made as a collective. This western concept of thinking is a requirement for successful implementation of TQM.
- And lastly, it provides a record of the discussion and how they arrive at their conclusions. It provides documented traceability so that if questioned by auditors they can demonstrate that their changes and implementations were based on sound and reasonable scientific principles and knowledge.

The structure of the action plan is based on what I have perceived as real-life problems encountered in China from my research and observations. They address weaknesses that exist amongst the scientists and technicians that Chinese universities are outputting and address the key differences in the quality of personnel between East and West. In this manner, the action plan will force the staff member following the guidebook to challenge these observed weaknesses:

1. **Problems:** Problems you have in your area that you have encountered.
2. **Proposals:** Your proposals for improvement within the laboratory.
   a. Be specific and concrete.
   b. Include an implementation plan, with a time schedule and minimum and optimal implementation targets.
   c. Refer to any forms, charts or tables that you would use, and include samples in an appendix.
3. **Obstacles:** Obstacles to implementation in technician attitudes, company organisation and culture, etc., and how these might be overcome.
4. **Resources:**
   a. The resources required: funds, equipment, materials, person-hours, expertise, etc.
   b. The resources available within a company.
   c. Any resources that would have to be found outside a company.
   d. Alternatives that could be used to cover any shortfall in resources.
5. **Assessment:** Ways of assessing the results of implementing these proposals. Set acceptance criteria and establish reasonable and acceptable limits.
6. **Benefits:** The benefits your proposals would bring to production TQM within the laboratory. Benefits should be thought of in terms of labour savings, cost savings, reduction of waste, improvements in product, etc.

Including all six of these steps in the action plan gives no other option but to think outside the proverbial "box" and stimulate the creative and investigative thought patterns.

Interactive test: Each chapter has a multiple choice test that allows participants to check for themselves how well they can recall the contents of specific texts, or of the whole chapter. If you do not score well on the tests, then you are encouraged to reread the section again until you are both familiar and comfortable with the concepts and principles.

Regulatory and preferred reading references: This section at the end of each chapter presents the relationship of the content to either domestic or international standards. These references supplement the concepts and ideas that are presented within the chapter and further enhance one's knowledge base.

5. Deciding to introduce TQM

A institution's or company's senior management makes a firm commitment to introduce TQM within the facility. It is a major undertaking, as it is TQM, which actually breathes life into the cGMP (current Good Manufacturing Practices) and in my industry the GTPs (Good Tissue Practices) that are the regulatory standards. TQM is the implementation of these standards in real life situations and therefore is the governing principles for the total operation of a company. Without TQM, attempts to implement GMP or GTP guidelines will ultimately fail because a company does not have the required infrastructure to support them and employees lack the necessary understanding to make "quality" a reality.

When a facility decides to adopt TQM it cannot be done half-heartedly. It requires a 100% commitment by everyone, from the CEO, all the way down to the lowest levels of employment. It only succeeds when everyone participates, everyone contributes and when everyone feels as if they personally can make a difference.

For this reason, the order of this guidebook will begin by focusing on the most senior position within an institution or company and explaining the CEO's role in Total Quality Management. This is because in real-life, quality does start at the top and filters down throughout all the lower levels. If this is not the case then it will never be successfully implemented. Following the examination of the role of the CEO, the text will then focus on the roles of managers other than the CEO, and then subsequently work its way down through the layers of quality employees, laboratory managers, supervisors, technicians and general employees. That is Total Quality Management!

Table of contents

An outline on how this guidebook should be used in an educational training format within a company and the requirement for qualified personnel to act in the capacity of trainers to ensure that training reaches all levels of employees within a company.

The introduction will look at how TQM has changed the thinking in the industry with the primary focus on the end user being the customer/patient. Therefore it is essential that the quality systems are designed to meet the patient's needs and continually changes to meet those needs in a dynamic process. This involves continuous improvement and constant learning by the employees and technicians. TQM focuses on preventing problems before they occur, and developing the inspection techniques to make that possible. Unlike other quality programmes, TQM involves everyone, from the CEO of a company down to the night cleaners. Everyone has a role in providing quality to the end product.

SECTION ONE
Governance and leadership in TQM

The full implementation of TQM requires the commitment of the Chief Executive Officer (CEO) and senior managers at a company. The CEO must take charge personally, providing a vision of where a company is heading, and the leadership to achieve this vision. This requires that the CEO along with the senior managers, defines the company's philosophy, and develop long-term and mid-term plans based on this philosophy. Plans must then be translated into actual management policies. Deployment of these policies down through the organisation is under the control of the CEO. Therefore, one of the first steps in any implementation of TQM is for the CEO to actually present both a Vision and Mission Statement for an institution or company.

Chapter 2. Chief Executive Officer: Ensuring quality

The Chief Executive Officer has the primary role in ensuring that quality is maintained throughout every level of an institution or company. This involves a number of activities, the most important of which are presented in this chapter. It is imperative that the CEO not only encourages but enforces the undertaking by employees of the reading of this guidebook as well as the undertaking of all the follow-up exercises.

Chapter 3. Laboratory Managers: Managing systems

All that the Laboratory Managers do will have an impact on quality, but several functions are particularly important in ensuring a high level of quality in their own departments and in a facility as a whole. The functions included in Chapter 3 have to do with establishing, implementing and monitoring work systems in the laboratory, while those in Chapter 4 present ways of supporting the roles and contributions by the technicians. Because a facility may encompass multiple laboratories and other facilities spread across a country, it is imperative that all Laboratory Managers operate to the same standards and system in order to achieve an overall TQM environment.

Chapter 4. Laboratory Managers: Managing people

This chapter presents key actions that Laboratory Managers can take to maximise the contribution of the technicians to the success of a facility. It is essential that the Laboratory Managers implement all of these actions in order to be successful. The strength of any institution or company is its workforce and if technicians and personnel are not properly managed then subsequently it cannot be expected that a facility can achieve TQM.

Chapter 5. Facilities management: Laboratories and equipment

Managing facilities and equipment involves carrying out regular inspections; dealing with any problems and making sure they do not happen again. Deciding which forms of maintenance to use and keeping records of maintenance are all part of TQM. Establishment of engineering teams, with the qualifications to perform such tasks is essential to the implementation of TQM at a company.

SECTION TWO
The Quality departments

Chapter 6. Quality Operations: Implementation of Quality Systems

Each institution has a Quality Operations Unit consisting of QA, QC and the Quality Operations Manager that meet regularly for the purpose of improving the quality of the work, as well as continuous training in Quality issues. Quality Operations activities are at the core of TQM and play a major role in creating a dynamic atmosphere in the company workplace.

There will always be problems in work processes within the laboratories. Being able to identify them, report them through the proper chain of command and then take immediate action to stop any damage, while finding the root cause in order to prevent them from happening again is essential. This chapter presents systems that will help the technician to recognise and deal with problems. (Chapters 8 and 17 will provide detailed guidelines on using statistical methods to solve problems by analysing and interpreting data.)

There are many problems that cannot be solved simply by examining equipment and machinery. Data has to be collected, usually over a period of time, and then analysed and interpreted. When data has been collected, the statistical methods and tools presented in Chapter 8 will help to analyse and interpret it. Performance of trend analysis through statistical process control is a fundamental requirement of the TQM in the laboratories.

Quality inspections are essential to ensure that the technicians, products and performance have the desired quality features that its customers/patients want. The importance of proper audits, inspections and being able to identify potential problems are key to Quality Assurance's basic functions. In order to perform these audits properly, QA must be independent and not restricted in any manner of its investigations.

The quality of the education and training that a company provides for its staff determines the quality of the products and services provided. A company needs to approach training systematically, and implement it in a continuous framework of constant upgrading and improvements. Employee education is an investment in the future of any company

SECTION THREE
Working in the laboratory

Standardisation is an essential tool for maintaining and improving quality within an institution or company and its contract service providers (hospitals). Writing of Standard Procedures as written descriptions of the best way to do a job, carry out an operation, or complete a process, increase the probability of a successful outcome. These standards can also refer to the specifications of a product. The concept of standardisation appears in many different chapters. In this chapter, we will be dealing with operation standards. It is essential that the facility recognises that there is a balance between writing good SOPs and having too many SOPs. Understanding that TQM is about having quality SOPs, written to a uniform standard and having the lowest number possible and still operate effectively, is essential.

Chapter 12. Precision measuring equipment: Metrology and calibration ... 258

The purpose of measurement control is to ensure that the metrological equipment used to measure specific working ranges are calibrated and qualified to ensure they remain within acceptable limits of precision and accuracy. This ensures that the conditions in which the cell products are manufactured retain their quality characteristics, thereby meeting the regulatory standards. It is a matter of ensuring all measurement equipment is certified long before you need to use it.

Chapter 13. Disposal and storage: Waste and surplus materials 272

A workplace that is neat, orderly and well organised is always more efficient. The details in this chapter present a number of actions to be undertaken in order to achieve this. It has been proven that personnel actually perform better in a clean and organised environment.

Chapter 14. Healthy and clean laboratories 291

Everyone working within a facility is to work in a comfortable and healthy laboratory environment. This also proves to be a more productive environment. There are five sets of actions that can be undertaken to keep cell therapy workplaces healthy and comfortable and to avoid contaminating the environment within and around the laboratory.

Chapter 15. Safety and managing risk in the laboratory 307

Because of the potential risk of working with contaminated cell lines and infectious agents, laboratory safety is crucial in cellular immunotherapy. Proper management of laboratories can eliminate the opportunities of infection and deaths can be prevented. The severity of any infectious agents can be greatly reduced. There are key actions that can be undertaken to improve safety in a company laboratories. Technicians will appreciate in knowing that a company executive is concerned about their health and welfare and in turn will work much more efficiently.

SECTION FOUR
Production, materials and development

Chapter 16. Production control 329

Production control is the management of the production processes within the laboratories to ensure that a company produces immunotherapies of high quality that the market wants, in the right treatment regimen, and ready for infusion at the correct time.

Chapter 17. Process control

Process control ensures that the manufacturing processes used to produce the immunotherapies at a company are of the required quality in a continuous and stable manner. There are several mechanisms that will be discussed for maintaining process control.

Chapter 18. Auditing external suppliers

The quality of the immunotherapies relies on the raw materials and reagents that are provided by suppliers. These materials will have a major impact on the quality, efficacy and safety of the immunotherapy products.

Chapter 19. Post-treatment follow-up

A facility's responsibility for its treatments does not end when the therapeutic regime is completed. The success of an institution or company depends, above all, on whether the customers/patients are satisfied with the therapy they receive and consider it to have been of a benefit to their overall health status and improvement. No matter how good the quality and inspection systems are, as far as the client is concerned, it is whether the treatment worked or not, even to a limited degree which counts. This is why it is essential to have a good after-treatment service in order to follow-up on the degree of success, noting both adverse and positive effects that may have occurred, and whether or not further treatments are required.

Chapter 20. Product development

Product design and development is the process of creating a new therapeutic treatment based on modification of existing process or the introduction of new processes. It involves identifying a market need, creating the therapy to meet this need, and then testing and improving the immunotherapy. It consists of a series of activities: research, analysis, design, engineering, and modelling, and then testing, modifying, and re-testing until the therapeutic product and regimen is perfect. Design and development is usually carried out by a project team, with members from both outside and inside an institution or company. This chapter presents detailed procedures for managing the process of product design and development.

In conclusion

Suggested answers to self testing multiple choice questions

Preamble:
Guidelines for teachers and trainers

1. Introduction

Managing Quality in Biotechnology is a practical teaching and training manual for implementing Total Quality Management (TQM) primarily through self-guidance and at the individual's own learning pace. In order to ensure the training processes are completed, it is necessary that a company has trainers whom will assist and follow-up on the progress of technicians, dealing with any difficulties that personnel may have had with some of the chapters. In the chapters, the text provides clear guidelines for improving quality over the full range of management systems and practices but it is anticipated that some students, employees and technicians may still encounter some obstacles in their self-training progress. Any time that the individual encounters such a problem, they should raise the issue immediately with their designated teacher or trainer.

Trainers provide the necessary framework to enhance the technician's ability to learn through the following assistance techniques:

- They discuss what the technicians are doing at present in the area dealt within the text: What problems have they faced, solutions tried, and any successes achieved.
- They discuss how the technicians can incorporate the concepts from the text and bring improvements to their functional areas.
- They assist when required in the technicians preparing an action plan for implementing their conclusions, to be presented to the decision makers at the company.

A trainer may be any employee with a sound experience within the industry and basic moderating abilities, but does not necessarily need to have an extensive knowledge of TQM. I prize a trainer more for their competence and common sense when dealing with specific areas of company operations above all else. Once again, common sense is the key characteristic requirement for implementing TQM within the biotechnological field.

2. Methodology

This trainer guideline is solely intended to present a general methodology, leaving each trainer to develop and adapt their own skill-set to the situation, culture and required style as they please. If the trainer should be unfamiliar with TQM, then it is recommended that they should read the short "Introduction to TQM", to get a general idea of the basic concepts and understand their responsibilities.

Getting started – Orientation: Begin with orientation type questions. Their purpose is to get technicians focused on the theme of the chapter. In particular, help employees to reflect on their own work situation in relation to the theme. The trainer is encouraged to create relevant questions of their own to present to the technician in order to aid their training.

It is not easy for many people to speak openly and frankly, so give them a bit of time to think about each question in further detail before they give their responses.

Trainers should encourage technicians to talk of their own expectations from the chapter: the benefits they had hoped to gain from it for themselves, for their laboratory and for the company. Briefly note these down on a flipchart or whiteboard and see if they have been met or else can be achieved by the time the employees are finished studying the entire chapter.

Reading and discussion: The reading and discussion of any text forms the key basis of activities in the training process. Basically, if they do not read, they are not going to learn. With most of the chapters, trainers should concentrate on one or two paragraphs at a time before moving on to the next. Make certain that the student or employee has the general picture, is clear on any minor problems or questions they may have. Go through the paragraphs together until they have a solid understanding.

Ask participants to reflect on their own experiences and then ask them how they would apply these new ideas from the text to deal with the issues discussed regarding those past experiences.

The MADMEN questions

As a trainer, you may find that the discussions are best treated as a group activity if you have several students or technicians to train and all are at the same chapter in the book. The groups should be limited to only about five or six members in order to remain manageable. One person in each group should be designated to take notes. Where there are multiple groups, one group summarises its conclusions in a presentation to the other groups and then receives valuable feedback.

Have the MADMEN requirements or questions posted where they can be easily seen, in order to keep everyone focused on the key issues that they have to identify and resolve during any fruitful discussions.

The trainer does not necessarily need to take an active part in any of the discussions, unless it is slow to start, in which case the trainer should prompt the participants a little. The primary role as trainer is to encourage participation, and to monitor the discussions to ensure that they are going in the right direction – discussion can often be held back by participants going off on a tangent about their own experiences, spending too much time on insignificant details, or dealing with irrelevant issues. As the trainer it is imperative to keep all individuals within a group focused on the core issues.

Participants should finish these discussions with clear ideas for improvements in their laboratories and/or facilities, the obstacles that lie in the way, and how these obstacles can and will be overcome.

Action plan: Students or technicians either individually, or in groups should prepare an action plan following the discussions, with the intention to present these plans to the decision makers in a company. After the interchange of ideas during the discussion, with the MADMEN questions all answered, the concepts that go into the action plan will usually bear little resemblance to the guidelines in the text, but rather are the participants' specific conclusions regarding these guidelines as they pertain and apply to their own situations at the company.

Not every issue raised in the guidebook is relevant to the situation at every company, and therefore it is expected that technicians will adapt some, or add some new ideas of their own.

The 6-point structure, shown below, will provide a useful framework by which technicians can prepare their action plans.

3. The 6-point structure

As previously mentioned, any action plan should deal with the six major areas of problem resolution, which will be overseen by the trainers as the technicians discuss their solutions. Each trainer will focus on whether or not the technicians have dealt sufficiently, accurately and effectively with each of the points listed below, which in turn correspond to the issues raised by the MADMEN questions.

1. **Problems:** Problem has been accurately identified and described.

2. **Proposals:** Are the proposals realistic and achievable?
   a. Trainers ensure that training sessions are specific and concrete, and those technicians who have received training could apply what they learned on their jobs.
   b. Include an implementation plan, with a time schedule and minimum and optimal implementation targets. Trainers must have some concept of time frames and whether or not these will be acceptable to senior management.
   c. Refer to any forms, charts or tables used, and include examples whenever possible. Trainers should assist in making the document clear and easy to follow.

3. **Obstacles:** Any obstructions to effective implementation resulting from technician attitudes, company organisation and culture, etc., need to be eliminated. Trainers must be aware of political sensitivities within a company and know how to deal with them.

4. **Resources:**
   a. The resources required: funds, equipment, materials, manpower, expertise, etc.
   b. Identify the available resources within the company.
   c. Identify those resources that would have to be contracted from outside the company.
   d. Identify any alternatives that could be used to cover any shortfall in resources.
   Trainers must ensure that this is not just an impossible wish list and that the technicians have identified resources that are pre-existing or within a company's budget to make some of their action plans actually happen.

5. **Assessment:** Ways of assessing the results of implementing these proposals.
   Trainers should encourage technicians to examine the pros and cons of each solution and recognise that for every positive point, there may be a negative one as well. Approval is based on having more positives than negatives.

6. **Benefits:** Discuss the benefits any proposals bring to the technician's scope of work. Trainers must help decide whether the benefits are real or imagined.

4. Prepare your training programme

Study the materials. To prepare the programme trainers should first:

- Make themselves thoroughly familiar with the selected chapters and texts.
- Do the tests themselves in order to assess how well they have absorbed the content of the material.
- Instruct technicians to submit in advance a form in which they identify what they want from the training and any specific problems they are encountering. Give special thought to these problems, and perhaps discuss them with senior management in advance.
- Select graphic figures: Consider carefully which graphics to include. Trainers must decide how useful to the training purposes each of them is. Do not use them simply because they look pretty. If they are not appropriate to a specific learning section, then do not include them as they will only serve to confuse and discourage the trainees.
- Decide on the training duration: How many sessions will be necessary and how long and how frequent they should be? This will be determined largely by the complexity of the chapters, the methodology for training used and the competence and experience of the technicians.
- Keep in mind that *Managing Quality in Biotechnology* is focused on preparing for the implementation of TQM, and not simply conveying knowledge about it. The process of discussing the ideas in the text and how to implement them can therefore take quite a long time. In some texts, one paragraph alone will be enough for one training session. Likewise, the preparation of useful action plans can take quite a while.

Timing is affected by whether training is on an individual basis or by having small-group discussions, and whether groups work on their action plans during the training session rather than between sessions. The general rule of thumb is to assume they will take far longer than you anticipated!

Decide when and how to use the multiple choice tests: Some trainers may ask the technicians to do the interactive tests after each text or after they complete the entire chapter, on their own, or they may conduct an actual closed book examination session. The decision of test implementation should be left to the trainers.

5. Evaluate the training programme

When each training session is finished, obtain feedback from the technicians and employees. This is invaluable for planning future programmes and sessions. This can be done simply by giving them about 15 minutes to write answers to the following questions:

- What did you find most useful in this section or chapter?
- What did you find least useful?
- What benefits has it brought you, your laboratory or the company?
- What changes would you recommend?

An Introduction to TQM

1. Today's challenge

The challenge that institutions teaching biotechnology and companies face as the 21st century gets underway, is to succeed in a global economy where competition is fierce and where customers/patients are becoming increasingly demanding of quality and safety. Therefore, both institutions and companies must adopt a modern, global perspective. They must supply therapies and services that are competitive in both price and quality, and meet both the international as well as domestic regulations and requirements. To maintain a competitive edge in such an environment they must continually improve the quality of their products.

With close to 500 facilities offering immunotherapeutic treatments in China, a company must demonstrate that it is different from all the others and Total Quality Management (TQM) has proven itself as an effective way of managing and continuously improving quality by raising the standards and making it hard for competitors to meet the standard you attain. Although I have written this guidebook from an insider's perspective of the biotherapeutic industry, it is equally applicable to any other biotechnological industry.

What is Total Quality Management? What indeed do we mean by quality? Quality may be simply defined as meeting client requirements. Considering that the customer/patient for immunotherapies is expecting improved health and longevity, then such requirements are a tall order. Total Quality Management, as the name indicates, regards the continuous improvement of customer/patient-oriented quality as both requiring active management and involving the entire company – and often suppliers and clients as well. TQM can be described in practical terms as customer/patient focus, continuous improvement and teamwork. TQM is seen as a dynamic economic effort by companies to adapt and survive in a constantly changing, ever-toughening environment.

2. The background to TQM

Born out of management practice, and with roots in production, statistics and quality control, modern quality management began in the USA in the mid-1920s and has had a profound impact on modern business history. It was not however widely applied in the West until the 1980s, some two decades after it had taken off with remarkable success in Japan.

A core objective in the prevention of abnormalities, defects and deviations, in ensuring that all products and services conform to customer/patient requirements, is the reduction of variation. Biological products manufactured under the same conditions and to the same specifications almost always vary to some extent. These variations may be great or small, and come from the main components of the production process: equipment, people, method or materials – essentially, equipment deteriorates from the day you buy it and raw materials are never exactly the same, especially when dealing with an individual's own cells. The critical question is whether these variations have any effect on quality. The answer to this question however is often unknown, because the variations are usually not measured. TQM provides the necessary tools, which allow us to measure these variations and to take corrective action as required.

3. The central role of management

TQM must first of all have the active commitment of senior management at a company if it is going to succeed. Senior management has to take personal charge, providing vision, forceful leadership and clear direction. They must translate this vision into detailed, long-term planning, often for a period of up to five years or more.

This includes:

- Developing a clear long-term strategy for TQM.
- Formulating and deploying an annual policy and ensuring everyone understands it.
- Ensuring that objectives, targets and resources are agreed among all those responsible for putting the policies into action.
- Building quality into designs and processes.
- Developing prevention-focused activities.
- Planning how best to use quality systems, standard procedures, tools and techniques.
- Developing the organisation and infrastructure to support improvement, and proper allocating of the necessary resources.

Senior management must also establish, as evidenced by their own behaviour, an appropriate style of management throughout the institution or company, a style that can give not only direction, but also encourages personnel to take on more self-initiative.

4. Personnel involvement

At the heart of TQM is the recognition that any institution's or company's most valuable resource is its people. The involvement of everyone in achieving continuous improvement in quality is essential. The customer/patient is seen as the end receiver of the immunotherapeutic product and health service, which is the final stage of a series of processes involving many different technicians and employees. Everyone who participates in each of these processes contributes to the final quality of treatment that the customer/patient receives. Equally important is that everyone sees themselves as a valuable contributor in the overall innovation and improvement of the therapy and products available.

Furthermore, within this series of processes, called a "process chain", the work output of one process is the input subsequently for another. A weakness in one process will have a direct effect on the next. Understanding this is critical, since this process chain crosses departmental and functional boundaries. From the TQM perspective, the need for everyone's active involvement and the inter-relatedness of the several processes in the process chain, have major implications for management at all levels of a company, and ultimately determines the culture of a facility.

For human resources management two behaviours are clearly indicated: learning to listen and listening to learn. That means empowering technicians and other personnel to communicate both upwards in the hierarchy and laterally across department boundaries. Historically, communication in companies was always directed downwards, with the objective of keeping staff informed but having little to say in this top-down management style. That style is not acceptable in the TQM model. Those involved in the work processes will have valuable knowledge about on-going and developing problems and in the interest of continuous improvement, they need to be able to communicate their suspicions and knowledge to superiors freely without the fear of repercussions or not being taken seriously. Similarly, there must be constant communication between those involved in the related processes in the full process chain so they can all work together in improving the quality of the final product.

This change in management style results in an overall change in the culture of a company and will ultimately lead to an environment where low-level management and staff:

- Find that their views are sought out, listened to and acted upon.
- Have a clear understanding of what it is required of them, of how their processes relate to the business as a whole, and how they affect other departments in the institution or company.
- Are given practical scope through such devices as suggestion schemes, and especially through teamwork (e.g. quality operations meetings), to participate in achieving innovation and improvement.
- Are motivated to contribute to the continuous improvement of their own work output.

5. Tools and techniques

Central to the implementation of TQM is the gradual introduction of tools and techniques with a problem-solving focus. Techniques such as process mapping, whereby flowcharts are used to show all the steps in a process with the aim of revealing irregularities and potential problems, is not unlike work-study flow diagrams. TQM tools recommend those that are simple to use, those that most employees can be trained to use, and those, such as Statistical Process Control (SPC), that will require some specialised training.

Statistical Process Control can be used to measure variation and to indicate its cause. Some variation is tolerated in the output of processes. However, all variation has an identifiable cause and therefore can be reduced. Knowledge of the causes of variation is a powerful tool for the instituting of quality performance into a company.

The typical TQM method is the PDCA (plan, do, check, act) cycle, also known as Deming's Wheel or Deming Cycle, which will be discussed in the next paragraph.

6. The PDCA cycle

The PDCA cycle, developed by Dr. W. Edwards Deming, is an invaluable strategy for improving any situation, from solving a tiny process problem to introducing TQM itself throughout a company. It consists of four easy to remember steps:

- **Plan:** Gather data regarding the problem, identify the causes, decide on possible solutions or countermeasures, and develop a plan with targets, and tests or standards that will check whether the countermeasures are correct. This should be done systematically and thoroughly.

- **Do:** Implement the countermeasures that have been decided upon.
- **Check:** Check the results of the implementation of the countermeasures against the standards established in the "Plan". If the countermeasures do not work, begin the cycle again with "Plan" and repeat.
- **Act:** If the countermeasures are successful, standardise them and put them into regular use. They then become normal practice.

The PDCA cycle is essentially a means to achieve continuous process improvement. The PDCA cycle is not based on the idea of "get it right first time" as was the old standard, but rather on the fact that a manufacturing process will rarely get anything completely right the first time; perhaps not even the second or third time. That being the case, then the PDCA cycle must be continuously repeated if quality is to be a realised goal. With each application the improvements made must be standardised and become the base for further improvements. Each cycle is a new layer like rolling up a steady incline, never quite reaching the top.

Technicians can follow a plan-do-check-act cycle, exercising a degree of self-management in their job performance that not only contributes significantly to a company's pursuit of continuous improvement, but also improves the quality of their own working lives. Each time the cycle is put into practice the performance bar is raised as they climb that incline.

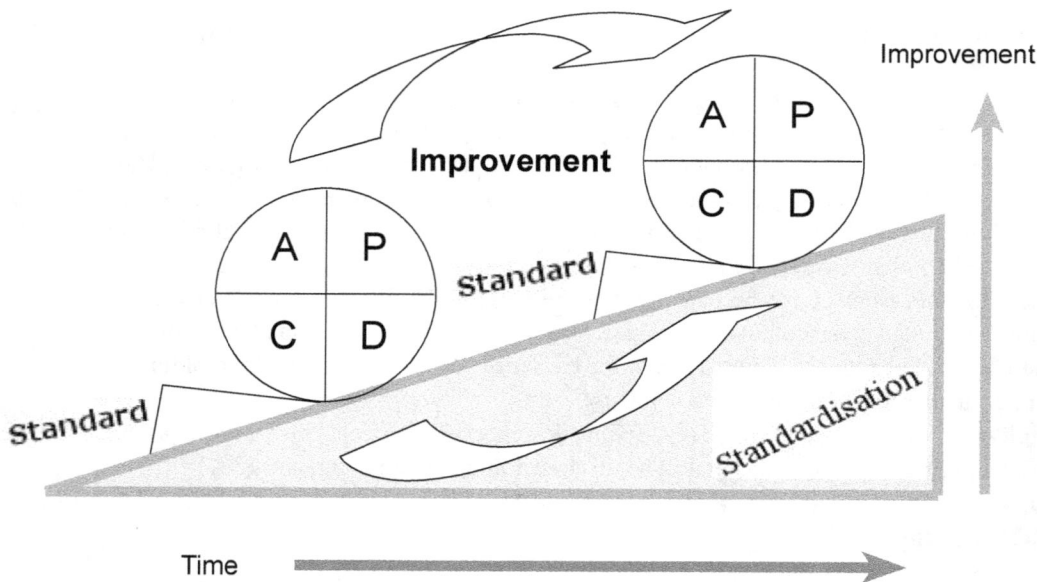

Deming Cycle

7. Teamwork

Identifying the root causes of problems is not always an easy task. In addition to the tools and investigative methodologies required, there is an important role for personnel to work together on shared problems in order to find the correct solution. Developing a team mentality in the workplace is not always easy either, as it relies on the differing personalities, goals and attitudes of the individual staff. People can and will be a root cause of problems.

8. Education and training

In order to achieve TQM at a facility, a full programme of education and training is essential. Training is needed to develop the practical skills of applying TQM tools, education to effect the changes in behaviour and attitude and to ensure understanding of what is involved in the ongoing pursuit of continuous improvement.

9. cGMP and GTP

TQM has a close correspondence with the International Cell and Tissue standards, the set of universally recognised standards of good management practice which ensure that the organisation consistently provides products and services that meet with the customer's/patient's quality requirements within the biotherapeutic industry. These regulatory standards define the requirements of the quality management system that can be applied in any immunotherapy organisation. TQM will assist a company in a very practical way to meet the requirements of cGMP and GTP accreditation.

10. The prognosis

Implementation of TQM is not easy and it is by no means to be attempted in a quick and haphazard manner. Improvement should be viewed as a slow incremental process which requires participation of everyone at a company. It must be accepted that this is an unavoidable challenge that companies working in the biopharmaceutical sphere must now face. The design of this book makes TQM everyone's responsibility and will not only initiate technicians on the path of self-enlightenment and self-implementation but is designed so that all other departments and their respective employees, whether senior or junior in position at an institution or company can actively engage in the process.

SECTION ONE
Governance and leadership in TQM

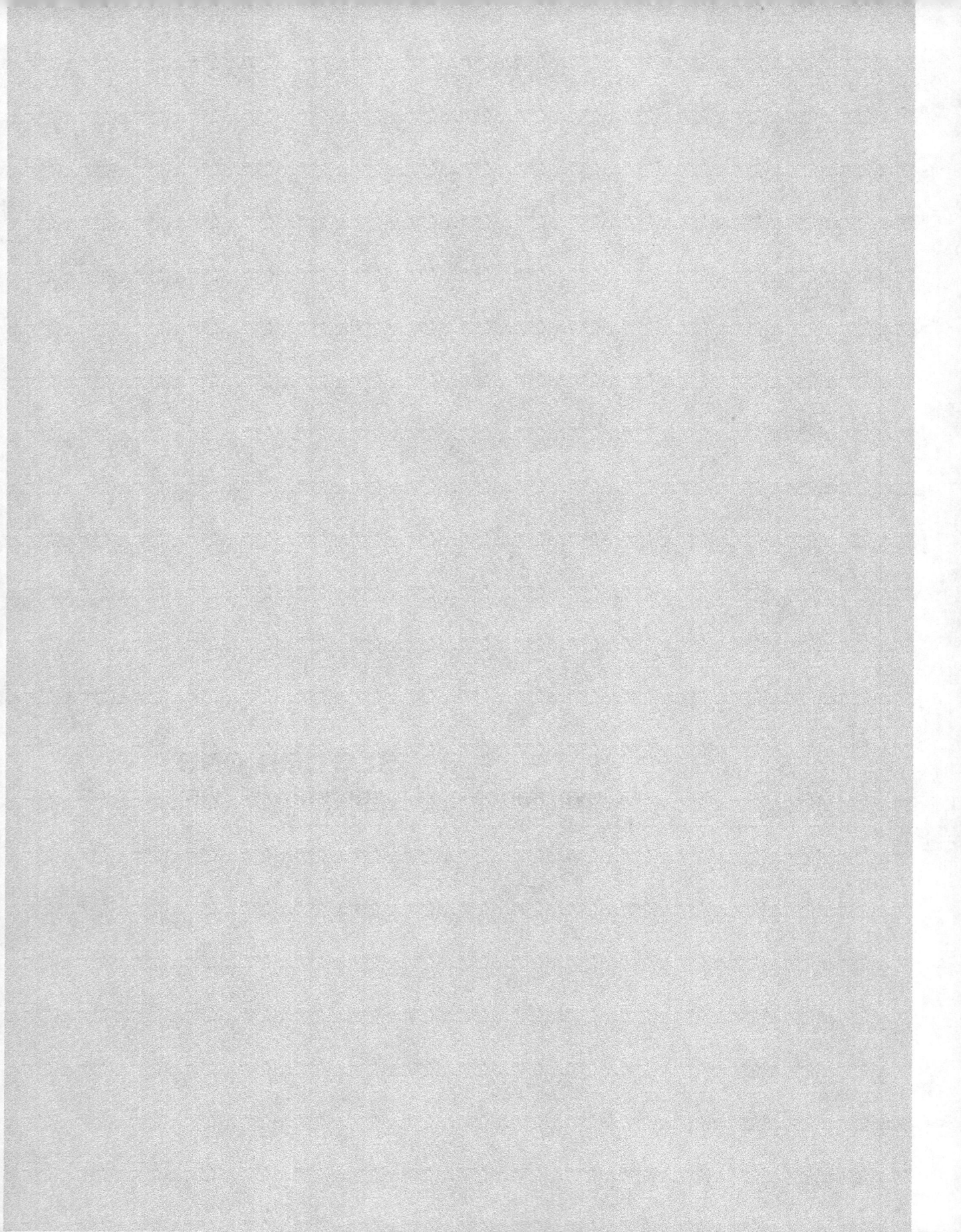

Chapter 1: Chief Executive Officer

Managing
Policy

The full implementation of TQM requires the commitment of the Chief Executive Officer and senior managers. The CEO of an institution or company must take personal charge, providing a vision of where as an organisation they are heading, and the leadership to realise this vision. This requires that the CEO, along with the senior managers, define a corporate philosophy, and develop long-term and mid-term plans based on this philosophy. These plans are translated into annual management policies, and these policies are deployed down through the organisation. This is known as policy management.

1.1 Policy management: An overview

For successful policy management, it is necessary to first define a corporate philosophy, and then prepare long-term and mid-term management plans. From these plans, it is decided what is to be achieved by the end of each year, and the concrete measures required in order to do this. These become the annual management policies, which are communicated to the different departments, making sure that they are implemented in a coordinated way and to an agreed upon time schedule. Good policy management is essential if an organisation is to successfully develop new products, improve quality, reduce costs, and strengthen its operations.

a. The first step is the institution or company defines its business philosophy: What are its aims, what is important, what are its values? On the basis of this philosophy, the CEO develops the long and mid-term management plans.
b. At the beginning of each fiscal year the CEO establishes the annual management policies for that fiscal year: Their targets for the year, and the concrete measures needed to achieve them.
c. Identify only policies that will be given priority status.
d. Concrete guidelines with numerical targets for important items such as treatment numbers, design and development, production, personnel administration, and services need to be established.
e. Set specific targets and methods for maintaining and improving quality in each department.
f. Identify control items: Measurable items used to judge whether or not tasks assigned to departments or sections or individuals have been carried out, including methods of evaluating this; if tasks have not been carried out as required, appropriate action should be taken.
g. Set the annual targets.
h. Set the strategic measures to implement the policies.

1.2 Prepare mid-term and long-term management plans

Long-term plans are normally for five years minimum, and mid-term plans for three years. They are essential if an organisation is to achieve all that it is capable of achieving. They should be based on the business philosophy, on a good understanding of the management strategies, and on a sound grasp of the corporate strengths.

    a. At each organisational level the policies will be based on:
        i.   The long- and mid-term outlook.
        ii. Results from the previous fiscal year.
        iii. Relevant business conditions.
    b. In this process of policy deployment, it is essential that information is exchanged through the different levels, and that general agreement is reached.

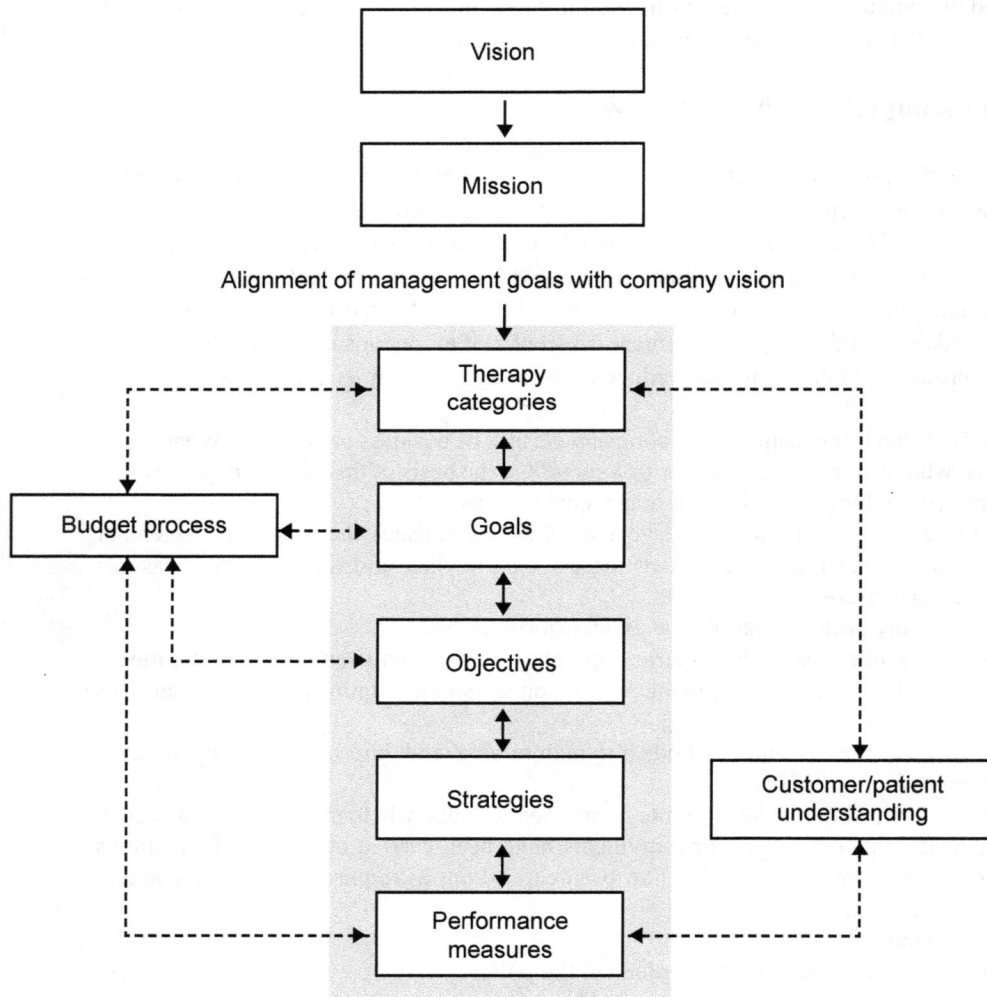

The management plan

1.3 Establish annual management policies

At the end of each fiscal year, senior management is to agree on their understanding of the long-term management plans, review the policies of the past fiscal year, and draft a plan for formulating annual management policies for the new fiscal year. This draft plan is discussed at a top level management meeting, and consensus is reached. Draft policies for the new fiscal year are prepared, and are then approved by the CEO.

a. Consider how a procedure for establishing annual management policies could be introduced in an organisation:
   i. Who would be involved in each of these steps?
   ii. What information would they need?
   iii. What meetings might they hold?
   iv. What current questions and issues, about both the present and the future, would they need to look at?
   v. What difficulties might be met and how could they be resolved?
b. The annual management policies must be deployed down through the organisational levels, from higher level to lower level departments and laboratories. Those at each level will establish their own policies, based on these annual policies, and prepare plans to implement them.

1.4 Deploy policies and prepare implementation plans

The annual management policies must be deployed down through the organisational levels, from higher level to lower level departments and sections. Those at each level will establish their own policies based on these annual policies, and prepare plans to implement them.

a. These policies are sent to the various departments in order to achieve the policy deployment.
b. The department managers examine the policies they have received, reflect on the work they have to do within their own departments, and formulate their own policies – their annual targets and their concrete plans to achieve them.
c. Implementation plans ensure that the policies deployed through an institution or company are implemented uniformly.
d. When department managers receive the annual management policies, each manager should take the following steps:
   i. Acquire an understanding of the management policy and establish their own concrete policies based on application to their respective areas.
   ii. Examine ways to harmonise their departmental policy with an organisation-wide policy, avoiding any conflicts.
   iii. Establish how they will evaluate their departmental activities.
   iv. Produce written implementation plans.
   v. Implement these plans in the form of departmental, laboratory, and individual activities.

vi. Hold advance discussions on items that have crossover with other departments and laboratories to foster cooperation and avoid conflict.

vii. The managers deploy their policies (their targets, goals and implementation plans) to the departments and individuals below them, who then implement them.

1.5 Control the implementation of policies

Check regularly that policies are being implemented as planned: that departments, laboratories, and individuals are carrying out the tasks assigned to them. If policies have not been implemented properly, it is important to decide whether the implementation plans were impossible to carry out, or the plans were carried out according to the schedule but failed to achieve the desired target.

a. Implementation has to be controlled. Assign employees to use control graphs and other reference materials to inspect, on a weekly or monthly basis, if:
   i. The measures in the implementation plans are being implemented as instructed.
   ii. These measures are having the anticipated responses and effects.
   iii. If any abnormalities appear (i.e. any unexpected results of implementing the plans), countermeasures should be taken quickly and the results reported to senior managers.
   iv. If necessary, modify the implementation plans through change control.

b. Plans for controlling the implementation of departmental policies should include:
   i. Achievements and problems from the previous fiscal year.
   ii Concrete targets for the coming year.
   iii. The schedule for their implementation.
   iv. Those appointed to be in charge, the individuals responsible for performing, and any cooperating departments or other staff involved.
   v. The anticipated outcome of the plans.
   vi. Use the "5Ws and lH" (who will do what, where, when, why and how) to present these points clearly in the plans.

1.6 Reflect on the policies at the end of the year

Examine the annual management policies at the conclusion of each year. This process of year end reflection will provide essential feedback for future improvements. There are two types of year-end reflection: reflection based on results, and reflection based on the processes.

a. Possibly at the middle and definitely at the end of each fiscal year, unit managers, department managers and CEOs make an analysis of:
   i. The status of policy implementation.
   ii. The achievement level of the annual policy items.
   iii. The status of the organisation's general quality management.

b. At the end of each fiscal year, review what has been implemented during that year.

c. Prepare summary reports to be used as a basis for the improvement of activities.

1.7 Carry out a CEO diagnosis of policy implementation

To manage their policies CEOs must have a sound grasp of the mechanisms of policy management and they must display leadership. Unless they do so, no organisation-wide changes will take place, and policies will end up unachieved. They must step out of their office and visit the various divisions, hospitals and laboratory facilities where they should hold frank discussions on policy deployment and promotion. This is referred to as the CEO implementation of policy management and is performed through close contact with personnel of each unit.

a. CEOs must keep the following points in mind:
   i.   It is implementation at the front line that puts higher policies and plans into effect.
   ii. Plans become more concrete as they move down the organisation ladder of a company.
   iii. Interdepartmental and intersectional problems should be resolved in a cooperative manner.
   iv. Use the "5Ws and IH" method to specify how the targets and deadlines are to be met.
   v.   Make communication and checking methods clear.
b. Once the implementation of policies begins, the CEO needs to check regularly that they are being implemented as planned: make certain that departments, laboratories, and individuals are carrying out the tasks assigned to them.
c. If policies have not been implemented properly, the CEO will have to decide whether the implementation plans were impossible to carry out, or the plans were carried out as instructed but failed to achieve the target.
d. There are two ways of appraising implementation:
   i.   Appraisal of the results.
   ii. Appraisal of factors: specific activities and processes being implemented.
e. Use the PDCA cycle to carry out the appraisal as explained below:

P - Plan: Set up a concrete implementation plan, as described previously, with evaluation items and evaluation criteria; decide how to put it into practice and who will be in charge of each item; prepare a schedule and check that everyone who needs to know about it is informed.

D - Do: Implement the plan exactly as it has been drawn up. Do not implement it partially or haphazardly.

C - Check: Use the evaluation items that were chosen at the planning stage to review and evaluate the results of implementation.

A - Act: Analyse the results that have been evaluated and take concrete countermeasures. If necessary, incorporate the results into a new PDCA cycle. Once the desired result is achieved, standardise the new procedures.

Analysis by the CEO focuses on two areas:

a. *Evaluation of policy management*

    CEOs, other top executives, department managers, or section chiefs assume the role of evaluators. They periodically carry out a planned inspection of policy deployment, of implementation measures, and of the achievement of targets at the end of each term and during company-wide quality month. They also evaluate how well the plans suit the present conditions and note any improvements in the implementation and appraisal methods. They point out problems, suggest corrective measures and answer questions about any departmental or sectional difficulties.

b. *Evaluation of cross-functional management*

    It is impossible to get an accurate grasp of such matters as new product development, quality assurance, and successful management by only carrying out an evaluation of a narrow range of activities within one company function. Cross functional activities or processes (which cross the boundary between two or more functions) have to be evaluated from a special company-wide perspective.

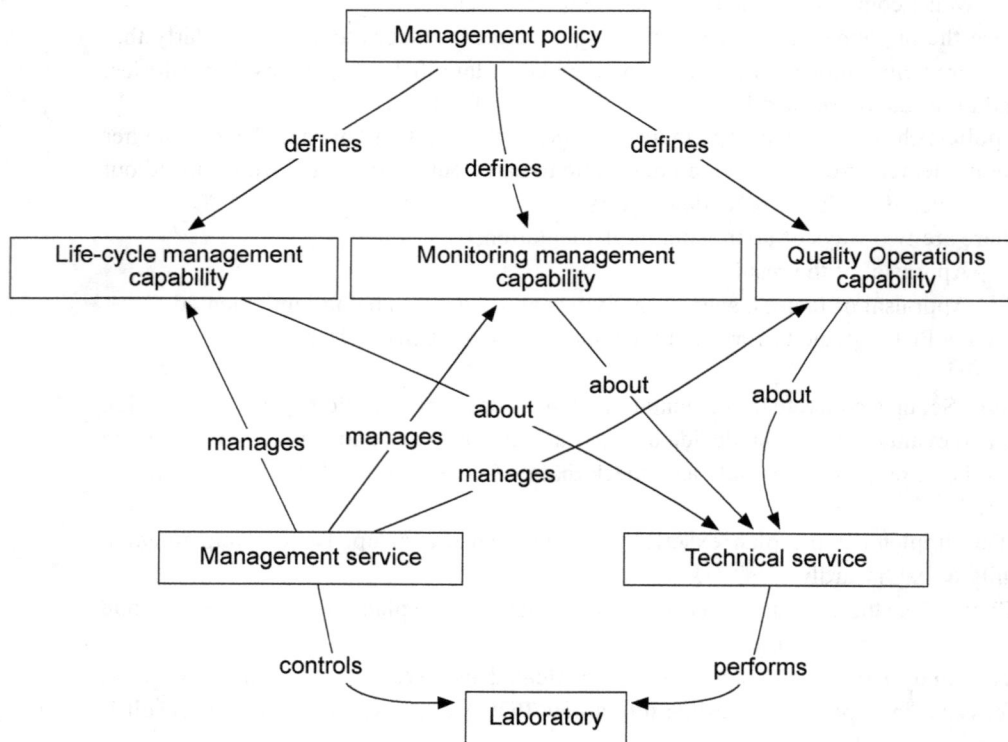

The implementation web

The monthly appraisal

Ideally, the CEO and managers should carry out an appraisal each month. The basic monthly appraisal measures control items and uses control graphs to judge whether the targets have been achieved. This corresponds to the C and A (check and act) parts of the PDCA cycle. Detect problems, identify the causes and take measures to deal with them.

The sources of policy implementation problems will normally be found in:

a. Changes in the external environment.
b. Failure to implement specific items.
c. Failure to carry out the implementation plans properly.
d. Delays in carrying out the implementation plans.

Approaches to the problems:

a. When failed implementation poses problems, investigate the causes, and analyse the root causes and eliminate them.
b. When problems exist in the implementation items, investigate why these were originally included in the plans, how they were approved, and identify and eliminate the root causes.
c. When problems exist in the implementation plans, then revise the plans.
   Document immediately any actions taken. For reference purposes, write up analysis of the policies that were not successfully implemented, and the actions taken to deal with them. Use follow-up forms, charts with implementation plan management tables and PDCA sheets. Assessment should avoid being vague in content.

Control items

Control items are measurable and should be used to judge whether tasks assigned to departments or laboratories or individuals have been carried out successfully. If not, then appropriate action must be taken.

Three important control items in appraising policy implementation are:

a. Accuracy of target values:
   Policy management helps to put the policies of senior managers into effect. The policies and target values themselves are therefore control items. They must be checked to ensure that they are realistic and achievable.
b. Deployment problems:
   It is essential to identify and solve any problems that occur immediately. The question of whether problems have been properly identified and resolved become important control item themselves.
c. Progress ratio:
   The progress ratio is the percentage of a task or process that has been completed in

comparison to the expected results that would be obtained if the task or process was to be completed fully. Therefore, in this respect, it is a measurement of successful implementation. In other words, if the task has only been 50% completed, then it would be expected that 50% of the desired results would be achieved. That would be a 1:1 ratio but if only 25% of the desired result is achieved, then work:results has a 2:1 ratio and this indicates a serious problem.

It is impossible to check everything. Select important, high contribution implementation items as control items. Always review the correctness of the control items that have been selected. If they are no longer suitable, then revise them.

Note the distinction between control items and control characteristics:

a. "Control items" are measurable items used to judge whether or not tasks assigned to departments, laboratories or individuals have been carried out; if not, then determination of "why not" should be conducted.
b. "Control characteristics" are characteristics used to judge whether machines, materials, facilities, equipment, processes, and systems are in the necessary condition to perform properly through specific, pre-acknowledged checks.

Control graphs

Use control graphs to measure the progress of implementation plans. These are usually time-event graphs (graphs that show the examined event over a period of time) in which action limits are entered. The action limit is the limit represented by the highest or lowest value accepted by quality control. If the actual values fall outside these limits, then a corrective action is required and the cause of the error in the process must be determined. Data providing the achievements for each control item over time is plotted on the graph. Any data that falls outside the action limits indicates a failure to achieve the implementation scheduled for that point in time. When this happens action must be taken.

Procedure for dealing with abnormalities

The term "abnormality" is often used in TQM to refer to either existing problems, or an indication of a potential problem. It can also refer to any unexpected outcome as in a deviation. In the context of policy implementation an abnormality refers to any result that was not planned or anticipated. In this context, it could be a planned action that was not done at all, or was performed improperly, as well as having an unexpected result.

There are three important actions to take when abnormalities occur:

a. Emergency measures: eliminate the obvious effects of the abnormality.
b. Recurrence measures: eliminate the suspected causes.
c. Core measures: eliminate the identified root causes.

To detect and deal with abnormalities:

a. Specify the control items and prepare a corresponding number of control graphs.
b. Collect data over a specific time period and plot it accurately on the control graph. Compare the plotted data with the action limits. Look for any trends in the plotted points that could indicate a potential problem.
c. Look for any signs of abnormalities, and confirm as soon as possible that they actually are abnormalities.

To do this, reduce processes down to their basic component processes.

a. Never ignore any abnormalities. They do not go away on their own. Assess them accurately. Every abnormality has a solution.
b. Enter data about any abnormalities on forms which can then be used as abnormality reports and issue these reports to all concerned parties. Record every event accurately.
c. Take any emergency actions required right away to prevent any further harm or damage to the product or process.
d. Analyse processes using defined problem-solving procedures and skills in order to identify the underlying causes. Training in problem resolution skills, especially for managers and team leaders, and laboratory technicians is essential.
e. Take any actions required to eliminate the underlying causes that were identified in the preceding step.
f. Record all actions taken to deal with the abnormalities in an abnormality report. This report will include any emergency actions undertaken to deal with the immediate problems as well as the recurrence-prevention measures taken to deal with the identified underlying causes.
g. Analyse the abnormality reports at least once per month to check that the actions taken have been appropriate. Confirm their outcomes, any subsequent problems detected, as well as the corresponding improvements that have been made. Every action taken and its outcome should be detailed separately.

Once the facts of the current situation have been analysed:

a. Then prepare a review report.
b. Leave space in the intermediate review or report for additions of the latest description of the current situation.
c. Select those problems from the reports that are considered of utmost importance and make them the focus of actions for future plans.

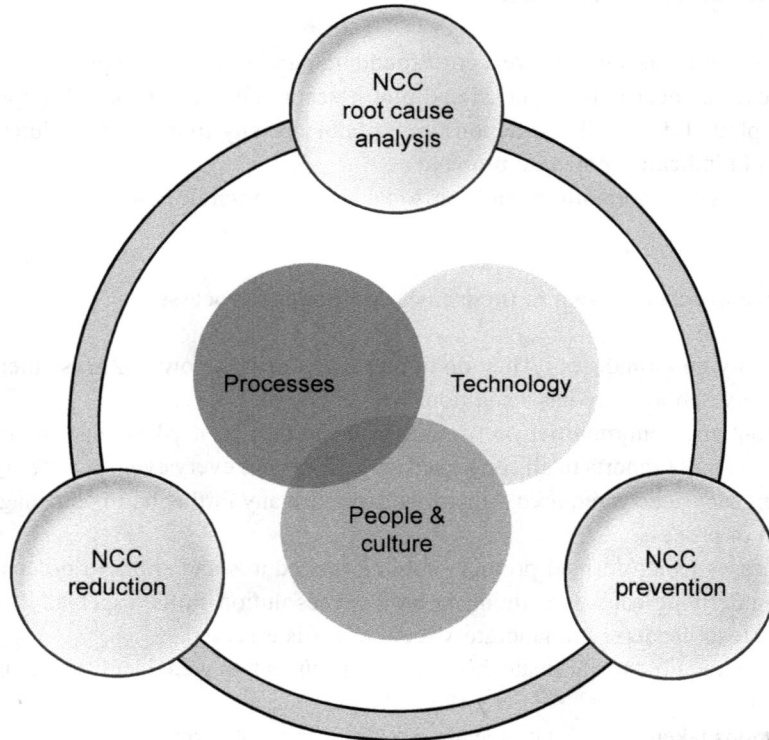

**The core causes of Non-Compliance or Conformance (NCC):
Methodology, machines and manpower**

Follow-up

After the evaluation and confirmation by the CEO, the suggested measures are to be implemented systematically.

Normally the following actions will be performed:

a. Technical experts:
   i.   Write up the problems and the measures to improve them. These documents are sent to the departments and laboratories with the abnormality and they are put into action immediately.
   ii. Write up all comments, questions and answers that arose during the discussions to identify and resolve the problems and these are also distributed to the affected departments and laboratories, as well as to all other related departments and laboratories to serve as reference data.

b. Affected departments and laboratories:
    i. Map out improvement plans for all the items pointed out in the report, and provide their action plans to the Quality Department for approval.
    ii. Carry out the approved action plans.
    iii. Submit interim reports with items that have not been fully improved within three months to Quality Assurance for an extension.
    iv. Write completion reports once items have been fully corrected.
c. In the next evaluation the CEO follows up on the items pointed out in the previous report to incorporate into the annual review.

Recommended reading

1. Burnett, S., Benn, J., Pinto, A., et al. Organisational readiness: Exploring the preconditions for success in organisation-wide patient safety improvement programmes. *Qual Saf Health Care*, August 2010; 19(4):313-17. [PubMed]

2. Langley, G.J., Nolan, K.M., Nolan, T.W., et al. *The Improvement Guide: A Practical Approach to Enhancing Organizational Performance.* San Francisco: Jossey-Bass, 1996.

3. Marshall, M., & Øvretveit, J. Can we save money by improving quality? *BMJ Qual Saf* 2011; 20:293-96. [PubMed]

4. Øvretveit, J. *Does Improving Quality Save Money? A Review of Evidence of Which Improvements to Quality Reduce Costs to Health Service Providers.* London: The Health Foundation, 2009.

5. Schouten, L.M.T., Hulscher, M.E.J.L., Everdingen, J.J.E.v., et al. Evidence for the impact of quality improvement collaboratives: Systematic review. *BMJ* 2008; 336:1491-94. [PMC free article][PubMed]

Self testing multiple choice questions

1. The first step in policy management is that a company:
   a. Develops its long-term plan.
   b. Develops its mid-term plan.
   c. Establishes its business philosophy.

2. When department managers receive the annual corporate policies the first thing they do is:
   a. Ignore relating them to the problems that they themselves have to deal with.
   b. Decide on their own policies and targets.
   c. Return the policies to the CEO with their comments.

3. Control graphs are used to check if:
   a. Countermeasures are creating the abnormalities and deviations.
   b. The measures in the plans are being implemented as prescribed.
   c. The measures in the plans are not producing any results.

4. At the middle and end of each fiscal year, departmental managers, laboratory managers, or CEOs make an evaluation of:
   a. The status of non-policy procedures implemented.
   b. Policy without investigating general quality management.
   c. The number of abnormalities.

5. The actions that have been implemented are reflected on at the end of:
   a. Each month.
   b. Each half-year.
   c. Each year.

6. Long-term plans are normally for:
   a. One year.
   b. Three years.
   c. Five years.

7. The long-term plans should show your major strategies with control items such as:
   a. Expectations of number of patients receiving a particular therapy.
   b. Converting existing equipment to perform unintended uses.
   c. How much you are going to charge for the treatment.

8. If a company chooses to prepare both mid-term and long-term plans, then the long-term plans will present _____ for marketing, therapy production volume, plant and equipment purchases and maintenance, and personnel costs.
   a. Basic targets.
   b. Extremely detailed targets.
   c. Full cost estimates.

9. The procedure for establishing policies for the fiscal year includes:
   a. CEOs do not necessarily agree on the long-term management plans or review them with department and section managers.

b. Reflections on the policies of the fiscal year just finishing are written-up and discussed but are not used for making new policy.

c. A draft plan for policies for the new fiscal year is presented at a top-level meeting.

10. The actions to be taken before the annual management policies are finalised include:
   a. Close out all plans from previous years even if they failed meeting targets.
   b. Draw concrete purchases for new equipment that may be needed in the future.
   c. Consider any relevant recent changes in the cell-therapy business environment.

11. When lower level departments and sections receive the annual management policies they will establish their own policies based on:
   a. The long and mid-term outlook.
   b. Business models taken from other companies.
   c. Results from all previous fiscal years.

12. When departments and sections have received the annual management policies they should:
   a. Avoid discussion of items that are strongly related to other departments and sections.
   b. Produce written implementation plans.
   c. Implement these plans as directives from the top without departmental, sectional or team input or activities.

13. The implementation plans should include:
   a. Only the achievements from the previous fiscal year.
   b. The cost of preparing the plans which must be recovered prior to implementation.
   c. The schedule for the implementation of the plans.

14. Points to be kept in mind with regard to policy deployment include:
   a. Only those plans at the top can be considered concrete and put into effect.
   b. Plans become more concrete as they move down the organisation levels of the company.
   c. Interdepartmental problems should be resolved through debate of the issues.

15. The two types of appraisal of the implementation of policies are appraisal of:
   a. People and results.
   b. Results and factors.
   c. Factors and people.

16. The monthly appraisal corresponds to:
   a. The P and D part of the PDCA cycle.
   b. The C and A part of the PDCA cycle.
   c. The P and A part of the PDCA cycle.

17. The source of policy implementation problems may lie in:
   a. Changes in the external environment.
   b. Focusing on the implementation of specific items.
   c. Moving too fast in carrying out the implementation plans.

18. The assessment of abnormalities and deviations should include:
    a.   Written analyses of policies that were not successfully implemented.
    b.   Any relevant forms except PDCA sheets, which are too general.
    c.   The reasons why the implementation was impossible from the onset.

19. One of the most important control items is represented in the question below:
    a.   Are policies and target values exceeded?
    b.   What is the failure ratio?
    c.   Are problems identified during the policy deployment stage?

20. An appraisal of the implementation of policies should check:
    a.   All items.
    b.   As many items as possible regardless of status.
    c.   High contribution implementation items primarily.

21. Control _____ are used for judging whether machines, tools, facilities, equipment, processes, and systems are in good condition.
    a.   Items
    b.   Characteristics
    c.   Features

22. The _____ of the implementation plans may be used as action limits.
    a.   Contents
    b.   Targets
    c.   Characteristics

23. Match the terms with the definitions:
    i.   Emergency measures   ①   eliminate the causes.
    ii.   Recurrence measures   ②   eliminate the root causes.
    iii. Core measures   ③   eliminate the outward effects.

    a.   i③, ii②, iii①
    b.   i②, ii③, iii①
    c.   i③, ii①, iii②

24. An abnormality is:
    a.   An unexpected outcome in a process.
    b.   A planned defect in a process.
    c.   An emergency situation.

25. The detection of abnormalities and the actions to deal with them include:
    a.   Specify the control items and prepare a corresponding number of control graphs.
    b.   Collect data every year and plot it accurately on the control graph.
    c.   On first occurrence it is not necessary to record but is if it happens a second time.

26. In the detection of abnormalities, the initial action to deal with them is:
    a.   Take any emergency action required right away.
    b.   Record any actions taken to deal with abnormalities in abnormality reports.
    c.   Begin an investigation to eliminate the underlying causes.

27. The primary purpose of reflection is to:
    a.   Provide feedback for future improvements.
    b.   Remove any remaining problems at the end of the fiscal year.
    c.   Find out who was responsible for any failures to implement policies and penalise.

28. Items that can be reviewed at the end of the term include:
    a.   Activities implemented during this term.
    b.   Implementation plans for the next term.
    c.   Non-closed investigations of failed target achievement.

29. The main point to keep in mind regarding reflection is:
    a.   Pay particular attention to changes in the economic environment.
    b.   Provide a lot of space for term reports of the current situations.
    c.   Select important problems from the many current issues and make these a focus of actions to be taken during the next term.

30. The CEO should make visits around a company to find out if:
    a.   The implementation of policies is producing substantial effects.
    b.   All the managers and section heads accept the policies without question.
    c.   Departments and sections are implementing activities based solely on their own policies.

31. To manage the implementation of their policies CEOs:
    a.   Must have a detailed knowledge of all that is going on in their company.
    b.   Must have a sound grasp of the mechanisms of policy management.
    c.   Must display leadership but do not need to know about operations of the company.

32. The items to be evaluated by the CEO come in two broad categories:
    a.   Evaluation of policy and cross-functional management.
    b.   Evaluation of production and policy management.
    c.   Evaluation of production and cross-functional management.

33. Those who receive the CEO evaluation should:
    a.   Explain their policies, their implementation methods, and the effects of these.
    b.   Use QC stories to explain why they do not have concrete data.
    c.   Present their regular business reports without comment on policy implementation.

34. In his or her evaluation the CEO should:
    a.   Confirm whether or not company policies are being deployed and implemented properly.
    b.   Ignore how work methods have been changed to achieve better targets.
    c.   Keep managers but not general employees fully informed of the TQM approach.

35. The CEO should approach the evaluation with the view that:
    a.   If people are not carefully controlled they will do as little work as possible.
    b.   Improper practices when diagnosed in the workplace should be highlighted.
    c.   Evaluation is useless in most cases.

36. After the evaluation by the CEO, follow-up activities should be carried out to ensure that the measures suggested are:
    a.   Carried out immediately without examination or discussion.
    b.   Examined carefully but not necessarily implemented.
    c.   Implemented systematically.

37. Documents written up by the Technical Experts describing the problems and the measures to improve them are sent to the involved departments where they are treated as:
    a.   Instructions.
    b.   Recommendations.
    c.   Requests.

38. Departments and sections receiving the Technical Expert documents:
    a.   Report their improvement action plans to the Quality Assurance department.
    b.   Submit interim reports with items not fully improved within six months.
    c.   Write completion reports once items are fully improved, therefore interim reports are not necessary.

39. The CEO follows up on the status of items pointed out in the Technical Expert report:
    a.   At regular intervals.
    b.   Twice yearly.
    c.   In the next evaluation.

40. At the beginning of each year the CEO identifies:
    a.   Policies which failed and removes them from the list of priorities.
    b.   Policies that will be given priority status.
    c.   All the minor policies from previous years that will be elevated in priority.

41. When writing the SOP on establishing annual management policies:
    a.   Assign one individual to carry out all the steps.
    b.   Evaluate who needs to be involved at each step.
    c.   Assign the task of identifying individual involvement to a later time.

42. In the Evaluation and Implementation Web, management policy _____ Quality Operations capability.
    a)   Manages.
    b)   Controls.
    c)   Defines.

43. Evaluation items are:
    a.   A measurement system to make evaluations during the Do phase of the PDCA cycle.
    b.   A judgment on whether activities have been successful in a non-appraisal manner.
    c.   Specific items to be evaluated during the Check phase of the PDCA cycle.

44. When implementing the Plan in the PDCA cycle, do so:
    a.   Casually.
    b.   Partially.
    c.   Completely.

45. In the Evaluation and Implementation Web, the Quality Operations capability:
    a.   Is defined by management policy.
    b.   Is about performance within the laboratory.
    c.   Maintains the life-cycle management capability.

46. When breaking down a process into component processes to deal with abnormalities:
    a.   Ignore any abnormalities unrelated to the current problem.
    b.   Do not take any action until the analysis is fully completed.
    c.   Confirm the effects of any actions taken.

47. Technical experts write up:
    a.   Responses to the minutes from the meetings.
    b.   The problems and measures to improve them.
    c.   The problems which it then sends to the departments to respond with appropriate measures.

48. Core measures eliminate:
    a.   The root cause.
    b.   The outward effects.
    c.   The secondary causes.

49. The business philosophy includes:
    a.   What is important to the CEO.
    b.   The goal is to make money.
    c.   What are a company's values.

50. Mid-term plans are formulated:
    a.   Every six months.
    b.   Every year.
    c.   Usually every three years.

Chapter 2. Chief Executive Officer

Ensuring quality

2.1 Establishing the company's quality policy

The success of a company depends on delivering therapeutic products and services of the highest level of quality. One of the primary functions of the CEO is to establish the quality policy, to involve the employees in realising it, and to ensure that any related corporations, clients, etc. are also aware of it.

a. The CEO should present the annual quality policy to the company employees, its suppliers and to the customers/patients.
    i. The CEO should prepare a clear statement of the basic quality policy that governs the company.
    ii. The CEO should hold meetings for questions and answers in order to explain the policy to the employees and to those affiliated service providers, making certain that everyone understands it.
    iii. Employees in marketing and client services must understand the quality policy, since they are the face of the company.
    iv. Employees should carry cards with the basic points of the quality policy for quick and easy reference.
    v. The CEO should promote Quality Operations training sessions and other study meetings to teach concrete methods for deploying the policy.
    vi. The CEO should communicate the quality policy clearly to customers/patients in order to gain their confidence to use your treatment services.
    vii. The CEO should seek to create the kind of workplace where everyone is motivated to put the quality policy into practice.
b. The CEO should raise the quality consciousness among the employees.
    i. Quality is a way of life. Unless it is expressed in a manner easily grasped by employees such as slogans, mottos, etc. they will not take it to heart.
    ii. The CEO should encourage employees to think up their own mottos for quality improvement, and give recognition for those that are suitable for display and becoming part of the quality slogans.
    iii. The CEO should host frequent meetings in each facility and laboratory to announce the most recent achievements in quality improvement and recognition.
    iv. CEOs must actively participate in these meetings, and thank employees personally. But most importantly, the CEO must be seen to be sincere and stay in the meeting for its duration.
    v. The CEO should combine the improvement proposal system with other quality

improvement activities in the workplace. This can be further enhanced by having a reward system for outstanding contributions.

vi. The CEO should encourage employees to participate in national events as a way of increasing both their quality consciousness and providing them with recognition on a much larger scale.

c. The CEO should involve employees in making quality improvements: encourage regular Quality Operations Meetings and training sessions.

i. Quality Operations (Q-Ops or QO) held-meetings are an effective way not only for employees to raise their quality consciousness, but also to participate in putting it into actual practice.

ii. Quality Operations training sessions should occur regularly to try to improve the quality of general work. Give time at these meetings for employees to identify problems in their workplace, with product quality and support them in finding a solution.

iii. The CEO should introduce quality control concepts and techniques at QO meetings and use them to find solutions. QO meetings should be a cornerstone of quality improvements.

iv. The CEO must be aware of current quality activities and have the achievements publicly displayed with a reference to the impact or savings to the company.

v. The CEO should encourage employees to put forward suggestions for improvement in their areas covering workplace safety, and quality and process improvement. There are no poor suggestions, all are appreciated and adopt the good ones that are provided.

To introduce a suggestion scheme:

a. Establish the suggestion scheme as an officially approved company system.
b. Appoint a company-wide officer to promote the suggestion scheme.
c. Assess suggestions at least once a month, award any prizes, and implement the good suggestions.

2.2 Defining the organisational structure

The CEO is responsible for the overall organisation of the company. This involves several key actions:

a. Draw up a company-wide organisation chart. Outline the organisational hierarchy of the company from senior managers to the lowest employees in each department. Quality Assurance (QA) department requires a copy of this for the Quality Manual.
b. A company-wide organisation chart should show the functions and responsibilities (by position, not name of an individual) of everyone from senior managers to the last employee in each department, and the reporting routes that connect them.

To prepare an organisation chart:

a. The CEO instructs the Human Resources (HR) department to draft an organisation chart. HR does so, obtains approval of the draft chart from the board of directors, and makes the chart available to all the different departments.

b. The departments give HR their feedback. HR then revises the chart, gets it approved by the CEO, and issues it with the CEO's approval stamp and the date of issue.

    i. The chart shows clearly the titles of the positions in each department.

    ii. Since people come and go in any company, names should be on another semi-official organisation chart.

    iii. Any organisational changes affecting the chart should be communicated quickly and accurately to all employees, and a record kept of these changes.

c. Remember that the QA department assures quality to customers/patients on behalf of the regulatory authorities. This department must be independent of the manufacturing, engineering and research departments and is therefore supervised directly by the CEO on the chart.

d. The format of the chart should be one that anyone can easily understand, so that it can be presented to stockholders, auditors, and customers/patients, and used as public relations materials.

e. Define clearly the tasks to be carried out by each department.

    i. The tasks of every department in a company must be clearly defined: task criteria must be established that specify the scope of the task and the procedures to be followed.

    ii. Common tasks should be identified and shown how they cross over between departments.

To draft task criteria:

a. Management clarifies in writing the basic policies which will determine the task criteria for each department.

b. Management then details in writing the specific tasks that relate to the criteria they have established for each department.

c. Each department checks that its specified tasks match the actual tasks performed in its workplace. Job descriptions must then be written to correlate exactly to the specified tasks.

d. If the department is not performing the tasks outlined, or is performing other tasks which are not specified, then it should submit a change control report requesting a correction to both the task descriptions and the departmental responsibilities to support what is actually done.

e. The scope of the tasks should be specified in writing in a manner that avoids any confusion.

f. Any work-related matters that are needed to implement the tasks must be written down as work documents, instructions, SOPs, and/or management procedures.

g. The procedures for carrying out current checks of position details, and periodic reviews and revision of task criteria should be determined.

h. The task criteria should be simple and easy for employees to understand.

When deficiencies are found in the task criteria they should be revised and updated:

a. The personnel concerned consult with each other and submit a change control form. Once approved, quickly carry out revisions. The revisions are then signed off by the related departments and distributed.
b. Department managers quickly communicate the revisions to subordinates to ensure that they know what exactly their tasks are.
c. Establish rules for the periodic review and, when necessary, revise task criteria; always keep written records of changes.
d. Define clearly the responsibility and authority of each employee.
e. Check not only for errors but also for omissions of tasks.

Each employee must know exactly what authority and responsibility they have. Management should draw up rules which will clearly define these for each job in every department.

By definition, authority is where someone has been officially given the right to do something, to make a decision, or to give instructions. Whereas, responsibility is defined as the given duty of carrying out a task or seeing that it is carried out. It is important to understand the differences of these two terms when examining the quality of roles.

Authority and responsibility often go together but not always.

a. The rules for job authority should be reviewed periodically.
b. The rules for job authority document the scope of responsibility and authority for each job for every employee in every department of the company.
c. Authority describes how the employees doing each job should be managed, and prepare plans and define responsibilities in case of an emergency.
d. Management should check from time to time whether the rules for job authority are being enforced correctly. The results of this investigation should be distributed to the personnel concerned so it can be corrected if necessary.

Note in particular that:

a. The authority for control of money should specify the maximum amount that the authority covers.
b. The authority for imposing penalties should be minimised in order to maintain and foster good working relationships, which would be jeopardised by someone being seen as the "bad guy".

How quality depends on the CEO

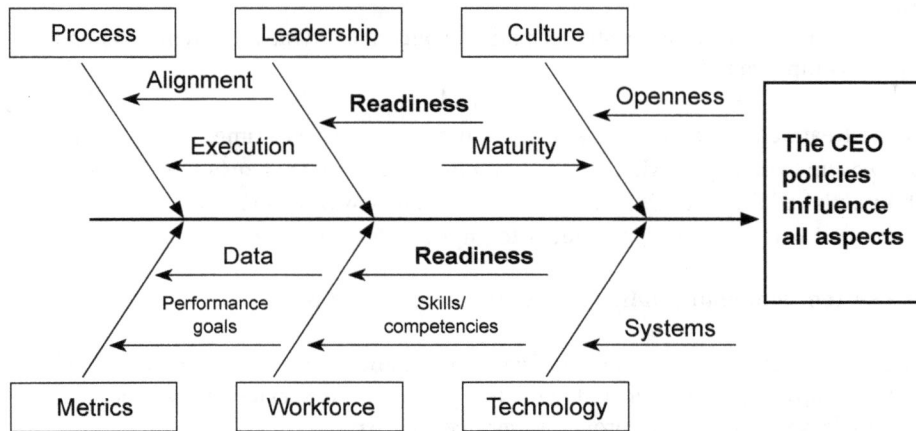

Success in adopting "best-practices" requires the CEO sets the culture, policies and procedures of the company.

2.3 Critical issues

Identify the critical issues facing the company, and put someone in charge of dealing with each of these. These issues will arise particularly in the planning, design and development of any new therapies, and in the production, inspection, marketing, and servicing of existing cell therapies.

    a. Set objectives for such areas as customer/patient satisfaction and service systems.
    b. Set objectives for reducing the percentage of ineffective treatments.
    c. Set a numerical target for reducing customer/patient complaints.
    d. Set a numerical target for increasing the efficiency of after-treatment service.
    e. Set a numerical target for improving quality in manufacturing and process inspection.
    f. Demonstrate clearly how inspection quality in the laboratory affects quality in marketing the products to the customers/patients.
    g. If necessary, improve the inspection methods and change the check items.

2.4 Meeting the customer's/patient's requirements

A company's quality policy will only be meaningful if it satisfies its customers/patients. There are three primary actions that must be taken in order to achieve this:

    a. Ensure that personnel at the company really understand what the patient wants. Normally this means the patient wants a safe and efficacious product that provides a significant degree of resolution to their illness.

b. Set up clear procedures for handling customer/patient needs. How clients are treated upon their arrival for therapy is just as important as the therapy itself. Ensure that there are Standard Operating Procedures in place which state exactly how this is performed.

c. Set up a system for analysing customer/patient concerns, any adverse reactions and subsequent complaints.

Therapies and treatments evolve, change and become replaced over time. It is essential that a company keeps up to date with what are the latest therapies and what customers/patients want. Find out how satisfied they are with their treatments, what new products should be looked at for development, and where the company needs to improve its after-treatment service.

Establish a system that continually updates the latest client needs:

a. Periodically survey how satisfied the customers/patients are with not only the treatment/therapy they received but with the personalised services they received. Evaluate their responses and improve the services in response.

b. Use statistical techniques to gather and analyse the post treatment information.

c. Calculate the percentage of income received by various categories, ie. type of treatment, disease treated, gender of patients, etc. Record this data every year and identify the trends. This will help with company decision-making as to which therapies should be focused upon and where trends in treatment categories suggest a company should expand or progress.

Complaints are inevitable. What is important is that a company handles them as quickly, responsibly, accurately and reliably as possible. Prepare an SOP that clearly describes the in-house procedures for handling treatment-related complaints. The CEO and senior management should periodically check what the customers'/patients' complaints are about, and the progress of any countermeasures taken to ensure that such claims will not arise again. Always be alert for the possibility of latent dissatisfaction as the result of adverse reactions, occurring long after the therapy. The body must be considered to be an biological incubator and therefore unnoticed consequences may not manifest until much later.

The continuous analysis of customer/patient claims will help a company to identify the causes of defects in the system or of the actual therapeutic products. This will allow improvement in the quality of the products, and, quite often a reduction in costs. The analysis also spots any long-term trends in market conditions and thereby provides an edge over competitors.

a. Normally it will be the QA department to handle complaints. Whichever department is in charge of handling complaints, it should use statistical techniques to compare and analyse data semi-annually, and should explain its long-term strategy to related departments.

b. The analysis of this data should be used at meetings to decide the effectiveness of treatments and make any decisions concerning product development.

c. After improvements have been introduced, confirm that the new results are acceptable, and record them. Provide any outcomes for review.

2.5 Technician and employee development

The environment in which a company does business is constantly changing. To ensure that the employees have the skills to respond flexibly to these changes, and to maintain the highest levels of quality in their work, it is necessary to have a good employee development programme. This should include management training, intrinsic skills training, and education in quality control. It is also necessary to encourage and support quality control education in all affiliated contract service provider companies.

To set up a company-wide education and training programme:

a. Designate who is in charge of the company-wide training programmes. This will usually be the HR department. The QA department is sometimes designated to look after quality control training.

b. Develop an SOP that thoroughly describes the training programme and how employees and technicians are evaluated on the successful completion of their training.

c. Draw up an annual training schedule for all employees, in consultation with different departments.

d. List the training requirements for each job and specify and control the requirements for each employee.

e. Examine the functions of the department in charge of employee development and training. Ensure that they coordinate programmes in such a way that department superiors can allocate time for the training of their employees.

f. It is expensive to use outside training specialists, or to send employees on external seminars, so a company-wide budget is usually allocated for this purpose and a company is selective as to who is sent off site.

g. Plan over the long-term to have in-house instructors teach both management skills and intrinsic technology skills, at least in the beginners' courses. They should also teach QC techniques and general quality operational knowledge.

h. National qualifications or industrial association qualifications are necessary for some jobs. More than one person should be qualified in each department as a backup. Therefore, both internal and external seminars for such qualifications should be included in the programme.

i. For training in intrinsic technologies, establish an in-house qualification system where there are four ranks: beginner, middle, upper and senior. Such a qualification should be given official recognition within a company.

   i. A beginner is new to the task, has read what is required but cannot perform the task.

   ii. Middle trainees can perform the task under direct supervision.

   iii. Upper trainees can perform the task without any direct supervision.

   iv. Seniors are able to perform the task, have a good knowledge base and therefore can train other employees in the performance of the procedure.

j. Establish a system where those who are trained and qualified to teach others are designated as Seniors and therefore leaders in the in-house qualification system.

Quality control education:

a. Employee development should also include quality control education. Everyone from top management to general employees needs to have a basic understanding of the fundamental quality control concepts and to be able to apply these in their work if required.

b. Provide a separate programme at each organisational level to meet the needs of that particular level. Those levels closer to the actual performance of the tasks may require a better understanding of the quality issues in order to perform their duties properly. Three to five levels in a company are optimal, derived from these groupings:
   i. CEO and Managing Directors.
   ii. Division managers, laboratory managers, marketing officers.
   iii. Department managers and section supervisors.
   iv. Technicians and technologists.
   v. General employees.

Quality training programme:

a. The programme teaches senior managers the correct understanding of quality control. They need to be able to play a positive leadership role in quality control activities.

b. The programme teaches middle managers the correct guidance procedure for quality activities and statistical techniques so that they can provide basic education in these.

c. The programme teaches technologists and technicians how to use quality techniques so that they can solve problems in their workplace.

d. The programme teaches general employees a basic understanding for quality that they can apply to their duties.

e. The programme trains in-house auditors to conduct internal quality audits.

f. The programme keeps records regarding the overall implementation of the programmes, and the educational results for each employee and each group.

Educate and support your contract service providers (CSPs) and vendors in quality control:

The final quality of the products will be determined by the quality of the input received from subcontractors and vendors for any materials and any cooperation received from them. It is necessary to ensure that their level of quality control is comparable to the contracting company. The company should provide their partner companies with continuous guidance and education in quality control.

a. Check their quality control system and insist that they carry out whatever improvements are required in order to be part of the contracting company's overall system.

b. When problems occur, instruct them to immediately inform the contracting company and to draw up concrete improvement plans.

c. Nominate someone to assist from the company in solving their problems and to monitor that any improvements they make are fully implemented.

d. Concerning vendors, audit prior to establishing a relationship and doing so only if they pass the audit. Re-auditing should be done every two years.

e. Send vendors a monthly written report on the quality of their products after receipt by the company. This report should give the quantity of acceptance lots, the quantity of rejected lots, the total quantity inspected, and the nature of the defects. If there are any significant problems, attach a letter requesting improvement, and instruct them to submit their improvement plans to the company's purchasing department.

f. Periodically, examine their quality control system and make any suitable recommendations.

2.6 Carry out routine and formal quality audits

If a company is to build up a good quality system, it is necessary to carry out regular quality audits to determine:

a. The quality activities are being carried out as planned.
b. The departments are achieving the planned results.
c. Activities are being carried out as effectively as possible.
d. Plans are suited to achieving their objectives.

To get full benefits from the quality audits, the CEO should hold regular meetings to discuss reports concerning the following items, and to outline improvements:

a. The establishment of annual quality targets.
b. The extent to which quality targets have been achieved.
c. The results from the internal quality audit.
d. The results of the assessment of the quality management status of subcontractors andother partner companies.
e. Out-of-control situations in processes and how any installation of any recurrence prevention measures is progressing.
f. Non-conforming products and how the installation of any recurrence prevention measures is progressing.
g. Reports on process changes.
h. Revisions of quality manuals, regulations, and operating procedures.
i. Any other quality assurance activities that concerns them.

The Quality Audit is a systematic, independent examination to determine whether quality activities and related results comply with the plans. It is also to determine whether these plans are being implemented effectively and whether they are suitable to achieve their objectives.

2.7 Conducting periodic internal quality audits

To have a good understanding of the status of a company-wide quality control system, you need to carry out periodic internal Quality Assurance Audits of each department.

QA visits each department and checks:

a. Whether the department understands the basic concepts of quality.
b. Whether it has implemented quality programmes in its daily tasks.
c. What has been achieved by each department in terms of quality.

The regulatory authorities prescribe that internal quality audits be conducted as follows:

a. Companies must carry out audits to assess the effectiveness of their quality system, and confirm that quality activities are carried out according to plan. Audits are normally performed by a team under the control of Quality Assurance. The results of the audit should be recorded.

b. Any issues, deviations, problems found on the audit are recorded as non-conformances, or out of specifications, on the appropriate forms and a Corrective Action/Preventive Action plan (CAPA) is generated by Quality Assurance.

c. Those responsible for the areas that have been audited must then take appropriate actions according to the CAPA. These actions should then be verified in follow-up audits and if completed successfully then the CAPA is closed.

d. Auditors must be individuals with no direct responsibilities for activities in the areas that they are auditing and therefore tend to always be Quality Assurance personnel.

e. Internal quality audits should be conducted by personnel who are familiar with the characteristics of the products and the realities of managing them particular to the work they are auditing. This does not mean that the QA auditor has to be an expert in the area they are auditing but must have an understanding of the general principles.

f. Audit reports are presented to senior management, primarily the CEO who then sees that the resources are made available to correct any insufficiencies identified in the audit.

g. Once the CEO has reviewed the internal audit then the system should be upgraded accordingly and improvements in quality should follow and then are signed off by Quality Assurance when completed.

h. At least once per year, Quality Assurance should provide a review of all audits and corrective actions undertaken to the CEO and senior management and this report is signed off.

i. The internal audit system is the key to improving the efficiency of the Quality Assurance system.

Internal quality audits, based on objective evidence and activating the PDCA (plan, do, check, act) cycle of the quality system, should confirm the following points:

a. The activities in the quality system are implemented according to plan.

b. That the quality system is functioning effectively.

c. All points raised in previous audits have been dealt with, and the effectiveness and efficiency of the system has been improved.

The Quality Audit System

Conducting the internal quality audit:

a. The audit team secretary (typically also from the QA department) sends advance notice to each department that is to be examined, and informs them of the items to be examined for that particular audit.

b. Each department then conducts a prior self-diagnosis for each of these items, and evaluates and documents this, and prepares the criteria, specifications, and process charts and reports that they will make readily available to assist the auditor/audit team.

c. When the auditor/team arrives, they first examines the department's documents. The person in charge at each workplace explains their self-examination on the basis of the documents. The auditor/team asks questions about the content of documents, and confirms report results.

d. The auditor/team conducts an on-site examination of the workplace, following the order as presented in the process chart.

e. The auditor examines and confirms whether or not quality control and quality assurance are being implemented as set out in the documents.

f. The auditor/team evaluates and scores the inspection results.

g. If any items are found to fall below a certain level, the auditor/team makes written comments and gives concrete reasons for the decisions. Corrective action requests are generated.

h. A copy of the audit report is filed with senior management for their signature and to make available any resources needed in order to undertake the corrective actions.

i. The auditor/team instructs the department to submit concrete proposals (usually within 30 days) and undertakings for resolving the non-conformances which it then reviews and if suitable, approves.

j. The auditor/team decides on a timeframe for the corrective actions to be completed.

k. The QA auditors monitor the corrective actions to see that they are being conducted as per the proposal and according to the timeframe. If there are any discrepancies, then these are written up as supplementary deviations that will require further corrective actions.

In the report the internal auditor/team should also make note of any positive or good points in regards to the inspection of the department.

2.8 Preparation of quality manuals

Quality manuals are the basic documents that a company will use to implement company-wide quality control. They contain the quality control rules, inspection rules, technical standards, product standards, operation standards, and other rules and standards for the assurance, maintenance and management of quality.

Company standards set out the correct procedures for the purchase, manufacture, inspection, control, and use of raw materials, reagents, equipment and products. It is essential to set up a high-level, internationally recognised, company standardisation system, managed by a quality assurance unit, to ensure that standards are established correctly for all the tasks performed in a company, and that standardisation becomes firmly rooted in a company psyche of the technicians and employees. The quality manuals must be in accordance with the GTP requirements.

To do so:

a. Establish rules that specify the procedures for writing, reviewing, approving, distributing, and implementing original drafts, and for documenting these in a way that is easy to access.

b. The first SOP written should be the SOP on how to write an SOP. This should be specified within the company Quality Manual so that it is evident that a company has an immediate understanding of the requirements for establishing a quality management system.

c. Document jobs and tasks in the order of the actual work flow and summarise all the relevant issues. Have these drafts reviewed by a review committee.

d. Be sure to harmonise department specific rules with rules of a higher level, such as regulatory requirements/government laws and company rules.

e. Establish rules for the periodic review of standards, and revise and archive those that are no longer appropriate.

f. Keep rules, working documents and SOPs as simple as possible, and the number of them as low as possible.

Procedures for preparing quality control manuals:

a. Each department provides documentation describing all its tasks, and the circumstances in which these tasks are carried out.

b. The department in charge of the quality system summarises the documented quality systems of each department into a description of a company-wide quality system.

c. A committee with members from different parts of a company prepares the first draft of the manual. This should be a practical document that can actually be used rather than an ideal one which sits on a shelf merely for show purposes.

d. A first draft is presented to the departments, who examine it to see if its implementation could result in problems. Departments then submit concrete proposals for improving the draft. This will ensure that the departments cooperate in achieving company-wide quality control.

e. The Quality Manual should specify:

   i. The interfaces of all the departments in a company (organisation chart).

   ii. The rules for all of those departments that are directly related to quality assurance. These include processing, quality control, etc.

   iii. The rules for departments sometimes not recognised as being directly related to quality such as HR, Finance, Administration, etc.

   iv. Interdepartmental rules and standards should be harmonised. In particular, values in design and research in cell therapy standards should be harmonised with values in cell processing and manufacturing and inspection standards.

   v. A company-wide Manual Review Committee should check whether standards and rules are harmonised between departments. It will coordinate between departments and ensure that nothing is overlooked.

   vi. It is important to create a simple quality manual which suits the actual situation and the range of tasks and activities performed.

2.9 Establish quality systems in non-production departments

Quality is not just for the laboratory departments. In order to have real company-wide quality systems it must also be introduced in other non-production departments, e.g. Marketing, Administration, and Research and Development.

Work in these departments may be thought of as intangible, blue-sky operations but in reality they should be continuously improved and refined just as much as the quality of products.

Establish numerical targets wherever possible, set an evaluation scale for improvement, and draw up an improvement schedule.

To establish quality control in non-manufacturing departments:

a. Get a good understanding of the actual work conditions. As far as possible, describe these numerically in terms of work procedures and the time required to complete specific tasks.

b. Since the work in non-manufacturing departments is often done individually, it can be difficult to solve problems or to set targets using a teamwork strategy. It is much easier to set common themes such as improvement of office work handling, or of methods of communication.

c. Work in the research and development department is typically new and non-repetitive. Implement QFD (quality function deployment) and quality research techniques in this department. See Chapter 20 for further details. (QFD: Analysis of the customer's/patient's requirements of quality, performance and function and the incorporation of these requirements into design and production in order to best satisfy the customer/patient.)

d. Evaluate the efficiency of office work by using a scale that can objectively measure such achievements as the standardisation of office work, or the reduction of office expenses.

2.10 Develop relationships with universities and research organisations

A company in its efforts to develop new products or materials, normally cooperates with outside research organisations and universities in joint research and development. Though these units are outside the governance of the company directly, there is still the expectation that they will work within a quality framework. In order to do so, it will be necessary for the company to audit the contract service providers and be satisfied that they perform in a manner similar to the company and maintain a quality management system that is in accordance with the rules and regulations of the cell therapy community.

Recommended reading

1. Bradley, E.H., Holmboe, E.S., Mattera, J.A., et al. The roles of senior management in quality improvement efforts: What are the key components? *J Healthc Manag*, Jan.-Feb. 2003; 48(1):15-28. [PubMed]

2. Øvretveit, J. *Leading Improvement Effectively: Review of Research*. London: The Health Foundation, 2009.

3. Parker, V.A., Wubbenhorst, W.H., Young, G.J., et al. Implementing quality improvement in hospitals: The role of leadership and culture. *Am J Med Qual*, Jan.-Feb. 1999; 14(1):64-69. [PubMed]

Self testing multiple choice questions

1. To present its quality policy, management should:
   a. Prepare a vague statement of its basic quality policy so staff can fill in the details.
   b. Give all employees cards with the basic points of its quality policy.
   c. Hold meetings where management tells employees how things are to be done, without question.

2. Management is responsible to increase the quality consciousness of:
   a. Their employees, service providers, and affiliated partners.
   b. Their suppliers, the public media, and shareholders.
   c. Their partners, shareholders, and employees.

3. To increase the quality consciousness of employees, managers should host frequent meetings at each worksite to:
   a. Announce company awards.
   b. Announce achievements in quality improvement.
   c. Announce national awards.

4. To actively encourage Quality Operation's activities, managers should:
   a. Contribute articles to national journals on quality control.
   b. Take up whatever suggestions are made whether suitable or not.
   c. Attend the beginning of each meeting and then leave after introductions.

5. To have an effective suggestion scheme:
   a. Treat the scheme as an unofficial company system to encourage more employees to make suggestions.
   b. Establish secretariats in each department to promote the scheme.
   c. Assess suggestions annually, and award prizes.

6. A company-wide organisation chart should show the functions and responsibilities of everyone from:
   a. Senior managers to junior managers.
   b. Senior managers to supervisors.
   c. Senior managers to the lowest employees.

7. The organisation chart should be drafted by:
   a. Senior managers.
   b. The HR department.
   c. The QA department.

8. Task criteria should be revised and updated:
   a. Annually.
   b. Twice yearly.
   c. Only when deficiencies are found in them.

9. The highest level at which a company-wide rules of job authority should be approved before being enforced is:
   a.   Managers.
   b.   Senior managers.
   c.   Board of directors.

10. Department-specific rules:
   a.   Should always be harmonised with company-wide rules.
   b.   Need not be harmonised with company-wide rules.
   c.   Should be harmonised with company-wide rules whenever possible.

11. Important issues referred to in the text on quality systems include those in:
   a.   Planning, design and development of new products.
   b.   The storage of raw materials.
   c.   Non-production departments.

12. To gain a really good understanding of marketing/patient recruiting issues:
   a.   Compare the results of the marketing department with that of the marketing departments in competing companies.
   b.   Compare the results of individual marketers with their counterparts in competing companies.
   c.   Compare the achievement of the whole company with that of competing companies.

13. When appointing someone to be in charge of an issue, establish and document procedures for handling:
   a.   Only abnormal situations that fall inside the designated scope of authority.
   b.   Any abnormal situations that fall inside/outside the designated scope of authority.
   c.   Only normal situations.

14. To get a good understanding of customer/patient needs:
   a.   Use information gained in the issues of adverse alerts by the regulatory authorities.
   b.   Use records of after-treatment service.
   c.   Personally try out newly developed therapeutic products in the market.

15. How to use customer/patient complaints information to improve products?
   a.   Use statistical techniques to compare and analyse customer/patient claims data for each six-month period.
   b.   Use data from the analysis of customer/patient claims at meetings on new products.
   c.   Establish a system to explain a company's long-term strategy to customers/patients to reaffirm that their complaints are not considered significant.

16. To have a customer/patient related focus in your quality policy set numerical targets for:
   a.   Improving contacts so patients know where to call to report a problem.
   b.   Reducing customer/patient complaints.

   c.   Showing how inspection quality in the laboratory affects perceived quality in the market.

17. The procedures for setting up a development programme include:
   a.   List the training requirements for each job.
   b.   Have all the personnel in a department attain national association qualifications.
   c.   Specify the training requirements for each individual employee.

18. Which of the following should be regarded as at the same levels for quality control education?
   a.   Section supervisors and departmental managers.
   b.   Managing directors and technologists.
   c.   Departmental managers and directors of marketing offices.

19. General employees should be taught the correct way of using QC techniques through:
   a.   Internal lectures.
   b.   On-the-job training.
   c.   External seminars.

20. Keep records of the educational results of:
   a.   Each employee.
   b.   Each group of employees.
   c.   Each member of management only.

21. To support quality control in your vendors send them a monthly written report on the quality of their products at acceptance showing:
   a.   The quantity of rejected lots only.
   b.   The quantity of acceptance lots only.
   c.   The quality improvements needed with acceptance and rejected numbers.

22. The CEO should hold regular meetings to discuss:
   a.   The establishment of annual quality targets.
   b.   Any changes resulting from internal quality audits that have a company-wide effect.
   c.   Revisions of quality manuals, regulations and rules.

23. Internal quality audits are aimed at confirming the following points:
   a.   Are activities in the quality system implemented according to plan?
   b.   Are only new issues but none of the points raised in the previous audit?
   c.   Are there any abnormalities occurring?

24. In the top management's internal quality control diagnosis:
   a.   The team arrives in each department without any warning to see what is really happening.
   b.   The department conducts its own diagnosis before the team arrives.

c. The department sends its own diagnosis to the team some days before they arrive and the auditors come to only confirm what they have written.

25. If possible the examination items used each year should be:
   a. The same.
   b. Different.
   c. Half the same and half different.

26. The quality control manuals should be first drafted:
   a. By each department.
   b. By the QA department.
   c. By a committee drawn from different parts of a company.

27. The manual should specify:
   a. The interfaces of all the departments in a company.
   b. The rules for production departments only.
   c. The rules for departments which are usually not recognised as being directly related to quality only.

28. To deploy quality activities in non-manufacturing departments:
   a. Set targets using a teamwork strategy.
   b. Set common themes.
   c. Focus on the improvement of individual employees.

29. To carry out work with outside organisations, it is imperative that a contract service provider (CSP):
   a. Has its own Quality Assurance so that a company does not have to audit.
   b. Establishes in advance how expenses should be shared.
   c. Shares a quality management system similar to that of the company.

30. To strengthen a company's international competitiveness:
   a. Consider safety issues of products from the perspective of the GTPs.
   b. Establish a system where patent, design and intellectual property are the priority.
   c. Assure international customers/patients that the company works to the highest domestic standards.

31. One of the common adverse reactions is:
   a. Death.
   b. Fever.
   c. Sterility.

32. Quality Operations training sessions should occur:
   a. Twice a year.
   b. Only when required.
   c. At regular intervals.

33. The organisation chart shows:
    a. The names of those employees in each position.
    b. The positions in each department but not the individual in that position.
    c. The reporting structure of the departments but no intradepartmental specifics.

34. If a department is not performing the tasks as specified in the generated tasks by management assigned to that department then the departmental manager should:
    a. Inform management only of what it does do.
    b. Submit a change control requesting corrections to the job descriptions and responsibilities.
    c. Ignore the situation as long as the department is doing the job.

35. Duplication of tasks by different roles in a company is:
    a. To be avoided.
    b. Is desirable as a backup situation.
    c. Normal company policy.

36. Authority is:
    a. The duty to carry out a task approved by management.
    b. The official right to do something, make a decision, or give instructions.
    c. Automatic approval by management.

37. Responsibility is:
    a. The duty to carry out a task approved by management.
    b. The official right to do something, make a decision, or give instructions.
    c. Automatic approval by management.

38. The CEO's policies influence:
    a. The workforce but not the leadership.
    b. The technology but not the culture.
    c. The processes and the leadership.

39. To analyse post treatment information, the company should use:
    a. A questionnaire.
    b. Statistics.
    c. Intuitiveness.

40. The department that normally handles customer/patient complaints is:
    a. Quality Control.
    b. Marketing and Client Services.
    c. Quality Assurance.

41. Implementation of a training programme requires:
    a. Hiring of external experts only.
    b. Development of an SOP to govern the programme.
    c. On-the-job training (OJT) only.

42. The top training level a company should implement is the ability to:
    a. Perform a task competently and be able to train others.
    b. Perform a task under supervision.
    c. Perform a task without supervision.

43. The quality training programme teaches general employees how to:
    a. Use quality techniques to solve problems.
    b. Self-audit their performance.
    c. Apply a basic understanding of quality to their duties.

44. When problems occur at a CSP technicians should:
    a. Continue working and see if it happens again.
    b. Immediately contact the company and inform it of the event.
    c. Solve it themselves and thereafter inform the company of the solution.

45. After the initial vendor audit, the vendor should be audited:
    a. Never as they have been approved.
    b. Every five years.
    c. Every two years.

46. The QA department should report directly to:
    a. Production.
    b. Management.
    c. The CEO.

47. In the global scheme, QA department actually works for:
    a. The company.
    b. The regulatory authorities.
    c. Themselves as they are completely independent.

48. Periodic audits by QA department are a good way to diagnose:
    a. Problems before they happen.
    b. A department's understanding of quality control concepts.
    c. If correct training has been implemented.

49. The CAPA is:
    a. Corrective Action and Preventive Action plan.
    b. Correct Application Planning Actions.
    c. Correcting Abnormalities of Production Activities.

50. Initial quality audits should be conducted by:
    a. Personnel from within the department being audited.
    b. Personnel with minimal knowledge of the departmental activities.
    c. Personnel familiar with the activities but not connected to the department.

Chapter 3. Laboratory Managers

Managing
systems

All that a Laboratory Manager does will have a direct impact on quality, but several managerial functions are especially important in ensuring a high level of quality departmentally and in the company as a whole. The functions included in this chapter have to do with establishing, implementing and monitoring work systems, while those in Chapter 4 present ways of supporting the contribution a company's technicians can make.

3.1 Implement the company's annual management policy

If the company is to successfully improve quality, and reduce costs, it must formulate and implement mid-term to long-term plans as discussed in the previous chapter, and develop annual management policies based on these plans. This is the responsibility of the CEO as indicated. The responsibility for the Laboratory Manager is to receive the CEO's annual management policies, examine them carefully, and implement them in the laboratory as departmental policies. To do so the Laboratory Manager requires the involvement and commitment of the technicians and technologists.

The Laboratory Manager should:

a. Discuss the implementation of the annual policies with all laboratory employees.
b. Ensure that they have a clear grasp of the objectives and methods needed to implement the departmental policies.
c. Work closely with the laboratory employees to solve any problems that arise in implementing the policies.
d. See Chapter 1 for more detailed guidelines on planning and implementing annual policies.

3.2 Clarify the extent of QA activities and clarify policies

The journey from the receipt of cells in the laboratory to the final delivery of the infusion product into the patient often goes through a lot of different steps requiring the support of various departments and contract service providers (CSPs). Each department is responsible for quality assurance activities at different stages along this path. It is important that the staff in the laboratory knows where his/her responsibility for quality assurance both begins and ends.

Essentially, the manager and the technicians must have a good understanding of the company's long-term and medium-term plans. When the manager receives the annual management policy from superiors, it is to be discussed fully with the laboratory employees. Together, they work out what is needed to implement this policy as departmental policies.

a. Evaluate the current and forthcoming work in the laboratories with reference to the work of the previous year.
b. Select development initiatives for the coming year that should lead to the implementation of the policies. An initiative may be a task, process or target that the manager decides to improve in some way.
c. Check carefully whether the selected initiatives fully satisfy the long-term and medium-term company plans. If they do not, then adjust them accordingly.
d. Laboratory Managers will discuss with the technicians how to implement these initiatives.
e. Review the job assignments of those who will be implementing them.
f. Plan and implement a training programme, based on the individual job skills of technicians.

Ensure that technicians and technologists have a good grasp of the specific objectives and methods needed to implement laboratory departmental policies:

a. Use quantitative values to define each objective. In other words, measure the objectively.
b. Encourage the laboratory staff to make suggestions about how the objectives could be better achieved and then summarise these suggestions.
c. Use graphs and charts to show how the achievements of the laboratory department contribute to the overall achievements of the company.
d. Devise a numerical score or index that will enable monitoring of the progress of policy implementation.

Problems often arise during implementation of the policies, so managers must work closely with technical staff to create a sense of cooperation with them:

a. Instruct laboratory staff to:
    i. Check and report periodically on the progress being made on each of the objectives.
    ii. Report back to their manager whenever there is a serious change in achieving the objectives or an unscheduled delay.
b. Do not wait for reports of problems to come through senior management. It is imperative that frontline managers go to the laboratories and check things out for themselves. Above all else, listen to what the technical staff have to say.
c. When problems are found, notify whoever is in charge. Motivate the technicians so that they aid in solving the problem. When assistance is requested from other departments, ensure that it is well coordinated between departments.
d. If a problem is considered serious, report it up the chain of command.

3.3 Define jobs, responsibilities and authority

Before a laboratory can be managed effectively, the company must clearly define the jobs, responsibilities and authority within each of its departments. When a company has done this for all the departments, then the primary task of the Laboratory Manager becomes setting the concrete policies and objectives regarding quality, quantity, responsibilities and costs for the laboratory, and to communicate these clearly to all his/her staff.

There are five primary areas where quality assurance activities are implemented in the laboratory, and where authority for their implementation must be defined through assigned responsibilities:

a. **Product characteristics** – The company must define who has the authority to decide the functions and the quality level of any cell therapy product, and who is responsible for examining/processing the results. This can be the Laboratory Manager or a designated technician within the laboratory.

b. **Process designation** – The company must define who has the authority to decide on the raw materials, reagents, facilities, equipment, processes and technical positions that determine which manufacturing process will be used in the laboratory, and who is ultimately responsible for examining/processing the results.

c. **Manufacturing** – The company must define who has the authority to manufacture the cell therapy product according to the process that has been designated; to maintain the facilities, equipment and reagents; to inspect and assess the raw materials and the in-process quality characteristics of semi-finished products; and who is responsible for examining/processing the results of all these activities.

d. **Final inspection** – The company must define who has the authority to inspect and assess the quality of each manufactured cell therapy; who has the authority to accept or reject such therapies and decide which of the rejected products can be reprocessed as well as decide which ones are to be disposed and how they will be disposed; and who is responsible for examining/processing and approving the results of all these activities.

e. **Packing and transporting the cell therapy for infusion** – The company must define which laboratory and/or technicians have the authority and responsibility for packing and transporting the finished products to the hospital CSP, how they will be transported and who will then infuse the therapy into the customer/patient when they are received by the CSP.

It must be remembered that "authority" is where someone has been officially given the right to do something, to make a decision, or to give instructions whereas "responsibility" is where someone is given the duty of carrying out a task or seeing that it is carried out properly. Authority and responsibility may often be used interchangeably but not always.

In order to define the job assignments, the Laboratory Manager should:

a. Specify all the jobs in each laboratory, including their own managerial functions such as planning, developing strategies, coordinating the functions of the laboratory, as well as promoting the company's capabilities.

b. Define the authority and responsibilities that go with each job.

c. Define the responsibilities and scope of authority of each employee.

d. Define the procedures to be followed in performing each job correctly. Ensure that only those procedures that are within protocols and SOPs are being followed. Any additional procedures will first be recorded as a deviation and this will subsequently require a change control document to be generated, in order to change the existing documents to incorporate the additional procedure(s). Should an additional procedure be incorrect and not intended, then the technician should undergo a retraining process which is documented.

e. Allocate specific functions and jobs to each unit within the laboratory.

f. Ensure that the jobs performed by the laboratory under their control are documented and that rules are established for assigning work within that laboratory. The process of assigning roles and functions should be contained within an SOP.

g. Ensure that tasks to be carried out by more than one technician or technologist will be coordinated. Instructions as to the specific responsibilities for each employee must be crystal clear.

h. Identify the quality characteristics of jobs using a numerical grading so that control charts and control graphs can be used to assess an individual's performance.

i. Decide who should draft all these aforementioned, required and suggested documents. Also, who should approve them and who will then implement them.

j. It must be kept in mind that the assignment of any task or job should only be made to an employee that is qualified for such performance, either by certification or demonstration on the job of fulfilling the requirements and that this is completely documented in his/her training log.

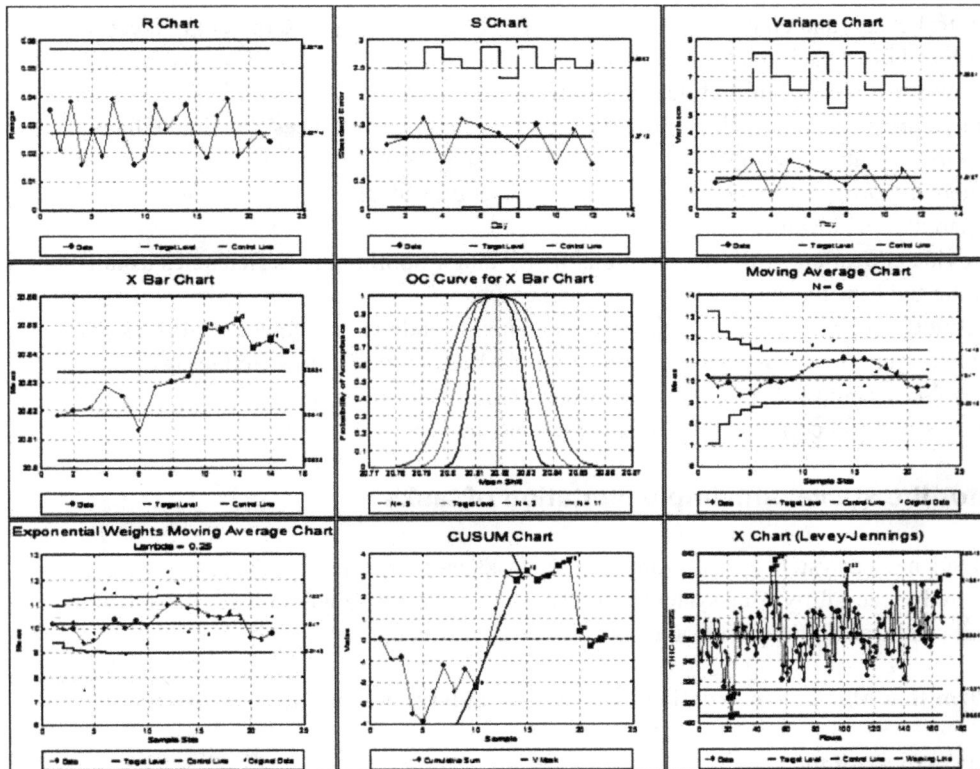

There is a wide variety of quality control charts and graphs that can be used.

3.4 Ensure that technicians understand and carry out tasks correctly

To ensure that employees understand and carry out their job functions correctly, the Laboratory Manager should:

a. Carefully adhere to strict guidance rules on job roles.
b. Communicate clearly and precisely any of senior management's instructions to technical staff.

Once a company has defined all the job functions, it is the primary task of the Laboratory Manager to clarify the laboratory's concrete policy and objectives on matters of quality, quantity, and cost. These must be communicated precisely and clearly to the employees, avoiding any ambiguity.

To do this properly it is essential that the Laboratory Manager must:

a. Know the extent of their authority with knowledge of how the laboratory interfaces with other departments, especially the next department in the chain which is most likely the hospital CSP.
b. Clarify the quality control points of the Laboratory Manager's job description.
c. Understand and control the quality characteristics of the jobs in the laboratory. Use control charts and control graphs to quantify these for visual ease of communication with technical staff.
d. Ensure that any abnormalities and defects occurring at designated control points are immediately identified. Provide a system where employees can immediately report any abnormalities and defects they discover by fostering an environment where reporting of such abnormalities is encouraged and there are not negative repercussions towards staff.
e. Provide education on those conditions where abnormalities and deficiencies usually occur and how to properly investigate the causes.
f. Actively support Quality Operations training and improvement activities by encouraging technical staff to participate in in-house training programmes.
g. Develop a technique or system by which confirms that technical staff have understood and correctly implemented the instructions.

3.5 Monitor the successful implementation of tasks

Managers need a system for checking that the tasks in their respective laboratories have been completed successfully. One system is to use check lists. These will contain those items of a task that can be easily identified and confirm the task is being performed well.

Confirm that all work instructions are easy to understand, and clearly specify:

a. The purpose of the task.
b. The quantity of work to be done.
c. The completion deadlines.

d. The procedure or methodology.

e. The materials and equipment required.

f. The specific standards to be followed (domestic/international).

g. How to report and treat any deviations or abnormalities.

h. How to properly report task completion.

Take a consultative approach:

a. Explain the objectives and improvement targets, and encourage laboratory technicians to give their opinion.

b. List the potential problems that are foreseen and discuss the preventive actions that will need to be taken if they occur.

c. Write a summary of any decisions made from the discussions, and monitor to ensure that they are carried out.

d. If procedures require revision, devise methods for doing so, and check the results of the newly implemented procedures with the technical staff.

3.6 Monitor the quality of work in the laboratory

There are a number of statistical and observational methods that can be used in order to monitor the quality of the work in the laboratory.

These include but are not limited to:

a. Preparation of charts, process capability index, etc. The use of numerical data and graphs shows trends in the production processes. It is extremely useful to look at daily trends over periods of a week or a month, or even longer in order to gain an appreciation of the actual situation.

b. Instruct employees to inspect key Critical Control Points (the specific features in a product or process that are inspected). Train them to report quickly when they notice any unusual noise, weight, colour, physical change, or vibration, of equipment or product, even if no abnormality is apparent from the numerical check data.

c. Check the work of technical staff periodically to see if they have deviated from the operation standards.

d. File a deviation report with Quality Assurance department for any operations and/or procedures that deviate from the SOPs or protocols. Following the impact assessment, if it is determined the deviation is an improvement, then modify the protocol/SOP accordingly using the change control system.

e. Ensure that employees write proper operation reports – these will include periodic reports on normal operations and emergency reports on abnormal operations. In order to do so properly it is essential that they:

   i. Understand the appropriate control points and evaluation methods, and report periodically on them. (Those points that can be examined readily to see if the process is continuous and stable.)

   ii. Report unusual events quickly on their own initiative.

iii. Realise that reporting only favourable items is the same as generating a false report and if uncovered then disciplinary action will be undertaken.

(See Chapter 7 Problem Solving, for more detailed guidelines on detecting and dealing with abnormalities.)

(See Chapter 8 Statistical Process Control, for more detailed guidelines on statistical methods.)

(See Chapter 11 Standardisation, for more detailed guidelines on using operation standards.)

3.7 Dealing with deviations

No matter how well the laboratory operates, things will always go wrong. That being the norm, then it may necessitate the generation of deviation reports. By definition, deviations are when a process or technical performance is not according to the methodology described in the SOP. A non-conformity is whenever the product itself is incomplete or does not function properly. An abnormality can involve performance, methodology, product, or any combination.

a. A non-conformity is any item that does not meet the required quality standard.
b. An abnormality in manufacturing is anything that appears in the production process which is different from what it is supposed to be, i.e. it does not meet the predetermined criteria governing the process.
c. An abnormality is not necessarily a deviation but it should always be investigated similarly.

What is critical is how the manager and laboratory staff respond to these.

There are four actions that the Laboratory Manager must take with deviations:

a. Respond quickly when a deviation is first reported.
b. Deal with the products immediately in quarantining them if necessary.
c. Initiate the investigation at QA's request to gather information about the causes of the deviation.
d. Check that the causes of the deviation are being properly analysed and acted on when discovered.

Employees should already know the lines of communication to follow when they are reporting any deviations they find, and should use short simple reports. More detailed reports can follow later according to the SOP on deviations and CAPA.

Laboratory Managers must respond quickly upon receiving a report of a deviation:

a. Ensure that Senior Managers have a clear summary of the situation and the actual therapeutic products involved.
b. Investigate how the deviation occurred.

c. Investigate collateral impacts or in other words, how far the deviation affects other processes and products in the same or other laboratories.

d. Upon impact investigation, know to what extent other departments are affected, and decide whether notices need to be sent to them.

e. Prioritise and organise the countermeasures to be undertaken once the causes are identified.

f. Upon review of the investigative reports, decide whether there is sufficient information to identify root cause or whether more information is needed.

g. Quickly send a deviation report to any related departments and to Quality Assurance department informing them of the deviation and asking them for their cooperation with concrete countermeasures. Check that they have received the report. If countermeasures are requested by QA, an instruction should be forwarded quickly to the person in charge of implementing the countermeasures if not the Laboratory Manager.

h. The deviation report to related departments should include the following:

    i. Issuing department (issue number, date of issue, issuing unit, and person issuing the report).

    ii. Sample/Patient ID, date the defect occurred and the manufacturing process in which it occurred.

    iii. Clear details of the deviation and related information.

    iv. The countermeasures that have been requested by QA.

    v. A list of all the departments to which the report is being sent.

    vi. Carry out a full investigation of the causes but remember that sometimes the first thing to do is to dispose or quarantine the rejected products.

    vii. Suspected impacts; this is not necessarily what is impacted but identifying the potential of being affected.

Rejected products as the result of a deviation have to be dealt with. There are a number of different actions that should be taken.

Depending on the circumstances and the nature of the defect, remedial actions include:

a. Put someone in charge of dealing with the affected products so that they will not be treated ineffectually.

b. Do an impact assessment on the affected product and determine whether it can be still processed, reworked, salvaged, or is to be disposed of.

c. When life threatening or safety-related deviations occur, authorise the Quality Assurance department to immediately stop the release for treatment of any related products, ensuring they do not reach the customers/patients and the CSP is fully aware of the situation.

d. If investigation of the cause reveals that there is no problem in market quality or risk to the customer/patient, then QA has the authority to allow the release of product and for treatment to continue.

e. When there are deviations that could lead to serious claims against a company, deal with the products first before fully investigating the cause. Then follow with reports on what the overall impact to the company might be. This secondary issue becomes a matter of concern for senior management and not the Laboratory Manager.

f. If the products have already been transfused into the customer/patient, assess the seriousness and urgency of the situation, consult the attending physician and the customer/patient, and take active measures to deal with the matter. Ensure that the patient is aware of all risks and provide any counteractive treatments or measure that might be beneficial.

To identify the causes of deviations, you will need to gather all the information you can. A systematic way of doing this is to ask the 4M questions – is it caused by man, machine, method, or materials.

Typical questions to ask are as follows:

a. Are the technicians/technologists following the operation standards?
b. Has any of the equipment, technicians, or materials been changed?
c. Were operators or inspectors replaced during the production process at any time? If not, was the method of operation, the inspection procedure, or the testing equipment changed?
d. Have any changes taken place in the manufacturing process due to changes in raw materials or reagents? Remember that raw materials and reagents from different suppliers may differ slightly even though it is claimed they are the same and that they conform to the same standard.

To check that the causes of the deviation are properly analysed:

a. Check the accuracy of the deviation report by going to the workplace and examining the product to see if the description in the report matches.
b. Use stratification (by deviation, by cell type, by production process, by operator, by equipment, etc.) and other statistical techniques to conduct the analysis. (Stratification is explained in Chapter 8.)
c. When the cause is in question, do reproducibility experiments to see if it is possible to reproduce the same deviation.
d. Once the cause or causes has been properly identified, plan and implement countermeasures to prevent recurrence of the deviation.
e. Check the outcome of these countermeasures. If the deviation no longer occurs and the problem appears corrected then submit a final report to close the associated CAPA.
f. If necessary, review and revise the technical and operational standards and any other standards related to them if the corrective actions require changes to the procedures. (Remember to generate change control documents.)
g. Check other cell processes that are similar to the deficient process and, if a similar problem is found in those, then review the entire production process and re-train the operators accordingly.

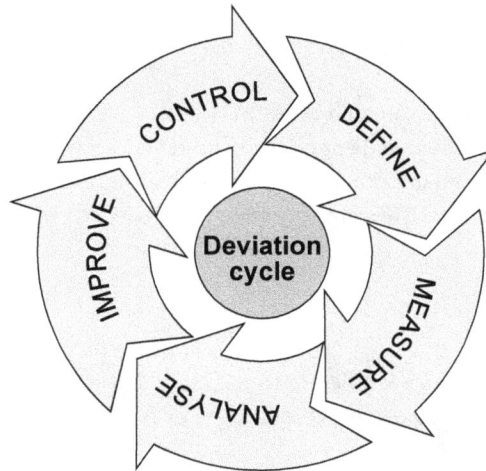

The deviation cycle

3.8 Be alert for further deviations post process revision

Production processes are sometimes revised in order to improve quality. This is a time when abnormalities and deviations can easily arise. An abnormality is anything that emerges in the production process which is not as it is intended to be.

There are three critical abnormalities that can emerge when you revise your cell production processes:

a. The quality does not change positively as planned.
b. The quality does change as planned but does not reach the target level.
c. The quality changes as planned, but other quality characteristics also change, producing unintended results.

To avoid the emergence of such abnormalities, you should have a standardised procedure for these revisions so that outcome can be regulated and remain steady from case to case.

This procedure should include the following:

a. Personnel:
    i. Specify those responsible for revising process conditions and those who will be in charge of this.
    ii. Train everyone involved in carrying out the revisions.
    iii. Ensure that those responsible are there when the revised process conditions are introduced, and that there is adequate supervision.
b. Check if:
    i. Revisions to process conditions are affecting other quality characteristics.
    ii. Revisions have changed costs.

c. Do not:
   i. Mix up the materials used before and after revisions, unfinished goods, and finished products.
   ii. Mix up the standards used before and after revisions.
   iii. Let the purchasing department and suppliers make revisions without permission.
d. Complete all other revisions that are related to the process condition revisions.
e. Test and inspect the finished cell therapy products and confirm whatever quality changes have taken place.

3.9 Keep work records

If the laboratory and the company as a whole are to maintain and improve quality, then it is necessary to know how well the laboratory is performing. For this purpose, it is essential that records of all events and occurrences are kept.

In particular the manager should maintain:

a. Achievement records that show how well the laboratory is performing against its targets.
b. Work records that will help to build up a picture of the overall situation of the company.
c. Daily job results records that will help identify potential abnormalities as early as possible.

Keep records of what your employees have achieved in the different production processes, and compare these with the plans.

Achievement records should include:

a. How much work has been done.
b. The percentage of deviations.
c. Records of quality characteristics.
d. The performance costs of the work.
e. The degree of success in meeting deadlines.
f. The laboratory's safety record.
g. The level of employee enthusiasm and morale.

When your laboratory's actual achievement differs greatly from what it planned to achieve, then corrective action must be initiated. Corrective action will be much easier to implement if all the data is converted into tables and figures. It is also a good idea to put control charts and graphs where everyone can easily see the level of work achievement and can compare this against the plan.

Work records are important to:

a. Give a clear picture of the present situation of the company.
b. Provide input for decisions about the future direction the company should take.

Work records include:

a. The names of products.
b. The name of the work and a description of the work.
c. The dates when work was started and completed.
d. The facilities and equipment that are being used.
e. The names of employees.
f. The working conditions.
g. The presence or absence of abnormalities.
h. The handling of any abnormalities, and the results.

Archive all records for a fixed period of time as established in the Archiving SOP and in fixed locations where they will not be damaged or lost.

Work records can be used to assess:

a. A company's process and manufacturing capabilities.
b. Its quality standards.
c. The control limits.

These assessments will then help with planning future processes and production, because they will show:

a. Optimal manufacturing conditions.
b. Factors that affect quality in particular laboratories.
c. The causes of abnormalities in the various laboratories.

3.10 Take a systematic approach to making improvements

Whether ideas for improvement come from suggestions, Quality Operations or elsewhere, managers should take a systematic approach to selecting initiatives for improvement, monitoring their progress and quantifying the results by using statistical techniques and introducing the PDCA cycle.

Use statistical techniques, graphs and tables to select initiatives for improvement, to monitor the progress of improvements, and to quantify the results. This will provide much more effective control of improvements, and will have the added benefit of providing employees with a concrete sense of what is going on in the laboratory, as well as demonstrating the advantages that improvements will bring.

a. Select improvement initiatives according to the level of improvement they can be expected to deliver. Take into account the balance between the time required for a solution, the costs, and the savings that the improvements are expected to bring.

b. Stratify the data that is gathered on the initiative from various angles, and look for evidence that will help to indicate what improvements are needed, and what measures should be taken. Train employees to use statistical techniques, so that they will be able to analyse the stratified data.

c. Set the improvement objectives, and decide on the concrete measures and the target values and outcomes needed to support these objectives. Summarise the data in control tables and charts to prove the objectives have been met or are in the process of being met.

d. Put a person in charge of each concrete measure that is to be taken, to determine the evaluation scale, as well as the intermediate and final check methods.

e. Evaluate the effects of any improvements in terms of money or time, since these are corporate objectives and support the continuation of the improvement process. Convert subjective evaluations, such as efficacy, safety, into a monetary value for this purpose.

As a reminder, the PDCA cycle is still one of the most effective ways to achieve improvements. The four stages as detailed previously are applied to improvements:

P – Plan: Set up a concrete improvement plan, choose criteria for evaluating it, decide how to put it into practice and who will be in charge of each item, prepare a daily schedule and check that everyone who needs to know about it is informed.

D – Do: Implement the plan exactly as it has been drawn up. Do not implement it partially or casually. Exact implementation is necessary in order to discover any defects there may be in the plan itself.

C – Check: Use the evaluation criteria that were chosen at the planning stage to review and evaluate the results of implementation.

A – Act: Analyse the results that have been evaluated and take concrete countermeasures and any other action that you think necessary. Ensure that the results are reflected in the next new PDCA cycle. Finally standardise the new procedures.

If the results are not satisfactory then change the plan, try it again, and if this time the results are satisfactory, then standardise the changes into the supporting documentation, and put it into regular use. The PDCA cycle can be used for the simplest of functions or for the most complex laboratory activity imaginable. It provides the manager with the means by which to continuously improve the level of quality in their laboratory.

3.11 Keep everyone informed of how the laboratory is performing

All employees should be kept informed of how each laboratory is performing within the company. This provides the opportunity for staff to see how their work area compares against others within the company as well as discussing which methods have proven effective.

Good graphic methods of displaying this information include:

a. Control charts, which will help everyone to understand the current situation, the presence or absence of abnormalities, and any transitions or changing trends in the quality characteristics of important products.
b. Transition charts that show changes:
  i. In actual defect ratios against the targets.
  ii. In daily performance against targets.
  iii. In safety records.
  iv. In the number of improvement proposals.
c. Charts that indicate production volumes of cell expansion against planned volume in a cumulative manner.
d. Charts that demonstrate the gap between actual progress and the progress expected according to the standards.

Any table or chart that indicates daily or cumulative changes against target values will provide a good picture of the present situation and will help with decisions about the future.

3.12 Establish report-writing procedures

Manager's reports have to be documented properly:

a. Use the proper, established company format in writing reports.
b. Keep records of communications with all other departments.

Review the efforts that have been made to improve quality over an extended period. Appreciate what direction they are moving in: for example, is the focus on solving problems that have already arisen (e.g. too many failed cell production packs) or on the future (e.g. improving product quality)? Then consider whether this is the best direction to continue.

When preparing reports:

a. Standardise the format as much as possible.
b. Use formats that are easy to fill in and can be completed quickly.
c. Itemise the contents in order of importance, putting detailed explanations if required into appendices at the very end.
d. To make it easy for the reader, wherever possible present information visually in the form of graphs or reference tables which allow visual review of the time sequence.
e. In addition to the specific content required by the format, include, where appropriate, your own comments and proposals, for example:
  i. Actions to prevent abnormalities recurring.
  ii. The disposal of rejected items.
  iii. The assumed causes and the effects on other departments.
f. Write reports in time and distribute them to the appropriate personnel.

When communicating with other departments put everything in writing and maintain records.

a. Clearly indicate the scope of the matters being discussed, and briefly state what is being requested, the objectives and the targets. Be sure to sign (or stamp) and date the document.
b. Itemise the contents in order of importance, putting detailed explanations and numerical information in appendices.
c. Keep and exchange minutes of any decisions that are agreed upon.
d. If written communication is insufficient, arrange for a meeting, and discuss beforehand what topics are to be raised.

Recommended reading

1. Clinical Pathology Accreditation (U.K.) Ltd. Standards for the medical laboratory, 2009.

2. Dewar, S. *Clinical Governance under Construction: Problems of Design and Difficulties in Practice*. London: King's Fund, 1999.

3. Institute of Biomedical Science. Guidance on the role of the clinical laboratory quality manager, 2008.

4. Quality managers and the clinical laboratory, *The Biomedical Scientist*, April 2002, pp. 378-79.

5. U.S. Food and Drug Administration. Guidance for industry: Quality systems approach to pharmaceutical CGMP regulations, 2006.

Self testing multiple choice questions

1. To implement company policy effectively in the laboratory:
   a. Evaluate the work in your laboratory with reference to the previous three months.
   b. Examine with your technicians whether the initiatives selected for development satisfy long-term and medium-term company plans.
   c. It is not necessary to review with your technicians how to implement initiatives.

2. To ensure that your technicians have a good grasp of the specific objectives and methods needed to implement the laboratory policies:
   a. Let your employees decide how the objectives could be achieved by themselves.
   b. Use graphs and charts to show how the achievements of their department contribute to the overall success of the company.
   c. Use qualitative values to define each objective.

3. When any problems arise during the implementation of the policy:
   a. Try first of all to solve them yourself without informing anyone.
   b. Call in the persons in charge, investigate the causes, and examine possible countermeasures.
   c. If a problem is serious pass it on to your superiors and let them solve it.

4. The primary areas where authority for the implementation of Quality Assurance activities must be defined include:
   a. Research and Manufacturing.
   b. Process Design and Final Inspection.
   c. Packing Manufactured Goods and Sales.

5. Job assignments are descriptions of:
   a. The responsibilities and authority of each department.
   b. The jobs and responsibilities of each department.
   c. The jobs, responsibilities and authority of each department.

6. To define the job assignments of each department in a company:
   a. Specify all the requirements in the department.
   b. Limit the number of the jobs.
   c. Identify the quality characteristics of the jobs as numerically as possible.

7. To clarify their policies and objectives and communicate these to their employees, managers should:
   a. Communicate with other departments and copy what they are doing.
   b. Understand and control the quality characteristics of the jobs in their department.
   c. Ensure that abnormalities will not occur.

8. To communicate your superiors' instructions to your employees:
   a. Relay those instructions exactly as told and let staff decide what they mean.
   b. Consider how these instructions fit in with the responsibilities and capabilities of your department.
   c. Ignore them if they do not correspond to your department policy.

9. To ensure that your employees understand and follow your instructions:
   a. Provide education and training in advance to ensure that they can follow the standards.
   b. Never give employees jobs that are even a little above their skills level.
   c. Let technicians develop a habit of reporting when a job is completed and only after they have actually completed several jobs so as not to waste your time.

10. When you are assigning jobs and giving work instructions:
   a. Always give instructions that are appropriate to the operator's abilities.
   b. Judge what they are capable of from talking to the other operators.
   c. Always carry out prior training before operations start, no matter their abilities.

11. In your work instructions clearly specify:
   a. The cost of each job so employees are aware of the price of failure.
   b. How to properly report the stages of the work performance.
   c. The various sites where employees can find the methods for themselves.

12. To take a consultative approach in communicating your instructions:
   a. Explain the objectives and improvement targets fully.
   b. Do not invite employees to give their opinion.
   c. Ask employees to write a summary of any decisions reached during discussion.

13. To set up a system that will enable technicians to recognise if the tasks they are responsible completed properly a manager must:
   a. Personally complete each task.
   b. Assume appropriate actions were taken to complete the task.
   c. Identify an item in each action that can be measured to show if it has been completed.

14. Management check items are indicators that can be checked to see:
   a. How well the manager has defined the tasks that he is responsible for.
   b. How well a management task has been carried out.
   c. How well the performance of the department compares with other departments.

15. A manager should select as priority management check items those factors that most affect:
   a. Employee morale.
   b. Cost reduction.
   c. Business performance.

16. Analyse factors that appear to affect management check items as early as possible so:
    a.   Deadlines will not be met.
    b.   Costs can be avoided.
    c.   Measures can be taken quickly to deal with them.

17. To monitor the quality of the work in your department:
    a.   Use numerical data and graphs to show trends in the manufacturing process.
    b.   Check workplace operations regularly to see how often employees are deviating from the operation standards.
    c.   Inspect all the designated inspection items personally.

18. Ensure that when employees are submitting reports they:
    a.   Understand the appropriate control points and evaluation methods.
    b.   Should only report the favourable items.
    c.   Report unusual events only if they are told to do so.

19. To ensure that quick and appropriate action is taken whenever defects occur:
    a.   Ensure employees know that all information must stay within the department.
    b.   It is not necessary to distribute a report to related departments informing them of the defect and asking for their cooperation.
    c.   Distribute a report to related departments informing them of the defect and asking for their cooperation.

20. Take the following actions when you are informed about a defect:
    a.   Ensure you have a clear idea of the situation and the actual products concerned.
    b.   Assess the defect financial cost as this will determine whether action is taken.
    c.   Wait until you have more information before doing anything.

21. A defect memo sent to another involved department should include:
    a.   The issuing department.
    b.   Demand that they immediately undertake countermeasures.
    c.   Assignation of blame for the defect.

22. When there are defects that could lead to serious claims against a company the first priority:
    a.   Deal with the products.
    b.   Investigate the causes.
    c.   Take countermeasures.

23. If products that may be rejected have already been transfused into the customer/patient:
    a.   Recall the products immediately.
    b.   Assess the seriousness of the situation.
    c.   Consult the customer/patient.

24. To gather information about the causes of the defect, ask the following questions:
    a.   Are operators following the operation standards?
    b.   Has a new manager been appointed?
    c.   What day of the week was it as performance is always bad on a Monday?

25. To check that the causes of the defect are being properly analysed:
    a.   Go to the workplace and examine the product.
    b.   Change the standards immediately.
    c.   Draft concrete plans for improvement of future product.

26. The likely result of the appropriate revision of processes is:
    a.   Quality changes as planned and reaches the target level.
    b.   Quality does not change.
    c.   Quality changes as planned, but other quality characteristics also change.

27. To prevent abnormalities when process conditions are revised:
    a.   Mix up the materials used before and after revisions.
    b.   Ensure only the new SOP is used after revisions.
    c.   Change the employees before and after revisions.

28. Achievement records should not include:
    a.   How successfully deadlines were met.
    b.   The level of employee morale.
    c.   The facilities and equipment that were used.

29. To ensure that everyone can see the level of work achievement it is best to
    a.   Ask the supervisors to tell everyone.
    b.   Display the achievement results in control charts and graphs.
    c.   Hold monthly meetings and announce achievement awards.

30. Work records include:
    a.   The department's production capability but not the work description.
    b.   The presence or absence of abnormalities in facilities and equipment.
    c.   The quality characteristics of the work results but not the work conditions.

31. Work records cannot be used to assess:
    a.   A company's quality standards.
    b.   A company's profit/loss statement.
    c.   Its process capability.

32. To use job result records to recognise abnormalities keep them up to date on:
    a.   A daily basis.
    b.   A weekly basis.
    c.   A monthly basis.

33. Use statistical techniques, graphs and tables to:
    a.   Select initiatives for improvement.
    b.   Monitor the progress of improvement.
    c.   Highlight lack of training and educational programmes.

34. In selecting improvement initiatives take into account the balance between expected improvements and:
    a.   The percentage of failures.
    b.   The savings that the improvements are expected to bring.
    c.   The techniques that can be used to select them.

35. Evaluate the effects of improvements in terms of:
    a.   Subjective evaluations.
    b.   Time lost.
    c.   Decrease in abnormalities.

36. Which is the correct sequence of the PDCA cycle when applied to work instructions:
    a.   Make changes to the operations, standardise operations, instruct subordinates to carry out operations, check the results.
    b.   Standardise operations, instruct subordinates to carry out operations, check the results, make changes to the operations.
    c.   Instruct subordinates to carry out operations, standardise operations, check the results, make changes to the operations.

37. A control chart will help everyone to understand:
    a.   Any trends that are taking place.
    b.   That abnormalities do not occur.
    c.   Which employee is not performing satisfactorily.

38. Methods of displaying control information include:
    a.   Real-time electric signboards that show the gap between actual progress and the progress expected in the standards so employees are reminded of failings.
    b.   Transition charts that show performance against targets.
    c.   Graphics that show changes in the number of suggestions for improvement.

39. When you are preparing reports:
    a.   Standardise them as much as possible.
    b.   Itemise the contents in order of cost as this is of primary importance.
    c.   Do not include any visual graphs or figures as reports are always written in nature.

40. When communicating with other departments:
    a.   Clearly indicate the scope of the matters being discussed.
    b.   If written communication is not enough then do not bother them again.
    c.   Do not discuss anything that will make your department look bad.

41. Quality Assurance should receive a deviation report:
    a. Only if the deviation deviates an SOP but not a protocol.
    b. Only if the deviation is not an improvement.
    c. Whenever any operation deviates from the SOP or protocol.

42. Whenever any operational or procedural abnormality occurs, then a deviation report should be filed with:
    a. The CEO.
    b. QA department.
    c. QC department.

43. A deviation is always the same as :
    a. An abnormality.
    b. A non-conformance.
    c. An intentional change.

44. When a deviation is identified, then the Laboratory Manager must undertake _____ action(s).
    a. One
    b. Four
    c. Seven

45. When compiling the countermeasures it is important that they be:
    a. Organised and integrated.
    b. Prioritised and organised
    c   Prioritised and archived.

46. When an investigation concludes and it is determined that there is no risk from the product, it can be released by:
    a. QC department.
    b. The CEO.
    c. QA department.

47. When identifying the root cause of an abnormality, a systematic way of doing so is to ask what is known as the ____ questions:
    a. 4Ms
    b. 4Ws
    c. 4Hs

48. When the cause of a deviation or abnormality is in question, then it is advised to perform:
    a. A repeat test in a contract service provider operated laboratory.
    b. A repeat test in a different laboratory operated by the company.
    c. A reproducibility experiment within the same laboratory.

49. The five steps in the deviation cycle are:
    a. Define, Measure, Analyse, Eliminate, Control.
    b. Define, Measure, Analyse, Improve, Control.
    c. Refine, Measure, Analyse, Reduce, Control.

50. Assessments from work records are useful in planning future processes because they can provide the _____ manufacturing conditions.
    a. Minimal
    b. Optimal
    c. Strictest

Chapter 4. Laboratory Managers

Managing
people

4.1 Ensure that laboratory employees understand and follow standards

A Standard Operating Procedure (SOP) is a written description of the best way to carry out a task or a process. It may also refer to the specifications of a product thereby establishing the acceptable limits for an immunotherapy product. If a company is to produce quality goods consistently and efficiently, it is essential that its employees understand and follow the standards thoroughly.

First of all, the technicians must be taught to recognise the benefits of following an SOP: they integrate and simplify operations and get rid of any inefficient processes by improving work methodology and thus provide a platform upon which to build future improvement.

Explain the work standards as set out in the SOPs to your technicians:

a. Ensure that technicians recognise the purpose of having work documents and operating standards.
b. Explain clearly to each technician the operating standards relevant to their work, and check that they fully understand and appreciate them.
c. Explain clearly each technician's responsibilities in using the relevant standards and why deviations cannot be accepted.
d. Ensure that the materials, reagents and equipment both specified and required by the procedural standards are available.

To ensure that the technicians follow the standards and rules:

a. Establish a control system to check that the work standards are being adhered to, clarify the procedure to those in charge and train them to use it properly.
b. Quickly find the cause of any non-conformance and act to prevent its recurrence.
c. Where an employee is at fault for failing to follow the standards, make it clear to him/her that the standards must be followed. Write up a deviation to file with QA department and impose retraining.
d. Create an environment in which everyone is willing to follow the Standard Operating Procedures without question unless they know of a method to actually improve the standard.
e. Standardise the skill levels of technicians and technologists so that they can see the importance of standards and rules in improving their skills.

4.2 Raise the employees' technical skill levels

Biomanufacturing technology is constantly advancing, requiring ever-higher skill levels. If you are to keep on providing quality products and services, you will have to continuously improve the technical skills of your employees to match the trends in the industry.

To do this:

a. Plan training programmes and encourage employees to participate.
b. Encourage employees to improve their work techniques on the job.
c. Exchange ideas with other departments.

To plan training programmes:

a. Review the jobs in the laboratory and the levels of knowledge and skills needed to perform them.
b. Plan long-term training programmes to raise technicians' skills to desired levels. The plans should include lists of those who will take part, and the content and duration of the programmes.
c. Plan and implement training for individual technicians. Keep a good balance of daily work and training. Use this plan to control individual training for each year.
d. Train technical staff to gain the specific qualifications that are required for certain procedures and operations. Special in-house proprietary technology should have a company-specific qualification system. Clarify the methods for acquiring these qualifications, e.g. doing examinations, education courses, in-house training, or on-the-job training.
e. Develop plans for each technician to acquire essential national qualifications, and implement training opportunities to help them to achieve these.
f. Encourage technicians by giving recognition to the higher skill levels that they achieve. Establish criteria for different skill levels, and present certificates that show the skill level they have achieved. Give recognition to those who work productively, as well as to those who improve their technical skill levels.
g. Train technicians to follow the work instructions when they begin working on a job for the first time, and whenever new operation standards and work instructions are introduced.
h. Keep records of the training courses that each technician has taken in their personal training files held by the Human Resources department. These records should show the date, the name of the instructor, and the content of the course as well as the level of achievement.
i. Maximise the training potential of Quality Operations activities. These can focus directly on issues related to the workplace environment, operation methods, and product quality. The theory and principles taught during these activities are essential for technical staff to comprehend the highly restrictive (rules and regulations) nature of the work they perform.
j. Create an atmosphere which encourages everyone to actively seek training in order to move up to a higher skill level with the company.

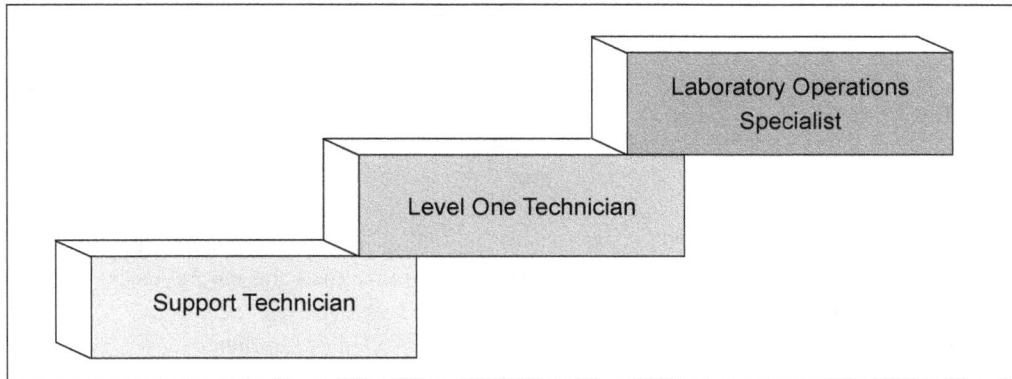

Career path of laboratory technicians

Encourage employees to improve their work techniques

As well as providing formal skill training for employees, you should also encourage employees to continuously improve their skills themselves by fine-tuning their work techniques, and developing their own original ideas for doing things better but recognising anything outside of the Standard Operating Procedure will require prior approval and change control documentation.

Exchange ideas

a. Improve the technical level of staff by exchanging the latest information with other departments. Such an exchange will also provide a sense of the skill level within a department by comparison to others.

b. Although with some skills it is important to know the company's level within the industry as a whole, it is also useful to compare at the laboratory level with other laboratories within the corporate system of affiliated contract service providers.

c. Employees should be sent to technically advanced training programmes to learn the latest skills, when such workshops become available. They then pass on what they learned to their colleagues upon their return.

d. Alternatively instructors from technically more advanced affiliated companies and consulting firms may be invited to visit the company to provide in-house training workshops.

e. Technicians should be encouraged to take skill-level certificate examinations organised by government or third-party accrediting organisations.

4.3 Delegating authority

Managers may make the decision to delegate some of their own authority to their employees as a way of upgrading and educating them. This can prove beneficial to the laboratory, since management efficiency is inversely proportional to the number of items being managed. This means, the less that a manager has to manage the better they will likely be in managing it.

Managers should delegate some authority for making decisions, giving instructions, consulting and giving advice as this is also part of succession planning.

Take the following steps:

a. First clearly define policy in the laboratory area that is being delegated.
b. Make certain that standards are established for the duties being handed over.
c. Ensure standards are established for dealing with any abnormalities that may arise.
d. Make certain that the authority to be delegated is not more than the employee is capable of handling.
e. It is imperative that the employee understands what is expected of him/her.
f. Require the employee to report on the progress and results regularly in order to assess his/her proficiency.
g. Managers must provide guidance throughout the delegation process.

Delegation is not a simple matter, and often proves much harder than originally anticipated. A manager must always remain responsible for whatever the subordinate does, but at the same time, must leave the individual free to perform the task without any interference unless absolutely necessary. Delegation often requires a fine balancing act by the delegator on knowing when to interfere and when not to.

4.4 Motivate employees

The attitude that laboratory technical staff have towards their work will have a major impact on the success of the company. A negative attitude reduces the efficiency of the production process and the quality of the therapeutic treatments that a company produces, whereas a positive attitude creates a dynamic workplace where productivity and quality remain high.

Managers should:

a. Be alert for signs of any negative attitudes, especially in regards to attendance and punctuality.
b. Take steps to motivate employees through positive reinforcement.
c. Try to maintain a balanced workforce.

Signs of a negative attitude are often expressed in how laboratory employees interact with each other and in their general appearance. Most commonly negative attitudes are evidenced by increased levels of absenteeism without prior notification and by poor punctuality.

These present two significant challenges:

a. First and most important is the immediate problem of ensuring that production is not disrupted by poor attendance and punctuality.
b. Secondly, the apparent failure to properly motivate the department personnel and the obvious need to improve employees' existing attitudes is recognisable.

In order to keep production of therapies in the laboratory on schedule, it is necessary to know the absentee rate in advance so that accommodation utilising other employees as substitutes for those who will be absent can be arranged.

Managers should:

a. Introduce a Request for Leave Form that employees must use in order to request leave in advance. Sufficient timeframe should exist between filling in the Request Form and the actual leave time.
b. Educate employees on the negative effect that unrequested absences has on production and the company as a whole.
c. On the longer term, train employees in several skills so that they can easily substitute for absentee technical staff should the need occur.

When employees are continually late:

a. Analyse the reasons for technical staff arriving late or leaving early, and take appropriate action to remedy the situation.
b. Educate employees on the effects that tardiness can have on the stability of production and the company as a whole.
c. Reward employees who are always punctual or do not leave early over a defined period of time, ie. one year.

A positive attitude among employees is often reflected in:

a. A willingness to take part in teamwork.
b. Confidence in acting on their own initiative.
c. Drive and energy.
d. A sense of responsibility.
e. Willingness to try to find improvements in their work.
f. Appropriate dress.

Several actions for employee motivation are:

a. Provide a good example:
    i. Laboratory managers can set for their technicians and employees a good example through their own positive attitude.
    ii. Ensure that all technical staff can visualise the value of any rules introduced prior to the actual introduction.
    iii. Encourage technical staff to openly come forward and discuss any troubles or concerns they may have.
b. Provide support:
    i. Support technical staff in raising their skill levels.
    ii. Carefully observe, through daily management practices, what technical staff are actually doing, and aid them if necessary.

    iii. Schedule regular meetings of staff so that they can improve their intra-communication skills.

  c. Encourage initiative:

    i. Ensure that your employees understand the full context of their jobs, and the relative importance of their role to the therapeutic product.

    ii. Encourage technical staff to discover problems in their work and present their own original solutions for approval.

    iii. Hold competitions that actually encourage technical staff to come up with innovative ideas.

    iv. Establish a suggestion scheme, and a system to evaluate their ideas and to see that the excellent ones are implemented.

  d. Challenge technician capabilities:

    i. Assign tasks slightly above the employee's ability level, whenever it is appropriate to do so without jeopardising the production process.

    ii. Select themes for improvement which interest and challenge.

    iii. Give technical staff appropriate guidance when they take on these challenges.

    iv. Hold meetings where employees can present their achievements to the company as a whole.

    v. Ensure that technical achievements are recognised in a company award system.

Contrary to the current situation in many companies around the world, one of the long-term ways of improving the attitudes of technical staff is to ensure that the company maintains an even distribution of employees of different ages and different years of service without a forced retirement policy.

This distribution should be balanced in the following categories:

  a. The total number of employees in age brackets.

  b. The number of employees of each sex.

  c. The number of employees in each functional area.

4.5 Involve technical staff in future improvements

The laboratory will be much more productive, quality will improve, and costs will be reduced if employees feel that they play a real role in achieving these goals. It is important that they do not see their functional role as simply carrying out instructions in a robotic manner. Encourage them to be both problem conscious and improvement conscious.

Managers can take two key actions:

  a. Set up a suggestion scheme.

  b. Support the activities of Quality Operations.

Encourage laboratory employees to make suggestions for improving quality, efficiency, and productivity, for reducing costs and for improving safety. Evaluate their suggestions periodically, implementing the good ones and develop a process of rewarding those who suggested them (not necessarily financial).

Take the following steps:

a. Set up a suggestion scheme that all employees can actively participate in.
b. Ensure that employees understand that the purpose of the scheme is to make work better, easier, safer, or quicker.
c. Set concrete goals for each workplace which should help stimulate suggestions.
d. Create a work atmosphere where staff willingly make suggestions by demonstrating a genuine interest in receiving their suggestions.
e. Evaluate suggestions fairly and quickly, letting each employee know what you think of his/her suggestion. When you have to reject a suggestion, tell the employee why.
f. Select those suggestions which significantly impact on improvement.
g. Display those suggestions selected where everyone can view and discuss them.
h. Implement accepted suggestions as quickly as possible.
i. Should implementation require changes in equipment or materials involved in an existing production process then change control documents will be required. No changes can take place until the change request is approved.
j. Once the suggestions have been implemented, and their positive effect has been confirmed, standardise them. Then integrate them into the production control system:
    i. Establish a procedure to control the new or improved process.
    ii. Inform all employees of the new control procedure.
    iii. Place an employee in charge of ensuring a smooth transition.

4.6 Employee involvement: Support Quality Operations activities

The second method referred to was supporting Quality Operations (QO) activities. QO arranges small groups of employees to meet, to train, and to discuss work problems and carry out improvement activities. Employees who participate in QO activities usually develop quality consciousness, problem consciousness, a willingness to make improvements, and a strong sense of quality management.

At the Quality Operations meetings, discussions following presentations should involve suggestions for improvement, which come from either individuals or groups. Once the QO presentations are underway, most suggestions will likely come from that department, but ultimately these meetings will lead to cooperation and therefore to more and better suggestions will develop from technical discussions. Once employees are aware of the contribution they can make to product quality, they will be motivated to look for ways of improving their own work processes. You will often find at the planning stage that the opinions of technicians in the workplace will be useful both in gaining an understanding of the problems and in implementing the plans.

Tips for evaluating and giving commendations:

a. Do not use monetary incentives to encourage employees if it can be avoided since monetary incentives are inadequate if trying to achieve stability in Quality Control which requires a genuine and sincere integrity.

b. Monetary incentives are often counterproductive since QO activities encourage self-motivation to be quality oriented for quality sake.

c. Any commendations relate only to the achievements of those who make suggestions (individuals or groups) and should not be part of the overall job performance rating.

d. Consider how much employees have studied and to what extent their skill levels have improved. It is important to evaluate employee progress.

e. Consider the autonomy, cooperation, and willingness of employees to participate. This applies not only to work time but time outside the normal hours of the company's operation.

f. Any evaluation must provide due praise where it is merited. Failure to provide recognition leads to the demoralisation of the employee.

g. The evaluation and commendation must be conducted in such a way that their colleagues can make the most of their skill set in future activities.

Quality Operations meetings and sessions are more likely to be successful when management is flexible and open and provides support to the activities.

Management should:

a. Encourage all employees to participate: make it clear that QO activities have an important function in the company.

b. Keep up to date with what the QO department is doing. Talk directly to the QO manager and keep records on how many meetings and presentations there are and how many actually attend. What concepts are they working on; how long will it take to solve them and what method of assessment is being used?

c. Comment on QO topics when requested and be quite clear about what management actually expects as an outcome from the activities.

Encourage the Quality Operations Manager to:

a. Keep members of the Quality departments motivated by encouraging suggestions, making the most of their sense of competition and their eagerness to improve, and show appreciation in what they have done.

b. Begin with simple, specific, immediate problems, or those common to all members, or pertaining to company policy, and then move on to more complex problems. It is best to begin with a problem that can be solved in less than three months, and later move on to more long-term issues so that participants do not lose interest.

c. Schedule meetings carefully in order to achieve maximum attendance: if meetings are held outside working hours, this should be at a time that suits as many members as possible.

Evaluate Quality Operations activities in terms of:

a. The number of sessions per month.
b. The number of hours per session.
c. The number of issues solved per year.
d. The number of presentations per year.
e. The participation ratio: the number of participants to the number of employees.
f. The positive impact from Quality Operations on employee performance.

Quality Operations activities should be ongoing. When one issue is solved, the members must identify and tackle the next one. Ideally, the QO should have a realistic goal in solving a set number of problems each year; ie. such as one major issue each quarter. In cooperation with the other Quality supervisors and QC group leaders, QO creates an atmosphere in which members want to partake in the activities:

a. To encourage groups to identify and tackle a series of new problems.
b. To encourage members to take on new jobs or roles in the QO activities.
c. To encourage promotional activities as an occasional change from the normal activities of lectures and conferences. These will also provide an opportunity to let others know of your achievements and for your employees to hear what they have achieved.

Once staff demonstrate that they are capable of managing numerous daily jobs themselves, upon their own initiative, managers should then transfer authority for these jobs to them. Such jobs would then be described as being self-managed. Self-management through staff groups will improve the quality of work, and at the same time will leave managers free to concentrate on more important issues.

Common themes for Quality Operations sessions:

a. Reducing the number of rejected therapeutic products, improving product quality,
b. Achieving consistent quality, toward fool-proofing the manufacturing process, and improving operations.
c. Establishing or improving customer/patient satisfaction: reduce the number of complaints.
d. Prevention of careless mistakes and errors on inspections, and the dissemination of incorrect information.
e. Performing systematic troubleshooting and preventing the recurrence of abnormalities.
f. Efficiency increase in production, using time more efficiently, improving the duration of processes, reducing delivery time and streamlining processes.
g. Improving operating procedures and the laboratory layouts, minimising downtime, mechanising processes, and using more technical tools.
h. Preventing equipment malfunctions, introducing automation, improving technical tools and equipment.
i. Automating the office, the production area, and reducing labour requirements.

Recommended reading

1. U.S. Food and Drug Administration. 21 CFR Part 11: Electronic records; electronic signatures.

2. U.S. Food and Drug Administration. 21 CFR Part 211, Subpart B: Organization and personnel.

3. U.S. Food and Drug Administration. 21 CFR Part 211, Subpart I: Laboratory controls.

(Note: CFR stands for Code of Federal Regulations.)

Self testing multiple choice questions

1. To ensure that the production process runs efficiently laboratory managers must:
   a. Restrict employees from being creative in their work processes.
   b. Restrict technicians understand the work standards.
   c. Hope that technicians follow the standards and rules.

2. To encourage employees to follow the standards managers should:
   a. Make the employee's responsibilities clear to both the employee and his/her supervisor.
   b. Create an atmosphere in which everyone feels they do not need to follow the standards and rules.
   c. Permit employees to set their own responsibilities.

3. Work standards are important because:
   a. They show which employees are inefficient.
   b. They improve work efficiency.
   c. They solidify a process so there is no need for further improvement.

4. Employee skills need to be improved:
   a. To achieve improvements in quality.
   b. To fill the workplace.
   c. Because of reversals in technology.

5. To educate and train your employees systematically in the skills they need:
   a. Review the jobs in the department and recruit only people that have those necessary skills so that an education policy is not required.
   b. Plan and implement education and training for individual employees.
   c. Develop plans for every employee to acquire national qualifications.

6. Managers should encourage their employees to improve their own skills by:
   a. Fine tuning their machines.
   b. Fine tuning their work techniques.
   c. Dressing better.

7. A department can raise its overall levels of skills by:
   a. Exchanging proprietary information with other companies.
   b. Sending employees to learn skills with companies that are technically more advanced.
   c. Exchanging managers with other companies.

8. The text suggests that:
   a. The less you have to manage the better you will manage.
   b. The more you have to manage the better you will manage.
   c. The more of you there are to manage the worse you will manage.

9. When delegating authority a manager must:
    a. Instruct the person to whom he is delegating to define their own policy in the area that is being delegated.
    b. Ensure that standards are established for the duties he is delegating.
    c. Ensure to still manage any abnormalities that might occur by being a constant presence.

10. A laboratory manager is ____ responsible for whatever his delegated subordinate does.
    a. Always.
    b. Sometimes.
    c. Never.

11. What impact can the attitude of employees have on the success of a company?
    a. A major impact.
    b. A minor impact.
    c. Little or no impact.

12. A manager may find indications of the attitude of employees in:
    a. The way they wear makeup.
    b. Their absenteeism.
    c. The way they interact with visitors.

13. Poor attendance and punctuality present the manager with the challenge:
    a. To ensure that production is not disrupted.
    b. To encourage employees to interact better with each other.
    c. To corroborate any doctor's notices accounting for the absences.

14. The attendance rate which ensures capability is the ratio of employees at work to:
    a. The number absent.
    b. The number needed.
    c. The number on requested leave.

15. When confronted with poor punctuality a manager should:
    a. Threaten to dismiss any employees who are often unpunctual.
    b. Educate employees on the effects that poor punctuality can have on production.
    c. Give the employees benefit of the doubt and excuse their lateness.

16. To motivate your employees to work to the best of their abilities:
    a. Ensure that they understand the importance of what they are doing.
    b. Do not interfere by observing what they are doing and offering them help.
    c. Never give them jobs that expand their ability level.

17. To motivate your employees you can:
    a. Arrange opportunities for employees to talk to senior management.
    b. Create an atmosphere where employees feel free to discuss problems with their superior.

    c.   Implement every suggestion for improvement that employees make.

18. A balanced workforce consists of equal numbers of employees of:
    a.   Different ages and sex.
    b.   Different ages and years of service.
    c.   Sex and years of service.

19. To get useful suggestions from employees:
    a.   Ensure that employees understand that the purpose of an improvement suggestion system is designed to reduce costs.
    b.   Dissuade employees from making any useless suggestions.
    c.   Show a real interest in receiving suggestions.

20. To encourage employees to make further suggestions for improvement:
    a.   Take a very long time to evaluate suggestions, perhaps even a year.
    b.   Limit the number of suggestions that can be made in each workplace.
    c.   Create an atmosphere in which members are willing to make suggestions.

21. Suggestions that have been adopted may take time to implement because:
    a.   They may require changes in existing equipment.
    b.   They may involve other companies.
    c.   They need the CEO's approval.

22. When suggestions have been implemented on a trial basis and their positive effects confirmed then the next step is:
    a.   Standardise them by introducing them immediately into the laboratory.
    b.   Write the change control document to request that all related standards be upgraded to include the suggested improvement.
    c.   Integrate them into the new control procedure as an unsigned amendment.

23. Once the change control documents have been approved by QA, managers and supervisors should:
    a.   Recheck the procedure to ensure that it was correct.
    b.   Standardise them by revising all SOPs, procedures and work documents.
    c.   Integrate them into the control system immediately and then do the paperwork.

24. Most suggestions for improvement will usually come from:
    a.   Individuals.
    b.   Groups.
    c.   Quality Operations activity meetings.

25. Quality Operations activities are more likely to emerge where managers:
    a.   Check closely on everything that is going on in their department.
    b.   Are flexible and open.
    c.   Take little interest in what is going on in their department.

26. For Quality Operations activities to have a full impact managers should:
    a. Make it clear to technicians that QO activities have an important function.
    b. Take charge and lead all the activities themselves.
    c. Remain at arm's length as they do not need to understand the activities.

27. The leader of a Quality Operations activity should:
    a. Only deal with the most challenging and hard to resolve problems.
    b. Show appreciation of what other participants are doing.
    c. Always hold meetings outside working hours.

28. To maintain the momentum of Quality Operations activities:
    a. Restrict participants to keep the same roles in all QO activities.
    b. Do not introduce and tackle new problems.
    c. Ensure managers fully understand and support the activities.

29. To make the most of QC statistical techniques:
    a. Use available data to get a full understanding of the facts.
    b. Train all employees in statistical techniques.
    c. Use statistical techniques to solve every problem.

30. Managers should transfer authority to employees when:
    a. They encounter a job they do not like to do themselves.
    b. The employees demonstrate capability of managing some of the manager's routine jobs.
    c. They have to go on frequent business trips and do not have time to properly manage the laboratory.

31. SOPs are beneficial because they improve work efficiency:
    a. By increasing the technical difficulty.
    b. And a basis of further improvement.
    c. By eliminating procedural steps.

32. Records of the training courses taken by employees are held by:
    a. The Laboratory Manager.
    b. The Quality Assurance department.
    c. The Human Resources department.

33. One reason that managers should practise delegating authority is because of:
    a. Succession planning.
    b. Replacement planning.
    c. Project management planning.

34. The number of employees in each functional area should be:
    a. Reduced.
    b. Determined by relative importance of the functional area.
    c. Balanced.

35. When managers reject a suggestion from an employee, they should:
    a.   Use it as an example for everyone else of what not to suggest.
    b.   Discard it immediately and say nothing further.
    c.   Explain to the employee why it was rejected.

36. When giving commendations or rewards to employees for their suggestions:
    a.   Only give a monetary incentive.
    b.   Avoid giving monetary incentives.
    c.   Do not give anything other than praise.

37. When holding a Quality Operations activity or meeting it is important to have:
    a.   Only senior management in attendance.
    b.   Maximum attendance from the entire company.
    c.   Only selective personnel who are identified as future leaders in the company.

38. When difficult problems are tabled at Quality Operations activities and meetings, they:
    a.   Stimulate interest and general participation.
    b.   Take too long to process and as a result interest in the activity is lost.
    c.   They quickly separate those employees that can think from those that cannot.

39. Evaluation of the Quality Operations activities should be done on the basis of number of sessions per month and:
    a.   Number of people not in attendance.
    b.   Number of hours per session.
    c.   The negative impact this has on attitudes.

40. Efficiency in regards to time includes the following:
    a.   Decreasing development time and minimising downtime.
    b.   Minimising down time and increasing holiday time.
    c.   Improving processing time and increasing development time.

41. Efficiency in regards to equipment includes the following:
    a.   Using more people provided with better equipment.
    b.   Introducing automation but not in the office.
    c.   Preventing equipment malfunctions and improving laboratory layout.

42. The stipulation regarding reagents mentioned in any SOP is:
    a.   Only reagents that are purchased new each time are used.
    b.   They are not past their expiry date.
    c.   They must all be from the same supplier.

43. When a technician commits a performance deviation, then he/she should be:
    a.   Severely disciplined.
    b.   Relocated to a different facility.
    c.   Undergo retraining.

44. When planning training programmes, ensure that they are:
    a. Short term.
    b. Long term.
    c. Never long term.

45. Normally, in most laboratories, the specialist technician is also referred to as :
    a. First tier.
    b. Second tier.
    c. Third tier.

46. Delegation is a:
    a. Simple matter.
    b. Not a simple matter.
    c. Relatively easy.

47. A negative attitude displayed by a technician will:
    a. Decrease efficiency.
    b. Create a dynamic workplace.
    c. Increase productivity.

48. The Request for Leave Form:
    a. Requires sufficient time to process.
    b. Should be submitted just before leaving.
    c. Is necessary only for long term absences.

49. Employees with a positive attitude often:
    a. Are responsible but lack energy.
    b. Have drive and energy and dress appropriately.
    c. Are eager to try improvements but lack confidence.

50. Managers should assign tasks that:
    a. Are slightly above the technician's ability level.
    b. Do not require any assistance or guidance.
    c. Are not interesting or challenging but are achievable.

Chapter 5. Facilities management

Laboratories and equipment

Managing facilities and equipment is key to any biotechnological company and involves carrying out routine inspections to ensure that the optimal state is maintained at all times. Policies must be in place for dealing with any problems and ensuring they do not happen again after they have been identified and corrected. In a GMP environment, the maintenance of both facilities and equipment must be carefully recorded and therefore deciding upon which forms should be used is essential to proper management. Subsequently, the proper handling, storage and archiving of these records becomes a primary concern.

5.1 Keep the workplace neat and clean

Often taken for granted, the initial step in maintaining facilities and equipment must be to keep the workplace clean and tidy. If the simple act of maintaining an orderly environment cannot be met, then it becomes obvious that the more difficult tasks of monitoring, repairing and maintaining equipment to achieve ultimate performance is unlikely to be met as well. Defects in a cluttered, unclean and unsafe environment are far more likely to occur.

It is not solely the responsibility of facility management to ensure that the workplace environment is kept orderly and employees must learn to look after their own work areas and equipment that they are responsible for. The only means by which the condition of facilities and equipment can be maintained properly is for these employees to use a check sheet with pre-determined check points that must be completed to a preset time schedule. These check points (also referred to as inspection points or items) will cover all the places, tasks, machinery and equipment in the employee's charge.

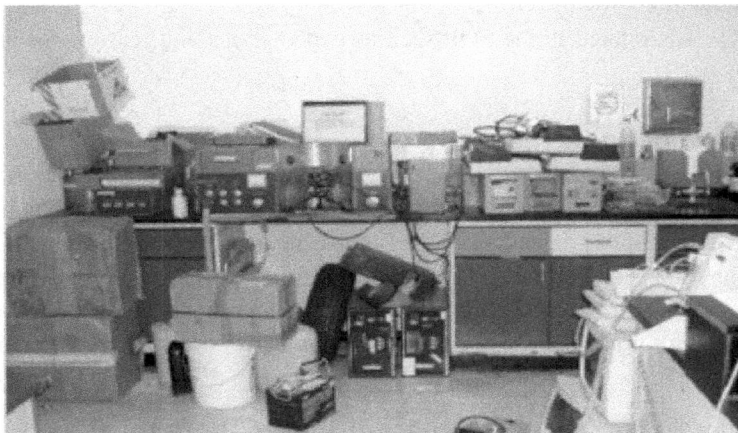

A typical cluttered laboratory workspace

5.2 Follow the user's manual

It is important that technical operators understand the functions of facilities and equipment and can use them properly. It is imperative that technicians be trained according to the manufacturer's manual, rather than to a company's established alternative practices. Therefore the SOPs for the proper maintenance of facilities and equipment should accurately reflect the contents of the manual and should not be contrary. Alternative practices will usually void any warranties provided by the manufacturer or supplier and should therefore be prohibited.

The manufacturer's manual (also referred to as the operation manual, or the user's manual) prescribes the optimum operation procedure for facilities and equipment, the tools and materials to be used, the proper handling methods, and any operational points that need special attention. Due to the size of some user's manuals, it becomes necessary that the SOP reflects only the key features of the operation thereby condensing the details, and avoiding non essential requirements which can always be referred to in the manual if the operator requires further reference information.

The SOP for the operation of equipment or facility should include:

a. The primary purpose for the operation of the facility or equipment.
b. The range or scope for the usage of the facility or equipment.
c. The essential characteristics of the facility or equipment.
d. Definition and explanation of any technical terms.
e. Identification of any parts (diagram preferred) and their names.
f. Any installation and assembly instructions required prior to use.
g. Methodology including the operational procedures and the means for making any adjustments or calibrations.
h. Identification of any abnormalities that can emerge when the facilities and/or equipment are in operation, which might be encountered and established countermeasures to be taken in such cases.
i. Any restrictions on training requirements and use of the equipment or facility. Special needs such as gowning, gloving, etc.
j. Any historical details related to the SOP such as explanations and reasons for past changes.

5.3 Carry out daily and periodic inspections

From the first day of installation, facilities and equipment begin to deteriorate with age and use. Therefore it is important to inspect facilities, equipment, machines and materials every day, looking for any abnormality, deviations or failures that might be present. These daily inspections are cursory but on a monthly or quarterly basis, a far more detailed inspection should be carried out, looking for problems that might still be in early stages of development.

Follow these inspection procedures:

a. Operators familiar with the facilities and equipment conduct normal daily inspections.
b. Use a check sheet on which the inspection items and methods have been identified in advance for ease of use and oversight prevention.

Checked items will typically include:

a. Any known parts or items that may become loose due to vibrations.
b. Specific sites where accumulations of debris or dust can occur.
c. Tubing, ports or common sites of blockages, obstructing objects, etc.
d. Connections subject to leakages of any fluids.
e. Electronic components, wires, or circuits exposed and easily damaged.
f. Components known to experience wear and tear (rapid deterioration).
g. Any system self calibrating or checking failures when turned on.
h. Proper operation of all switches.

Performing the inspection:

a. Use all five senses when performing an inspection.
b. Check gauges, printouts, display screens of those facilities and equipment that are monitored with measuring devices.
c. Report inspection results on the check sheet and get the manager or supervisor of the department to confirm and sign them off.
d. Process any abnormalities that are found according to the plan details provided in the section of this book on dealing with abnormalities.

5.4 Correcting abnormalities

Abnormalities need to be classified by their frequency of occurrence:

a. Occasionally
b. Regularly
c. Repeatedly
d. Chronically

Most abnormalities in regards to facilities are concerned with temperature, pressure, air quality and humidity. Controlling the environment properly is the single most difficult task in facility management. Breakdowns in heating, ventilation and air-conditioning (HVAC) systems are typically caused by leakages, the escape of gas, the release of vapour, and by soiling and blockage of filters caused by contaminants, waste and dust.

Include the following in any abnormality report:

a. Identify the section or parts of the facility or equipment affected.
b. The type and nature of the abnormality.
c. The normal parameters or acceptable limits.
d. Suspected causes for the problem.
e. Methods for processing the abnormality.
f. Initial containment and contingency plans.
g. Corrective actions undertaken or under consideration.

Any actions that are immediately undertaken should be minor in nature that do not place the product at risk. These are then reported after the fact. If the actions will impact directly on the product, then the impact is no longer minor and a CAPA needs to be generated with no immediate actions taken until such time that the CAPA is approved by QA.

If the cause of the abnormality cannot be identified then it will be necessary to suspend the operation immediately. This will be followed by disassembly and inspection of the facilities or equipment by qualified personnel. When a precision inspection of the disassembled equipment identifies the causes, take immediate measures to eliminate them from reoccurring. A full report of the corrective procedures to eliminate the abnormality is to be written and the CAPA is to be confirmed and closed by QA. This is followed by submitting a request as part of change control to edit the operating procedure and any other documentation that does not correspond to the changes that have been instituted.

5.5 Dealing with failure

A failure is when the facility or equipment ceases to function properly, either partly or completely. When a failure occurs it is important to re-establish correct operations as soon as possible. Protection of the product is the ultimate goal and if it cannot be achieved then operations should be shut down.

Three steps when dealing with failure are:

a. Find where the failure emerged.
b. Investigate the causes.
c. Process the failure.

Failures can be classified under:

a. The type of failure:
    i. Accidental. (Errors by technicians, technical glitches, etc.)
    ii. Trauma. (Bumps, knocks, physical abuse, etc.)
    iii. Degradation. (Time, erosion, exposure, etc.)
b. The cause:
    i. Single-cause failures.
    ii. Multiple-cause failures.

  c. The degree of loss of function:
    i. Partial failure.
    ii. Complete failure.
  d. The seriousness of the failure:
    i. Minor failure.
    ii. Major failure.
    iii. Critical failure.

When a failure occurs, the operators should complete a preliminary report, identifying the facility or equipment that caused the failure, the date, the type of failure, and an outline of what happened. The engineering department will use these preliminary reports to investigate the failure.

The engineering department should:

a. Detect the failed sections.
b. Assess the failure.
c. Classify the failure.
d. Investigate the cause or causes.

Use these processing methods:

a. When there are minor failures, repairs and replacement parts can be performed on the spot and the process usually continues uninterrupted.
b. When there are serious failures, a precision inspection needs to be conducted in order to determine the exact causes, to ensure that the appropriate countermeasures are undertaken.
c. Following implementation of emergency countermeasures it is necessary to perform a full evaluation of their long term suitability in order to prevent future laboratory shutdowns.

To prepare and maintain processing reports:

a. Report the causes of failures, the countermeasures taken, the date of their implementation and those responsible for the implementation, production-suspension periods, repair costs, the date of finalising countermeasures, those who examined the countermeasures, and actions to be taken in the future.
b. Archive reports as reference materials for similar problems that may be encountered in the future.

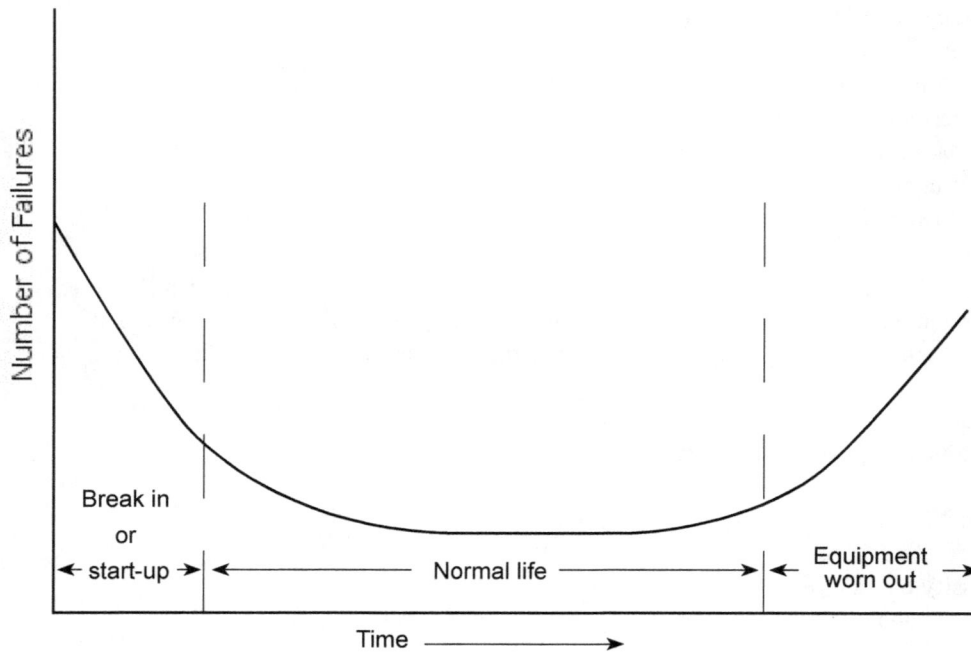

Typical equipment lifespan versus failures

5.6 Periodic maintenance

Periodic maintenance is the best way in identifying where repairs will be needed prior to the occurrence of a problem. Maintenance in this mode includes minor repairs, which are performed immediately with minimum cost, and the expectation that they will reduce the deterioration of the facility and equipment by prolonging their usage life.

The PERT method is often used to manage periodic repair work. PERT is an abbreviation for "programme evaluation and review technique". It is a schedule management method for large-scale repair plans and complex repair work. It presents the sequence of repair processes in the form of a interconnecting chain and creates a repair work schedule on the basis of the time required and the costs, e.g. the number of engineers required.

To draft a PERT Plan:

a. Identify necessary job activities: Confirm the contents of the job, and estimate the man-days (number of days x the number of employees) and the amount of materials required.

b. Prepare a job network or GANTT Chart: Present the number of job activities and their relationship to each other in the form of an event chain. Show the milestone points at which jobs can be started and concluded and arrange them in sequence. Jobs that can be performed simultaneously should overlap one another in the chart.

c. Estimate the work schedule: Enter the time required for each job on the chain, and make a final estimate of how long it will take to complete all the work. Identify the successive groups of jobs that will make up this final estimate of the completion period. These are known as critical paths.

d. Set up the network according to a time scale. Distinguish jobs that will take time from those that must be performed quickly. Make a total of the job loads (the number of employees needed and the material). Put all the different job loads on the time-scale GANTT or network chain.

e. Review and shorten the work schedule: Reassess the number of employees assigned to perform each job and the time allowed. Attempt to shorten the job schedule by changing the number of those employed on critical paths. Refine and reduce the schedule as much as possible.

f. Draft and accept the final work schedule: Prepare the work schedule through this trial and error process.

5.7 Manage the procurement of repair parts and materials

Periodic repairs may require the temporary suspension of facilities and equipment. The longer the repairs take the greater the losses to the company. One factor that can cause delays is not having available replacement parts or necessary materials. Orders should be placed well in advance with inventory levels maintained to ensure no shortages.

To manage this process efficiently:

a. Decide what repairs are routinely performed.
b. Place orders for materials and repair parts well in advance of schedule.
c. Choose an automatic purchasing order system for routine orders.

Select the equipment, tools and parts that require repair or replacement on the basis of their frequency of deterioration and breakdown. Factor these into the repair schedule.

Give higher priority to:

a. Those required in management inspection procedures or where an inspection audit is required by law.
b. Those that require preferential treatment because of:
   i.   The failure rate.
   ii. The size of production loss in the event of a failure.
   iii. Repair cost in the event of a failure.

Calculating priority:

Priority = production loss x failure frequency x repair cost

Fixed Period Ordering Method. Fix the ordering interval in advance, and determine the order quantity each time according to the current inventory levels and the quantity required. Keep in mind the time needed for procurement as well including shipping delays. Add a contingency time for any other possible delays.

Fixed Quantity Ordering Method. This method is not based on fixed ordering intervals in advance but instead on fixed quantities. The company orders a set number of units quantity (cost effective ordering quantity) when the inventory drops to a specified level. The re-ordering point is set at a level where the quantity needed for a period of time is sufficient and considers the time it will take to procure more, plus a small safety margin in case of delays. This cost-effective ordering method minimises storage and ordering costs because frequency is based solely on reaching that re-order level.

5.8 Prioritising maintenance

Optimal maintenance is achieved when priority is decided upon based on the importance of the particular facility or equipment.

To decide on importance:

a. The effects of a failure on safety, the environment, product quality, and loss of production.
b. The frequency of failure.
c. The cost of repairs and recovery.

The importance of the same facilities and equipment will not necessarily be the same at all companies. Therefore each company must prepare its own priority plan rather than rely on one that it may have obtained from another company.

5.9 Different forms of maintenance

Actual maintenance is composed of a large number of related activities, which include but not limited to inspection, adjusting parts, replacing material, and carrying out repairs. There are several different varieties of maintenance.

Deterioration activities:

a. Preventing deterioration, e.g. lubrication, adjustments, calibration and cleaning.
b. Measuring deterioration, e.g. regular inspecting and testing.
c. Post deterioration, e.g. periodic repairs and post-failure repairs.

Definitions of maintenance classifications:

a. Preventive maintenance – examination of facilities and equipment when designed or installed. It is aimed at preventing deterioration and failures from occurring.
b. Planned maintenance – incorporates periodic routine maintenance as ways of implementing preventive maintenance.
c. Periodic maintenance – refers to inspection and repairs at regular intervals.
d. Condition-based maintenance – constant monitoring of the condition of facilities and equipment. Also considered as part of planned maintenance.

e. Breakdown maintenance – getting facilities and equipment functioning again after experiencing failures.
f. Corrective maintenance – designed to prevent the recurrence of failures and is based on maintenance activity records.
g. Maintenance prevention – the facilities and equipment are designed to require less maintenance and to have a longer life span.

Modernisation and the increasing complexity and size of facilities and equipment, have increased the number of unexpected failures exponentially.

Special importance should therefore be given to:

Preventive maintenance, especially condition-based maintenance.

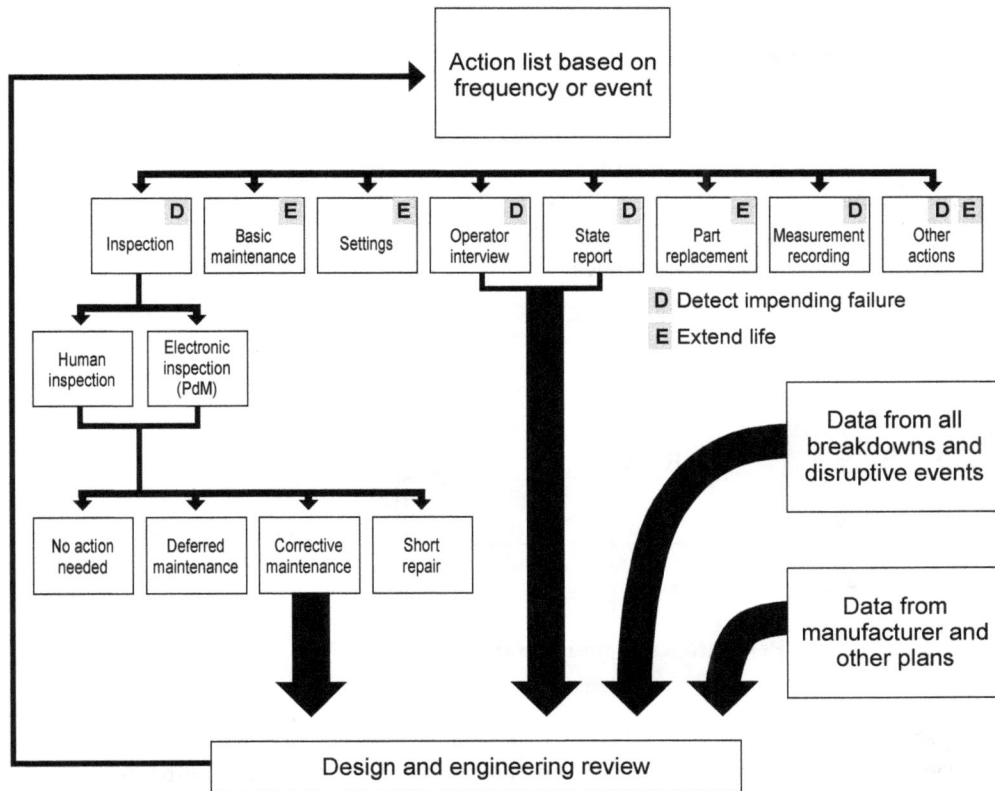

A flowchart of condition-based preventive maintenance

5.10 Decide what form of maintenance to use

The decision about the style and frequency of the maintenance to be used can be complex. Most companies perform periodic maintenance, condition-based maintenance and breakdown maintenance.

a. Periodic maintenance conducts its inspections and repairs at regular intervals and is the one that is most often put into practice by companies. It is carried out at intervals that suit the different facilities and equipment.

b. Condition-based maintenance requires constant monitoring of the condition of facilities and equipment. Most often it is selected for facilities and equipment which are under high risk and where failures would cause major losses and product is irreplaceable.

c. Breakdown maintenance is used to get facilities and equipment functioning again after failures and is usually chosen when maintenance costs are high so companies actually wait until a problem actually exists before responding. It is the most dangerous form of maintenance because it waits until a problem actually occurs.

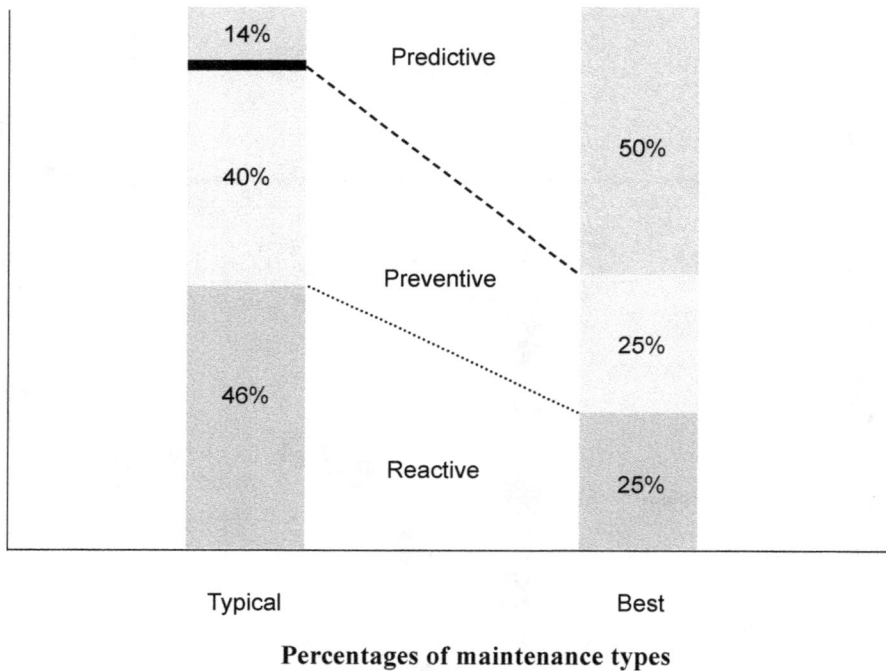

Percentages of maintenance types

Companies should strive to have predictive maintenance making up 50% of their maintenance programme. This requires increased monitoring as in condition-based maintenance, whereas break-down (reactive) maintenance should never be more than 25% of the maintenance programme. If this can be achieved then the company will have very strong management over equipment and facility failure.

Necessity for inspection: To determine the levels of need for inspection, it is necessary to perform an assessment based on an examination of the following seven points. If any of these seven assessments raise a flag then inspections should be done more frequently.

The seven points are:

a. The existence of legal and regulatory requirements.
b. The estimated suspension of operation period being prolonged.
c. The negative impact on quality.
d. The negative impact on safety.
e. The negative impact on the environment.
f. The inability to respond to disasters quickly and efficiently.
g. The cost of maintenance as compared to cost of repair.

Effectiveness of inspection: **Assess the effectiveness of inspections on the basis of facility and equipment by:**

a. Deterioration characteristics.
b. Facility and equipment life-spans.
c. Progression of deterioration.
d. Nature of the deterioration.

Need to disassemble: Equipment that needs to be taken out of action and broken down in order to inspect it properly will have fewer inspections as a company cannot frequently withdraw equipment from production without sustaining significant losses. Therefore, the more complex and involved the inspection of a particular piece of equipment may be, the lower the frequency of inspections. The relationship is inverse.

Frequency based on need to disassemble:

a. Negligible: It is possible to determine deterioration using normal measuring devices. Frequent inspections are therefore possible.
b. Maybe: It is possible to determine using diagnostic technology but not in all cases. Moderate frequency of inspection results.
c. Required: It is impossible to inspect deterioration without disassembling. Full inspections will be performed on an essential only basis and therefore at a low frequency.

Cost-effectiveness: Assess the manpower needed for inspection, as the more personnel required and the greater the cost/loss involved, the lower the frequency of inspection that results.

Classify facilities and equipment into three groups according to cost:

a. Overall maintenance cost = Maintenance cost + loss from failures.
b. Maintenance cost = Breakdown maintenance cost x frequency of failure
    + periodic repair cost x frequency of repairs
    + inspection cost x frequency of inspections.
c. Loss from failures = Cost of loss per failure x average failure frequency.

Select the inspection programme which can be conducted and maintained at the lowest cost without sacrificing quality of the final product.

Finally, determine the optimal maintenance style on the basis of the four criteria described above.

Use condition-based maintenance when:

a. Facilities and equipment need to be inspected a lot.
b. Inspection proves itself highly effective.
c. There is a high possibility of inspecting deterioration without disassembling.
d. Inspection will be highly cost-effective.

However, where it is impossible to inspect deterioration without disassembly, periodic disassembly inspections are necessary. When periodic disassembly inspections fail to reveal the necessary information, there is no choice but to conduct breakdown maintenance.

5.11 Condition-based maintenance

Carry out condition-based maintenance to continuously monitor your more complex facilities and equipment, especially those where failure could lead to large production losses. Use diagnostic equipment such as analysers, gauges and monitors to take continuous measurements. This will enable you to detect at an early stage any signs of deterioration or any incidents that could lead to failures. Having these monitors alarmed increases the reaction and response time to prevent a failure.

Keep inspection records up to date, especially those that are being measured manually on a daily basis, so that the status of facility and equipment deterioration can be easily determined from the records and actions can be taken immediately.

Diagnostic technologies:

a. Rotating machines that are often used in manufacturing require precision diagnosis using analysers.
b. Vessels, tanks and piping require diagnostic technologies that can identify corrosion and cracks.
c. Many kinds of thermal controlled devices require diagnosis to determine the degree of deterioration – thermocouples, radiation calorimeters and similar equipment are used for this purpose.
d. Systems based on artificial intelligence or computer controlled require programmes to analyse their proper performance.

5.12 Breakdown maintenance

There will always be accidental failures in any company. There are also some facilities and equipment that are so difficult and expensive to inspect that it is best to leave repairs until a failure actually occurs. When this failure does occur, repairs must be carried out immediately to minimise production losses, and also because sudden suspensions can have a major effect on other facilities and equipment within the production stream. Where equipment is essential to production and breakdown losses cannot be afforded, then a company must have a backup system available that can immediately take over following a failure.

Breakdown maintenance requires strict management, a constant state of preparedness, an efficient allocation of personnel, efficient and rapid procurement of materials, and superior management of the overall work progress.

Follow this procedure:

a. ***Investigate causes and examine countermeasures.*** When a failure emerges, investigate the causes and discuss countermeasures as soon as possible. If necessary, disassemble the equipment to assess failed components and the extent of the failure.

b. ***Draft repair plans.*** When causes are known and the materials, or parts for repair or replacement have been determined, prepare a repair timeline/schedule which identifies the necessary repairs that have to be done, and the time needed to do them.

c. ***Arrange personnel and material procurement.*** Assemble the staff needed for repair work. Make arrangements for procuring all parts and materials as quickly as possible. Depending on how urgent the situation is, consider using materials that may have been assigned to other products or projects.

d. ***Implement repairs and manage their progress.*** All repairs must be closely monitored to ensure they are to the planned timeline. This requires a superior management capacity that is both knowledgeable and efficient.

e. ***Report repair results and keep records***. Once repairs have been completed, record the results in a report to management for sign off at a senior level.

f. ***Manage an inventory of repair parts and materials***. It is necessary for key equipment, components, etc. to identify which parts have a frequency of failure and maintain replacements at all times. Order in advance.

g. ***Archive repair records.*** At some point, repairs of equipment and machines may come into question. Therefore, it is imperative that a detailed record is available if it should be requested on a regulatory audit.

5.13 Recurrence prevention

It is necessary to investigate the causes of failure or breakdown and take recurrence prevention measures to see that it does not happen again. Any and all corrective actions undertaken are fully documented so that immediate action can be taken the next time the identical failure or breakdown occurs by referring to these documents. The term corrective maintenance is used to refer to the total process of recurrence prevention, planned improvement of facility and equipment functions, extension of their life, and improvement of their reliability, maintainability, operational ease and safety.

To carry out recurrence prevention:

a. Classify the failures according to the nature of the breakdown such as damage to wiring, connection faults, physical abrasion, corrosion, etc.

b. Identify causes and results:

    i. Analyse the frequency of abnormalities: draw histograms and Pareto diagrams and conduct analysis.

    ii. Analyse the factors behind the abnormalities, using statistical methods.

    iii. If necessary, conduct detailed failure analysis: Failure Tree Analysis (FTA) and Failure Mode and Effects Analysis (FMEA).

c. Green-lighting, and other methods to generate lots of different ideas will help in determining what improvements can be made in order to eliminate future or potential problems. Think of the three R's when developing new ideas in order to **R**educe the faults, **R**efine the equipment or operational procedure, or if necessary **R**eplace the equipment in order to eliminate problems.

d. When new ideas are authorised and implemented, then it will be necessary to revise the Standard Operating Procedures to reflect these changes.

FTA is a technique used to analyse the root causes and ratios of failures by tracing back events and developing a tree-like figure. FMEA is a technique for analysing the types of failures of constituent parts (failure-phenomena) and their effects on higher items (e.g. facilities, equipment, machines, tools, parts, and systems). It identifies incomplete designs and latent defects. The effects of failure types are appraised using such criteria as degree of influence, failure frequency, and failure detection and repair difficulty.

5.14 Keep maintenance records

Keep records of all maintenance activities for future reference.

There are two main categories of records to keep:

a. Records of maintenance results:

    i. Periodic reports, e.g. daily and monthly failure reports.

    ii. Analytic reports, e.g. maintenance results by manufacturing facilities, and equipment, failure analysis tables and failure statistics by causes.

b. Appraisal of maintenance activities:

    i. Appraisal of individual maintenance activities, e.g. failure duration per case and the costs to the company per day during that failure duration.

    ii. Appraisal of the efficiency of maintenance activities, e.g. maintenance cost per production quantity as compared to the potential loss if such activities were not being conducted routinely.

5.15 Total preventive maintenance

Total preventive maintenance involves everyone within the company participating in maintenance activities. Through this means, a comprehensive maintenance system is established that covers the entire lifespan of facilities and equipment, involving everyone, from CEOs to frontline technicians, in the process.

It requires:

a. Participation by design, operation, maintenance, and other corporate departments and functions.
b. Making active use of statistical techniques to solve problems scientifically.
c. Small group activities where employees in the same workplaces form groups to resolve repetitious problems or develop new concepts and ideas for maintenance.

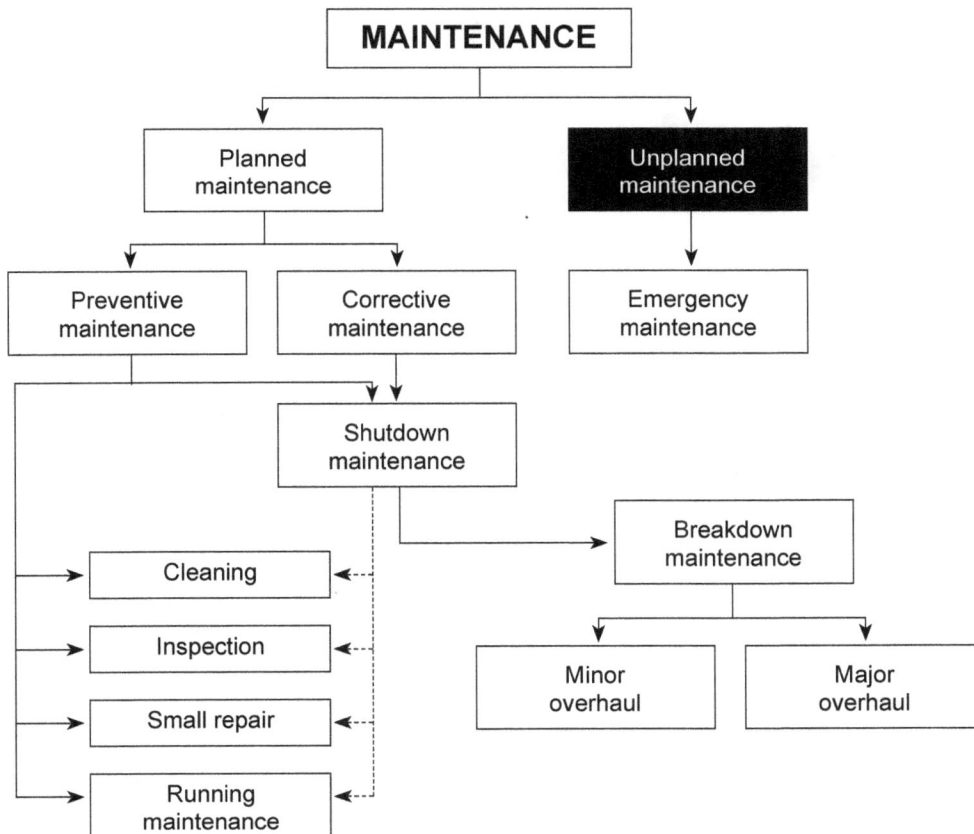

The maintenance cycle

Recommended reading

1. B.C. Buildings Corporation. Mandatory preventive maintenance standard to meet section 4.78 of the W.C.B. O.H. & S. Regulation, 1999.

2. Levitt, J. *Complete Guide to Preventive and Predictive Maintenance*. Norwalk, CT: Industrial Press Inc., 2003.

3. Mobley, R.K. *An Introduction to Predictive Maintenance*, second edition. U.S.A.: Butterworth-Heinemann, 2002.

4. Simmons, D.A., & Wear, J.O. *Hospital Equipment Preventive Maintenance Manual*. Kerala, India: Scientific Enterprises, 1969.

5. Wu, S.M., & Zuo, M.J. Linear and nonlinear preventive maintenance models, *IEEE Transactions on Reliability* 2010; 59(1):242-49.

Self testing multiple choice questions

1. The first step in maintaining facilities and equipment is to:
    a. Carry out regular maintenance activities.
    b. Keep the workplace neat and clean.
    c. Keep the employees motivated.

2. In a GMP environment, the maintenance of facilities and equipment must:
    a. Have employees looking after their own facilities and equipment.
    b. Be carefully recorded by whomever is responsible for the facilities and equipment.
    c. Be performed by regulatory designated inspection points.

3. Defects are more likely to occur:
    a. In a cluttered and unsafe environment.
    b. When employees carry out maintenance checks on their machinery.
    c. In a clean environment.

4. Technicians should be trained to use equipment according to:
    a. Best alternative practices.
    b. In-house modified procedures.
    c. The manufacturer's manual.

5. The daily inspection includes:
    a. Inspection at the start of operation.
    b. Rewiring equipment as necessary.
    c. Inspection only after the operation is completed.

6. The manufacturer's manual is also referred to as:
    a. The Standard Operating Procedures.
    b. The user's manual.
    c. The equipment guidebook.

7. In daily inspections use a check sheet on which the inspection items and methods are written:
    a. Before the inspection.
    b. During the inspection.
    c. After the inspection.

8. The selection of inspection items should be based primarily on the effect that the facilities and equipment have on:
    a. Cell therapy cost.
    b. Cell therapy quality.
    c. Lifespan of the cells.

9. Normal daily inspections should be conducted by:
   a. Members of the maintenance department.
   b. Designated inspectors/auditors only.
   c. Technicians and operators familiar with the equipment.

10. Details of the inspection should be entered on:
    a. The prepared standardised check sheet.
    b. The non-conformance report.
    c. The CAPA report.

11. The SOP on operating equipment should also include:
    a. Alternative methods when failure is imminent.
    b. Discussion points on non-decided issues with the equipment.
    c. Any restrictions on training requirements.

12. Equipment begins to deteriorate:
    a. After two years of being in use.
    b. From the first day of installation.
    c. Only after the warranty expires.

13. To decide on permanent measures hold discussions with:
    a. Managers, supervisors and operators.
    b. Managers, supervisors and those in charge of maintenance.
    c. Managers, supervisors, those in charge of maintenance, and technical staff.

14. Inspections should be performed using _____ senses.
    a. Three
    b. Five
    c. Six

15. When there are serious failures where there will be prolonged downtime:
    a. Carry out repairs at once.
    b. Replace parts on the spot.
    c. Conduct precision inspection to determine the causes.

16. There are _____ categories for classifying abnormality frequency.
    a. Three
    b. Four
    c. Five

17. The single most difficult task in facility management is:
    a. Controlling breakdowns of equipment.
    b. Controlling absenteeism by employees.
    c. Controlling the environment.

18. If the cause of the abnormality cannot be identified then it is necessary to:
    a. Suspend the operations immediately.
    b. Slow down the operation to provide time for identification.
    c. Continue as best as possible.

19. A failure is considered to be when a facility or equipment ceases to function:
    a. Partly.
    b. Completely.
    c. Both a and b.

20. The fixed period ordering method specifically involves:
    a. Fixing the ordering interval in advance.
    b. Ordering a fixed amount of material only the inventory falls below a certain level.
    c. Determining order quantity according to current inventory levels.

21. There are _____ steps to dealing with failure.
    a. Three
    b. Five
    c. Seven

22. There are _____ ways to classify failures.
    a. Two
    b. Three
    c. Four

23. Degradation falls under the classification of failure by:
    a. Cause.
    b. Type.
    c. Degree of function loss.

24. Equipment maintenance refers to activities aimed at:
    a. Preventing equipment getting older.
    b. Identifying and measuring deterioration.
    c. Recovering after deterioration.

25. Maintenance prevention comes under the heading of:
    a. Planned maintenance.
    b. Productive maintenance.
    c. Preventive maintenance.

26. Maintenance prevention means:
    a. Preventing problems occurring.
    b. Designing facilities and equipment to require less maintenance.
    c. Constantly monitoring the conditions of facilities and equipment.

27. Critical is a reference used under the classification of failure by:
    a.   Seriousness.
    b.   Type.
    c.   Degree of function loss.

28. As soon as failure occurs the operator should:
    a.   Complete a CAPA report.
    b.   Write a preliminary report.
    c.   Generate a non-compliance report.

29. Condition-based maintenance systems monitor the status of facilities and equipment:
    a.   Continuously.
    b.   When their condition deteriorates.
    c.   At regular intervals.

30. When a minor failure occurs:
    a.   Precision inspection needs to be conducted.
    b.   Repairs can be performed on the spot.
    c.   Countermeasures need to be introduced immediately.

31. The first step to be taken in breakdown maintenance is to:
    a.   Draft repair plans.
    b.   Arrange personnel and material procurement.
    c.   Investigate the causes and examine countermeasures.

32. The best way to identify where repairs may be necessary prior to a problem is:
    a.   Contingency maintenance.
    b.   Breakdown maintenance.
    c.   Periodic maintenance.

33. Recurrence prevention is also known as:
    a.   Corrective action.
    b.   Corrective maintenance.
    c.   Permanent maintenance.

34. PERT is an abbreviation for:
    a.   Prevention of error recurrence technique.
    b.   Programme evaluation and review technique.
    c.   Prevention expert requirement training.

35. Priority equals:
    a.   Production loss x failure frequency x repair cost.
    b.   Production loss x technician errors x treatment failures.
    c.   Production loss x company strategy x upgrade cost.

36. Condition based maintenance involves:
    a. Waiting for equipment failure prior to replacement.
    b. Constant monitoring of the equipment and facilities.
    c. Constant evaluation of maintenance activity efficiency.

37. Breakdown maintenance is the programme where:
    a. Repairs are undertaken after failure.
    b. Maintenance is based on activity records.
    c. Scheduled maintenance occurs at specified times.

38. Ideally _____ of a maintenance programme should be predictive.
    a. 10%
    b. 25%
    c. 50%

39. Ideally, preventive maintenance should only be _____ of the prevention programme.
    a. 20%
    b. 25%
    c. 33%

40. Total preventive maintenance involves:
    a. Having absolutely no failures occur at any time.
    b. The CEO being fully responsible for maintenance.
    c. Everyone in the company participating in maintenance.

41. To determine the levels of the need for inspection there are _____ points to consider.
    a. Five
    b. Seven
    c. Ten

42. Equipment that needs to be disassembled for inspection will:
    a. Have a lower frequency of inspection.
    b. Have a higher frequency of inspection.
    c. Not affect the frequency of inspection.

43. Equipment with rotating parts such as centrifuges require:
    a. Diagnosis based on sound differentiation of the rotating heads.
    b. Precision diagnosis using an analyser.
    c. Vibrational diagnosis techniques.

44. Pipes and tanks require inspections for:
    a. Corrosion and cracking.
    b. Discolouration and seams.
    c. Smoothness of the welds.

45. Equipment that needs to sustain a precise temperature requires inspection using:
    a.   Heat sensitive colour change tags.
    b.   A thermometer.
    c.   Thermocouplers.

46. In order to inspect computer programmes properly, software is required to:
    a.   Ensure operating code is performing properly.
    b.   Hack into the primary services of the software manufacturer.
    c.   Perform updates for any changes to programme.

47. Green-lighting is a useful tool when performing:
    a.   Recurrence prevention.
    b.   Breakdown maintenance.
    c.   Continuous maintenance.

48. Failure Tree Analysis (FTA) is a technique used to analyse:
    a.   Performance failure of technicians.
    b.   Effectiveness of the immediate repairs.
    c.   Root causes of failures.

49. Failure Mode and Effects Analysis (FMEA) is a technique for analysing:
    a.   Impact of performance errors.
    b.   Types of failures.
    c.   Best selection of maintenance programmes.

50. FMEA identifies:
    a.   Incomplete designs and latent defects.
    b.   Incomplete training and obvious defects.
    c.   Incomplete designs and obvious defects.

SECTION TWO
The Quality departments

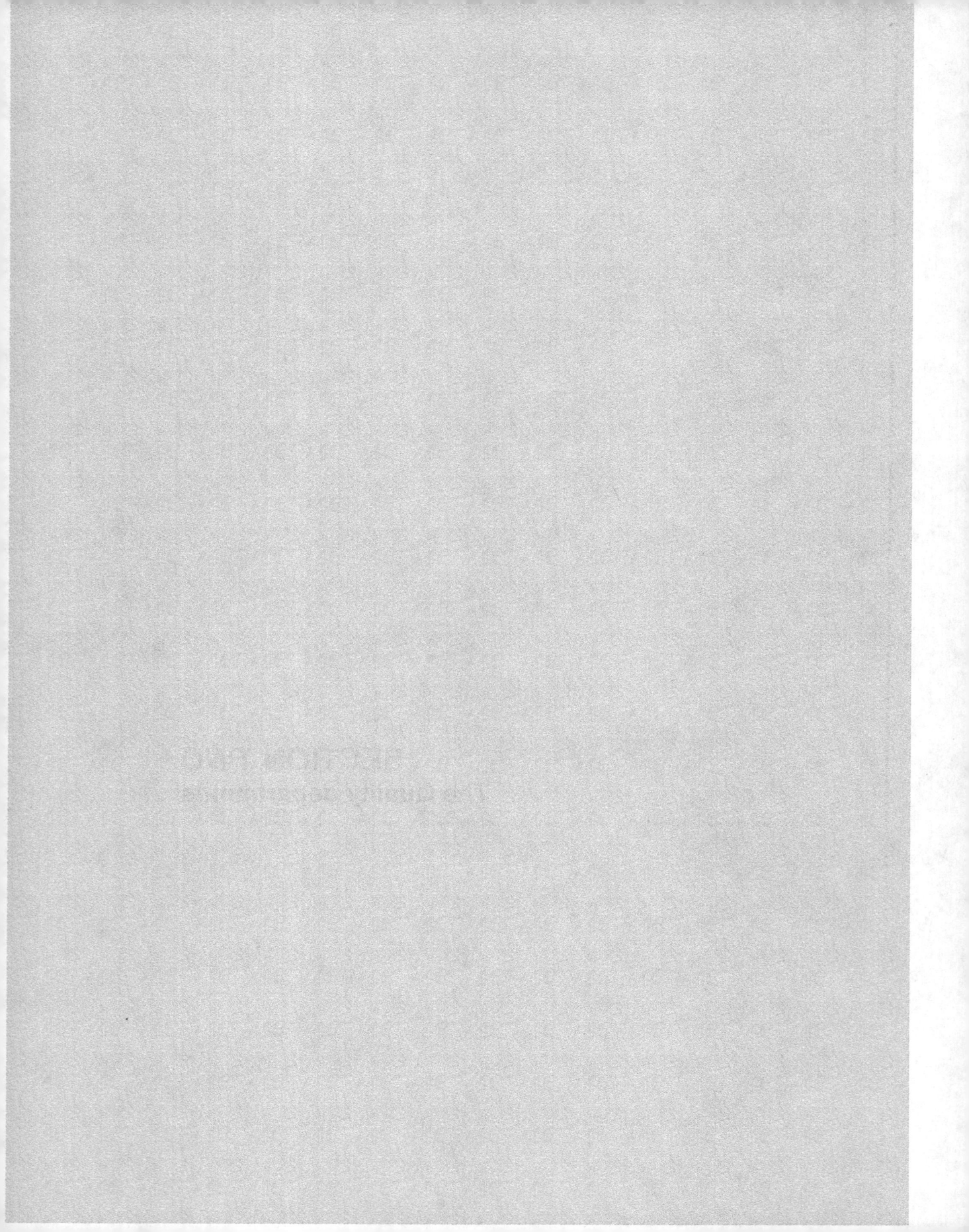

Chapter 6. Quality Operations

Implementation of Quality Systems

The Quality Operations Committee (QOC) is a small group of frontline employees from QO, QA and QC who meet regularly to try to improve the quality of the work performed at a company or institution. Quality Operations activities are at the core of TQM. They can play a major role in creating a dynamic atmosphere in the workplace. Therefore establishment of QOC is a priority.

6.1 Quality Operations Committee

The QOC normally takes a problem-based approach to improve the quality of work. The committee identifies needs, requirements and problems in the workplace, usually related to technical skills, implementation of QMS and product quality and these are referred to as "objectives". The committee sets about finding a solution to any of the objectives they have identified, using quality control concepts and techniques. It is the role of the QOC to try and be creative in seeking its solutions through the application of innovative QMS techniques.

Broadly, their agenda of QOC is to continually improve and maintain the quality of performance and products, and to constantly strive towards self-development as well as group development. Through the quality activities, they develop both a quality and problem consciousness, a willingness to make improvements, and an overall sense of quality management within a company. Ultimately these activities will lead to increased customer/patient satisfaction at the end of the quality chain.

The quality of any product or service is actually determined and achieved by the front-line employees. In the biomanufacturing industry this will be the technicians who prepare the protocols, procure the materials, test and produce therapeutic agents, and deliver these to the patient/customer in the form of treatment. During the treatment, the company operates within the framework of a service industry, and in that case quality depends on those contract service providers, such as the affiliated hospitals who provide the services, and in fact market the services to patients/customers. In the indirect departments (departments such as HR, Finance etc. that provide services to the entire company), quality is determined by their ability to provide services to other employees. Therefore it is the role of the QOC to improve the quality performance throughout all these departments and levels.

The QOC's problem-solving approach seeks to find and remove the root cause of problems through the four stages of the PDCA cycle – draft documents and protocols (**plan**), implement the protocols, SOPs, WIDs (Work Instruction Documents), etc. (**do**), confirm the results of the implementation (**check**), and carry out any necessary follow-up actions (**act**).

The QOC's approach, and quality management in general, is based solidly on facts. To do so, this means first getting all the facts, and then, wherever possible, converting those facts into numerical values. Once in numerical form it is much easier to analyse them objectively and accurately, and to arrive at a sound judgment.

This data-processing procedure involves:

a. Converting facts into numerical values whenever possible.
b. Distinguishing actual causes from the results.
c. Analysing results through stratification (where data is divided according to its sources, e.g. stratified by type of treatment, technicians, by equipment used, etc.).
d. Prioritising items for consideration.
e. Paying attention to the dispersion of the data points in relation to the acceptance limits or target values.

The methods that the QOC uses include:

a. Procedures for problem resolution.
b. The Corrective Action Steps.
c. Other statistical methods.

When employees start using these methods repeatedly, they find it easy to understand this well recognised approach to quality improvement.

Quality Operations activities also include the quality of working life. Most employees have a natural desire to develop their latent abilities and display them to good effect. Quality Operations gives employees the opportunity to fulfil this desire by gaining knowledge, solving problems, and achieving goals. Discussions at Quality Operations also help employees understand their co-workers better, to foster good relationships, and overall, to make the work place more pleasant, more cheerful and more dynamic.

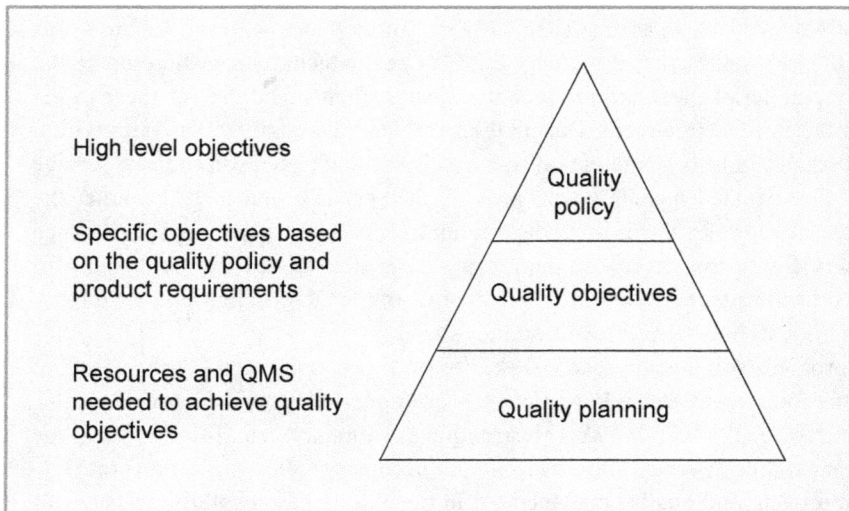

The planning pyramid

6.2 The human dimension to Quality Operations

Almost everyone has an innate desire for personal growth. Under the right conditions most people get a lot of satisfaction from improving their skills, and from using new skills together with their co-workers to achieve meaningful targets. Quality Operations provide the right conditions and opportunities for employees to achieve their desired personal growth.

The aims of Quality Operations activities are to:

a. Fully bring out employee's latent capabilities by expanding their knowledge base. Technicians have considerable ability and by providing a "quality" environment where they can continue to learn, their capabilities will continue to develop. Quality Operations activities provide a framework for learning and development.
b. Create a happy environment where members of the QOC respect each other, and allow each to display their abilities. Every viewpoint is expressed from a quality perspective, thereby improving the workplace environment because everyone has a sense of purpose. Other employees sense the spirit of cooperation and understanding amongst members of the QOC and this in turn encourages their learning and adopting of quality systems.
c. Contribute to the improvement and development of the entire company. Quality Operations activities become part of the corporate front line and are promoted actively from the level of the CEO downwards throughout a company. At the company, the CEO has entrusted the QOC with the task of determining the quality level of the products and services that will be provided to patients/customers. It is appreciated that in a quality work environment, all employees are entitled to exercise their abilities to the fullest, and as a result the company itself can improve and grow.

To make the most of the opportunities that the QOC provides, the employees and technicians should resolve to:

a. Bring out their potential abilities through self-development.
b. Act with good intent, transforming themselves into quality workers.
c. Seek opportunities to further their development as individuals and as a group.
d. Work together: ensuring that information is shared with everyone.
e. Encourage management to support the group in every manner.
f. Do their best to create a dynamic work environment.
g. Develop creative ideas to bring improvements to their work structure and environment and thereby achieve a dynamic working environment.
h. Cultivate quality consciousness, problem consciousness, and a willingness by all employees to seek improvements.
i. Make effective use of Corrective Action (CA) methods to resolve problems, prevent their recurrence, and to deal with potential threats that might be emerging.
j. Work towards the concept and goals of TQM.

6.3 Introducing Quality Operations activities at a company

In recognition of the diverse backgrounds and varied educations of employees at a company, it is necessary that the QOC introduces its quality operational activities in a sensitive manner that respects this diversity of educational and ethnic backgrounds, especially since the overall aim is to encourage the development of employees and not alienate them. Ensuring that the approach adopted by the QOC actually suits the working environment, is an imperative role of senior management.

There are three principal ways to introduce Quality Operations activities:

a. The first way is to introduce them simultaneously in all workplaces, such as offices, manufacturing facilities, testing and hospital departments. This is actively performed by the QOC and is a company-wide approach. By doing so, this promotes a sense of unity within the entire organisation as everyone shares the same learning experience and environmental change at the same time.

b. The second way is through formal training sessions. To achieve this, supervisors will need to be designated as departmental leaders in quality education and perform training activities. The training sessions are based on typical workplace procedures, examining common problems within their respective departments and using CA methods to resolve them. The departments compile their experiences into a quality presentation, and present their findings at a company-wide training conference. Through these training sessions the supervisors will gain the leadership confidence to perform other Quality Operations activities, to promote these activities amongst their staff, and to provide guidance when required. These training sessions should be ongoing, with the intention that they promote continual development at the departmental level.

c. The third way is to appoint demonstration groups made up of willing volunteers from within the company. These groups oversee Quality Operations activities on a department by department basis and gradually spread this training throughout the other workplaces. Any success achieved in one department will then serve as a role model for the other departments. After spending three to six months on problem resolution, the demonstration groups hold a meeting and present their achievements to a much broader number of people from across the company (managers, supervisors and top-level technicians). This encourages other departments to realise they will achieve the same results over time. Quality spreads to other departments by popularising the quality activities.

It should be noted that all three ways of introducing quality programmes require supervisors to constantly monitor the training activities and to provide assistance whenever it is needed.

6.4 How many employees in a company should take part in Quality Operations activities?

Although having all employees taken part in Quality Operations activities is not always practical as a company must continue to operate without interruption of its normal business functions. As training is a vital component of TQM, senior management must determine which departments and sections can opt out of which particular training at which particular time. This is a key decision that senior management must make and as a result, the decision will scale back participation within a company that must be compensated for in some manner at a later time.

There are three modes of participation in Quality Operations activities:

a. Participation by all the departments and sections: manufacturing, quality management and testing, facility and maintenance, human resources and accounting, purchasing, marketing, research and development, IT, hospital technical services, etc.
b. Participation by everyone at the same site: ordinary employees (including long-term and short-term employees and part-timers), and those employed by affiliated contract service providers, supervisors and department managers, all take part in the activities.
c. Participation only by those sections that actually perform quality activities: members attend meetings, express their opinions, and perform their assigned roles when they return to their departments and laboratories.

Participation by all employees within the company has a superior morale boosting factor by providing staff with an innate sense of unity. It also allows those in the same workplace to display solidarity and promote their combined range of abilities to other departments they normally do not interact with. The major problem is that in large companies, this approach is unwieldy as it can involve thousands of people, all gathered into a centralised facility that cannot properly accommodate them, which in turn becomes detrimental to the learning process and as a result has a subsequent demoralising effect.

6.5 Selection of leaders and their roles

The quality activity leaders/supervisors will be the driving force behind the activities and relies not just on those members of the QOC. It is important that the QOC properly selects people who can display this level of leadership, encourage others to cooperate in meetings, effectively gather ideas, and more importantly, create an atmosphere where everyone will feel free to express their opinion without fear of repercussion. Quality of the leaders is important if the training activities are to succeed.

Different ways of selecting leaders will be appropriate at different stages of introducing and establishing Quality Operations activities:

During the introduction of TQM to a company, supervisors are probably the most suitable leaders because of their inherent and established authority. These workplace leaders are at the forefront of this process, and thus quality management will reach their rank-and-file employees by simple diffusion. Yet, it will not become firmly established in the workplaces until such time that purpose trained leaders are made available to deal with the finer points. Employees must have a sense of active involvement in quality by management and this can be achieved through their frontline managers whom are part of the technician's daily routine.

Points to consider when selecting leaders:

a. Avoid selecting a leader with no leadership capabilities, which unfortunately do commonly exist in companies. How these people became supervisors or even higher in the first place is always a question. These leaders negatively impact and even impede training activities. They often have a negative attitude towards any change, thus become an impediment to the implementation of quality standards within a company.

b. Any activity leaders who have participated for two or three years in the corporate training processes and have a degree of understanding of quality methods, will be suitable leaders for TQM and in charge of other trainers because of their specialised experience and knowledge.

The primary role of Quality Operations activity leaders is to maintain the dynamics of the training sessions, to encourage staff to use their abilities to the fullest, and to support them with any difficulties they might encounter.

These quality leaders should:

a. Investigate what improvements technical staff would like to see in their working environment, identify specific problems, decide how to approach them, and select achievable targets.
b. Assess the basic qualities and skills of the staff they are responsible for, and as a result, assign them roles that allow them to put these skills to good use, thus create an atmosphere that motivates them.
c. Introduce venues by which staff can acquire the knowledge and skills needed to carry out the activities. Once sufficiently knowledgeable, then they should be encouraged to take on a training role themselves to disseminate the knowledge they have learned.

Technicians and employees should support their quality activity leaders by carrying out their assigned roles responsibly and diligently and by acquiring the skills and experience that will enable them to acquit themselves well, while continually improving the quality of their work and of their work environment.

Technicians receiving training should:

a. Provide active assistance to their leaders and participate in quality management.
b. Attend meetings and speak openly of their own experiences.
c. Carry out the roles assigned to them dutifully and in the required time frame.
d. Continually learn and adopt the technology and associated quality systems within a company.

6.6 Quality Operations meetings

Quality Operations activity meetings help members to work together towards achieving the same goals. Employees exchange ideas and information, while at the same time developing a spirit of cooperation and a sense of solidarity. But if these meetings are poorly managed, then the process will stagnate and members will actually become demotivated. Quality is a difficult concept for most people to adopt and sustain unless they are continually motivated and constantly apprised of its value. Historically within companies, the tendency has been to let quality lapse as people easily become distracted by issues they incorrectly assume to be more important. It is the purpose of the meetings to remind everyone at the company that there is nothing more important than the implementation and sustaining for a quality system.

If meetings are to be effective:

a. They should be planned well in advance.
b. All staff at some point should attend and be involved to avoid segregation and isolation of any department.
c. Roles should be distributed amongst employees so they are more actively involved: moderator, secretary, presenter, etc. so that active participation is a requirement.
d. The subject matter of each meeting should be detailed in advance.
e. Brainstorming sessions should be held to generate new ideas. Remember that there are no "stupid" questions nor are there any "stupid" responses in a brainstorming session.
f. All participants should be invited and strongly encouraged to give their opinions on any matter related to quality of product or performance.
g. Minutes should always be documented as a reminder of what was agreed upon and any roles that may have been assigned or volunteered for. The general rule under GMP is that if it was not documented then it never happened.

Planning meetings properly means it is necessary to decide on their duration, frequency, timing and place:

a. **Duration:** The length of meetings will vary according to such factors as workplace conditions, the agenda, and their frequency. However the average Quality Activity meeting should be about 60 minutes. Meetings that go on for longer than 2 hours usually have a demotivating effect and less than 30 minutes are considered ineffectual and unimportant.
b. **Frequency:** The number of meetings per month will vary according to their average duration. However, they should take place at least twice a month with once a week preferable so that everyone appreciates the importance of Quality. It is best to set the times and dates of activity meetings in advance to be held in the next one-month cycle and to adhere to this target. In order for this to be achievable, then senior management must fully support the agenda and make the time available for employees.
c. **Timing:** The times when meetings are held will vary according to work requirements and the other demands on Quality Operations members' time. When they meet during regular working hours, they should coordinate this with the normal production schedule to avoid conflicts. It is advisable to:
   i. Include QO activities in the monthly operational schedule.
   ii. Set up standing specific dates for meetings.
   iii. Use a SMS or messaging to determine when all members can meet.

d. **Place:** Meetings should be held in one of the following places:
    i.   The workplace.
    ii.   Conference rooms near the workplace.
    iii. Dining rooms.
    iv. Outdoor locations within the facility grounds.

Alternating environments help to stimulate and motivate personnel attending the activity meetings. These do not have to be exotic locales, simply some place a little different from the last.

6.7 Quality Operations Committee assemblies

After the Quality Operations Committee has completed an objective, it should re-appraise the methods that have been used for training, and to confirm that overall the training was well received by the staff and the company feels the goals were achieved. Members from the departments, training and sub groups should give feedback on the key points from their perspective and raise any questions regarding methods, or encountered difficulties, as well as raise any of their own creative ideas.

When presenting responses and perspectives the department spokesperson should provide:

a. A brief outline of any tasks the department carried out as a result of the training.
b. Any reasons for adopting a particular focus behind their identifying and selecting the problem to be dealt with, and the circumstances surrounding it, etc.
c. The situation they arrived at, using data to support their reasoning why it was considered significant.
d. The targets they selected and the basis upon which they were selected.
e. Any action plans they adopted, including the assignment of roles.
f. Analysis of the key factors using data to support the department's hypothesis and subsequent data to verify the hypothesis.
g. Details on the preventive measures decided upon and how they were implemented. The effectiveness of the recurrence prevention measures – compare the situation before and after the recurrence prevention measures, the level of target achievement.
h. The practices by which items and methods were standardised following the corrective actions that were implemented.
i. Any plans for follow up. A summary of the points needed for further examination has to be included. Identify the lessons learned from the exercise.

It is important to disseminate the information raised and provided by one department so other departments can also benefit. Quality Operations assemblies that are company-wide offer departments the opportunity to make presentations. By seeing how other departments or groups handle a situation, it provides confidence and reassurance and suggests a strategy to other departments that are encountering similar problems.

The lessons and benefits are multifold:

a. Presenters:
    i. Learn to express their ideas clearly, improve their communication skills, and gain confidence in giving presentations.
    ii. Become skilled at organising ideas and information so that people can easily understand what is being presented.
    iii. Experience the sense of satisfaction in their accomplishments.
b. Listeners:
    i. Learn about dynamic activities in other workplaces and expand their own scope of problem solving strategies.
    ii. Learn the practical applications of quality methods, creative thinking, and ways of organising information and ideas.
    iii. Learn that creative ideas that emerge from the Quality Operations Committee can be applied, achieve realistic results and are not merely a hypothetical exercise without any substance.

The Quality Operations Committee should assess the value and accomplishment of any company-wide assemblies.

QOC should pay attention to the following:

a. That any achievements being reported and examined focus on the context that they improved the work processes and performance. The QOC's perspective must not be the same as the corporate perspective which will be focused on profits and normally not entirely quality focused.
b. Any achievements reported are based contextually on the principle of quality first as a component of TQM. Everything else is secondary.
c. Consideration that willing participation is a critical factor for evaluation.
d. Consideration that all the managers, technicians and employees participated with a true sense of quality awareness. QOC has to determine that people were not merely going through the motions without a sincere commitment.
e. How presentations by the departments are related to the higher policies of the company without sacrifice in quality.

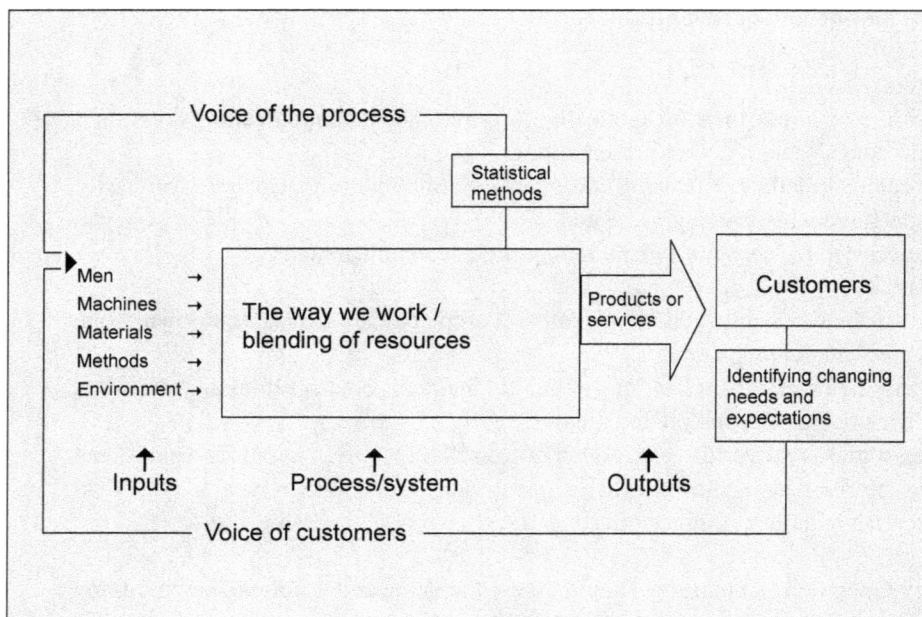

The Quality Operations information feedback loop

6.8 Evaluation of Quality Operations activities

Regular and appropriate evaluation of Quality Operations activities will help to motivate participants and to keep the programme proactive. It will identify where improvements are needed, and indicate any corrections that should be undertaken. There are two forms of evaluation: self-evaluation by employees, and evaluation by managers.

Evaluation should address the following questions:

a. Are Quality Operations activities in line with the action plans and targets?
b. How are the activities being conducted?
c. What abilities can be identified that the employees are developing?
d. Are the activities achieving satisfactory results according to expectations?

Self-evaluation by technicians and employees:

a. Self-evaluation after a problem has been resolved: Technicians and staff reflect on all the steps they have taken to resolve any problems, evaluate them, recognise any unsatisfactory points, and try to identify and correct the causes.
b. Self-evaluation at the end of the year: Activities are ongoing, so when staff have solved one problem, they should start planning to solve a new one. At the end of the year, the staff should review and evaluate all their activities in order to both assess and promote their accomplishments.

c. Limits of self-evaluation: Those employed as Quality Operations leaders can be expected to evaluate their activities willingly since that is their primary function and role within the company. The same may not be true for technical employees who see their roles within the laboratory in terms of performance and production of therapies. Their sense of accomplishment is in the output of what they produce, not necessarily in the application of theoretical and regulatory requirements. It is necessary that the importance of their quality roles are very much equal if not more important to those in the production and manufacturing category.

d. Self-evaluations tend to produce liberal ratings as no one likes to assess themselves badly. In order to establish a greater sense of objectivity, in addition to their self-evaluation, employees should be prepared to be evaluated by their managers and supervisors to see if they are in agreement. These secondary observations/evaluations will give employees a fuller picture of their implementation of quality requirements and point to any areas that may require increased attention and perhaps further training.

Evaluation by managers and supervisors:

a. Evaluate activities: Immediate superiors receive written and oral reports on quality activities. It is their role to evaluate these reports, and respond to their staff with appropriate instructions and advice as each step of the problem resolution procedure is completed. Their instructions should be appropriate to the level of their own technical skills.

b. Evaluate staff performance during Quality Operations and advise employees where they think improvements can and should be made.

c. Evaluate all Quality Operations activities for a specific term and the year-end report to the QOC. This is a key evaluation because of the emphasis on continuity of the quality framework within the departments.

Self evaluation of managers and supervisors:

a. After they have provided evaluations of their employees and their department, the managers should reflect on their own instruction methods and identify areas where they could make improvements.

b. Managers should reflect on any identified problems addressed by the QOC and re-evaluate their own day-to-day management procedures on how these issues were handled successfully or not depending on the QOC assessment.

c. Presenting a broad view of all the quality problems and issues they have been tackling, managers and supervisors should be able to report to the QOC the improvements they are introducing that will eliminate the root cause and rectify the problems.

6.9 Basic procedures for quality implementation

Once Quality Operations activities have been set up, there are a large number of procedures and rules to follow. These procedures involve a considerable commitment by the company and by the members of the QOC.

The basic requirements are that:

a. Members of the QOC all understand the concept and purpose of Quality Operations activities.
b. Members of the QOC actively exercise their function of leadership. This is supported by senior management that lend their authority to the QOC.
c. Members of the QOC have the ability to spread an awareness of the need for quality activities and are willing to participate in order to see it adopted by a company.
d. Members of the QOC have the ability to create an environment that encourages active participation from the employees.
e. At all times members of the QOC keep focused on the objectives of the activities and do not lose sight of the intended purpose.
f. Members of the QOC continually study quality methods and bring any innovations, regulatory recommendations, etc. to the quality activities where they transfer this information to the managers, supervisors and employees.
g. Members of the QOC steer the general quality and activity meeting to ensure they remain on topic and on schedule.
h. Members of the QOC carry out a self-evaluation of their own activities.

6.10 Quality Operations training

The goal of Quality Operations training is to achieve a level in the company where all employees are motivated, creative, problem-solvers, without reliance on the few members of the QOC to do all the thinking and implementation for them. Training must focus on the value of quality, on all associated programmes and activities within the company, on raising quality consciousness, and on the proper use of quality methods.

Training is achieved when employees:

a. Understand the basic concepts of the Quality Operations training programme:
    i. The objectives and value of a company-wide effort to promote quality activities throughout the workplace.
    ii. Quality perspectives and ways of thinking in quality mode.
b. Develop methods for quality implementation:
    i. Procedures for advocating quality activities.
    ii. The role of the QOC and its members.
    iii. How to participate in quality meetings and activities and contribute to the implementation of quality in the workplace.
c. Adopt methods for improving problem resolution:

    i.   Know the steps involved in problem resolution, the methods used at each step, and how to apply them to actual problem resolution in the workplace.

    ii.   Make use of tools: Pareto Diagram, Cause and Effect Diagram, Check Sheet, Histogram, Control Chart, Scatter Diagram, Stratification. (See Chapter 8 Statistical Process Control).

    iii.   Utilise the CAPA method of resolving problems.

    iv.   Use tools such as FMEA for the implementation of preventive actions.

Advocating methods of study:

a. For proper participation in Quality Operations activities:
    i.   Encourage staff to read: textbooks on Quality Operations and Quality Control, articles, and other publications.
    ii.   Listen to others and learn from them at Quality conferences, lectures, seminars, and similar forums.
    iii.   Learn from discussions at meetings held at the workplace, quality sessions, etc.

b. For advancing and managing Quality Operations activities:
    i.   Read articles, journals, books describing how to run meetings, and to select and present objectives.
    ii.   Learn from past activities, presentations, documented cases, meetings to exchange experiences and ideas.
    iii.   Learn from practical experience and bring that knowledge to the activities and meetings.
    iv.   Groups should develop innovative techniques in making improvements and share their ideas.

6.11 Adopt a quality plan

Activities presented by the QOC are designed solely for the purpose of advancing training and implementation of quality systems within the company. These activities and meetings are all part of the development of TQM. Therefore, an overall plan should be designed that prescribes the level quality that will be attained and explains the purpose, function and design of the activities, with identification of the activity targets, and concrete measures for achieving these targets. When preparing this plan, the QOC must look carefully at the present state of quality within a company and should appreciate the correlation with the vision and mission of the company. The presence of a quality implementation plan is essential in consolidating the quality activities and conveying an appreciation to the employees of the true significance of TQM.

To draft the plans:

a. Clearly define the type of activities to be developed, based on the fundamental principles of performance within a company, department, hospital and laboratory.

b. Consider problems that currently exist within the company, department, hospital or laboratory.

c. Decide the target level of the activities over a given period (short-term, mid-term or long-term), as well as the concrete steps to be taken in order to achieve those targets.

Points to consider:

a. It is a priority to have a good understanding of the present situation concerning quality and to be certain that any quality plans developed can be successfully implemented and inspected. Reality over dreams visualisation is required.
b. Ensure that the plans cover all the Quality Operations overview, including management, studies, and the resolution of problems in workplaces.
c. Ensure that the plans correspond with the operational structure of the company, department, hospital or laboratory.
d. Clarify the relationship between the plans and implementation of TQM within the company.
e. Establish realistic targets based on past achievements. For example, prior to introducing new activities, Quality Operations planners should examine past records, training courses, conferences, minutes of meetings, and reference materials, to see if the company has the capacity, capability and competence to achieve the new desired targets.
f. Identify concrete ways of achieving the targets and avoid the hypothetical. Provide a company-wide roadmap.
g. Establish management inspection items that can be checked to see if the target levels are being achieved, and define the methods for inspecting these.

6.12 CEOs must fully support Quality Operations activities

Quality Operations activities can only be successful if they are presented in the right environment. For quality implementation to succeed then management has to strongly believe in its relevance and importance to the company as a whole. CEOs, middle managers, promotional staff members, and personnel managers must all possess a good understanding of Quality Operations activities in order for them to create such an environment. Most importantly, the CEO must be the central focus that sends out a clear signal to everyone working at the company that TQM must be adopted and failure to do so will not be tolerated.

For successful TQM implementation, CEOs should:

a. Clearly indicate their belief and commitment to TQM and how they intend to implement these policies.
b. Appreciate the importance in introducing the quality activities in order to realise corporate principles, implement long-term strategies and further company-wide quality management (TQM).
c. Endorse a company's policy for introducing Quality Operations activities and announce it to all the employees.
d. Position Quality Operations activities clearly and pivotally within a company framework.

e. Take an active interest in the introduction of Quality Operations activities.

f. Support the promotion of Quality Operations activities by providing the resources and indicating the direction that this promotion could take.

g. Ensure that the Quality Operations activities are aligned with the particular regulations that a company adheres to by providing QOC with the necessary resources.

h. Ensure that company-wide training in Quality is mandatory for all managers, leaders and supervisors.

i. Ensure that an education budget is established and resource materials are provided for proper training and development.

j. Ensure that employees who are willing to do so are encouraged to take part in external activities (conventions, seminars, exchange-meetings to other companies and seminars).

6.13 Middle management roles

Middle managers must support a working environment in which employees are allowed to take on a significant role in assessing quality of their own workstation and fostering an environment in which staff willingly contribute their own quality ideas.

Middle managers should:

a. Practise management control activities for TQM as outlined by the QOC.

b. Create a supportive atmosphere for quality activities where employees willingly contribute.

c. Keep up to date with the status of Quality Operations and provide instructions and advice from the management perspective as the activities progress.

d. Get a good understanding of a company's policy on introducing TQM and communicate it accurately to their staff.

e. Get a good understanding of Quality Operations activities.

f. Support the promotional activities of TQM.

g. Implement any promotional plans through team efforts.

h. Inform the CEO of the status of quality activities and help them to appreciate the value of those same activities.

i. Inform their staff of introduction courses that CEO has decided upon, and of the related policies that their department will adopt.

j. Set up educational programmes to provide staff with the knowledge and skills they need to conduct Quality Operations activities.

k. Generate opportunities for employees to work together to practise and develop quality skills.

l. Evaluate Quality Operations activities during the in-process phase and afterwards.

m. Personally practise quality management activities. Middle management must recognise that the easiest way in which to introduce TQM practices to their staff is to be observed putting them into place themselves.

6.14 Promoting quality activities

Establish a company-wide structure to promote and facilitate Quality Operations activities. This should include a promotional committee and a promotional secretary. This organisation is different than the QOC which is concerned with introduction of quality training and implementation of the QMS for documentation. This promotional committee is concerned with setting up the infrastructure which facilitates and permits the QOC to carry out its functions seamlessly and without any resistance.

Since the promotional activities are conducted as a component of TQM, it is important to establish three systems:

a. A top-down system, by which corporate policies and TQM policies are communicated accurately to the staff.
b. A bottom-up system, by which the various problems that staff confront are communicated accurately upward to Quality and senior management.
c. A system by which the Quality Operations Committee receives instructions and assistance they need in order to conduct their activities correctly and energetically.

To set up the promotional organisation:

a. Establish a quality promotional committee, chaired by a top executive and staffed by department managers.
b. Establish a Quality Operations activities promotional secretary within the TQM promotional structure.
c. Establish an in-house registration, dissemination of information and reporting systems for Quality Operations activities.
d. Draft company-wide plans for promoting Quality Operations activities and the methods to ensure every staff member views the promotion.

The Quality Promotional Secretary:

a. Prepares the mechanism for getting Quality Operations activities started within a company.
b. Communicates with external parties and collects and disseminates information.
c. Conveys the policies and decisions of the CEO and the promotional committee to the staff.
d. Develops a good understanding of Quality Operations activities so that he/she is comfortable in discussing its role and function.
e. Assists the Quality Operations Committee in a secretarial capacity.
f. Provides whatever assistance is required to the QOC, and reports on the status of activities to management.
g. Performs any administrative duties for Quality Operations activities committees and meetings.
h. Prepares in-house reference materials, including handbooks and educational texts.

i.   Publishes Quality Operations activity news and other relevant public relations materials.
j.   Provides indirect assistance by offering advice to managers, and to QOC members.

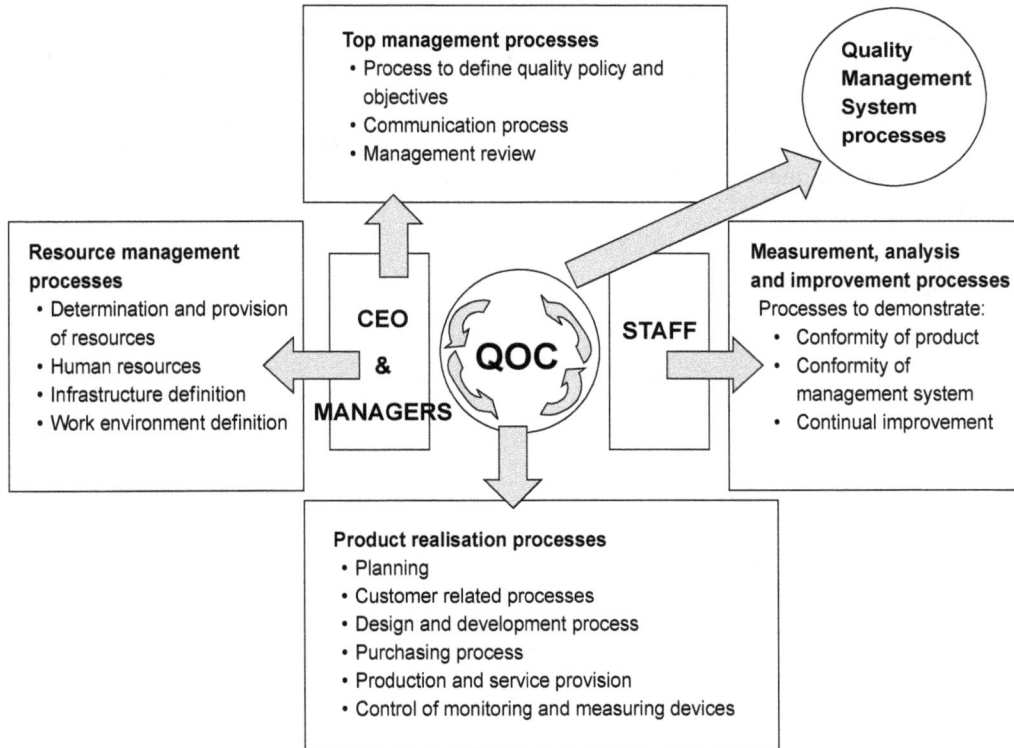

Communication flow through the company

6.15 Exchanges with other companies

When Quality Operations activities remain solely within staff's own workplaces and laboratories, their overall scope of work may become limited and short-sighted in the staff's perspective. When this happens they lose their initiative and quality subsequently deteriorates. Organising meetings or attending conferences periodically where staff will encounter their counterparts from other companies leads to a free exchange of ideas that is of mutual benefit. Fresh ideas from other Quality specialists overcome the stagnation and deterioration that companies are prone to.

Benefits of inter-company Quality Operations activities exchange:

a.   Provision of new ideas for activities, and for ways to carry them out.
b.   Provision of tips and techniques to help resolve management problems and revitalise activities.

c. Provision of a conduit and network for communication, thereby improving inter-company human relations.

The basic procedure for inter-company quality activities exchange-meetings:

a. Be aware of what information is not to be disclosed and confidential.
b. Discuss issues of a broad spectrum in an open framework that does not specifically identify problems within your own company.
c. Respect the privacy issues of other Quality specialists from other companies being aware that they are under the same restrictions on how much information they can provide.

Recommended reading

1. U.S. Food and Drug Administration. 21 CFR 211.22: Responsibilities of quality control unit.

2. U.S. Food and Drug Administration. 21 CFR 1271.160: Establishment and maintenance of a quality program.

3. WHO Technical Report Series, No. 902, 2002, Annex 3.

4. WHO Technical Report Series, No. 908, 2003, Annex 4.

(Note: CFR stands for Code of Federal Regulations.)

Self testing multiple choice questions

1. The criteria against which improvement in the quality of work will be measured are:
    a.   CEO demands.
    b.   Supervisor demands.
    c.   Customer demands.

2. PDCA stands for;
    a.   Plan, do, correct, act.
    b.   Prepare, do, check, act.
    c.   Plan, do, check, act.

3. Discussions at Quality Operations activities help employees to:
    a.   Understand their managers better.
    b.   Build better relations with colleagues.
    c.   Make the workplace more stressful.

4. The basic ideas behind Quality Operations activities are to:
    a.   Fully bring out employees' latent capabilities.
    b.   Contribute to the profits of a company.
    c.   Make the world a better place to live in.

5. To make the most of Quality Operations activities participants must resolve to:
    a.   Bring out their potential through self-development only.
    b.   Seek opportunities for development as an individual and as a group.
    c.   Critique any creative ideas their colleagues raise.

6. Promoters of Quality Operations activities should _____ check how employees feel about them before introducing them.
    a.   Always
    b.   Sometimes
    c.   Never

7. Introducing a Quality Operations activity should be done by:
    a.   Introduction at each workplace in turn.
    b.   Centralised meetings for all company employees.
    c.   Both a and b.

8. Of the three models of participation in Quality Operations activities the one not included is:
    a.   Participation by all departments and sections.
    b.   Participation by everyone in the same workplace.
    c.   Participation by those who belong to the same profession.

9. During the inauguration phase of Quality Operations activities the most suitable leaders are:
   a. Middle managers and supervisors.
   b. The cleaning staff.
   c. Persons with leadership abilities.

10. The five primary functions of Quality Operations activity leaders include:
    a. Train successors for QOC since QOC stays distant from training.
    b. Instruct the participants on how best to meet the profit expectations.
    c. Educate Quality Operations activities members.

11. Quality Operations activities participants are expected to:
    a. Follow the financial department's instructions.
    b. Study engineering technology and quality control.
    c. Carry out the roles assigned to them.

12. The five priorities in the workplace do not include:
    a. Quality.
    b. Costs.
    c. Leisure.

13. To conduct Quality Operations activities effectively:
    a. All employees should attend at some time.
    b. Only managers need to attend.
    c. Brainstorming sessions should be limited to senior management.

14. An average Quality Operations meeting should last:
    a. 10 minutes.
    b. 60 minutes.
    c. 180 minutes.

15. Which of the following components of a QOC presentation are in the correct sequence?
    a. Reasons for selecting a theme, action plans, establishing targets.
    b. Introducing the company, understanding the present situation, action plans.
    c. Action plans, examining recurrence prevention measures, analysing key factors.

16. Presenters must:
    a. Be born with the ability to publicly speak.
    b. Restrain from presenting any emotion.
    c. Keep their listeners engaged.

17. Advisors should keep their comments on presentations:
    a. To themselves so as not to embarrass the presenter.
    b. Strictly for senior management only.
    c. To open discussion so all other presenters and groups can benefit.

18. Following a presentation, managers, facilitators, etc. should pay attention to:
    a. The potential profits that a company could achieve from the new ideas.
    b. Whether all members participated with a sense of awareness.
    c. How presentations relate to higher policies.

19. Which of the following is most important to evaluate Quality Operations activities?
    a. Evaluation by CEOs.
    b. Self-evaluation by members.
    c. Evaluation by senior managers.

20. Evaluation by managers includes evaluation of:
    a. Ways to avoid implementing the problem-resolution procedures.
    b. The relevance of the QOC objective to their own department.
    c. Activities carried out by the promotional secretary.

21. The basic requirements for keeping Quality Operations activities going include:
    a. Meetings are always finished at the same time.
    b. Leaders and members keep the objectives of activities in mind as they carry them out.
    c. Having a conference hall off site at all times.

22. The seven QC tools mentioned in this text include:
    a. Pareto diagrams.
    b. Golf charts.
    c. Photos.

23. The study methods presented in this text for advancing and managing Quality Operations activities include:
    a. Read articles.
    b. Learn from experience.
    c. Both a and b.

24. The procedure for drafting plans for future Quality Operations activities includes:
    a. Vaguely define the type of Quality Operations activities to be developed.
    b. Consider the problems that exist within the company, department or section.
    c. Not setting any target levels of Quality Operations activities over a given period.

25. CEOs should:
    a. Remain distant from the introduction of Quality Operations activities.
    b. Clarify a company's policy for the introduction of Quality Operations activities.
    c. Try to grasp and take charge of the detailed problems posed by QOC groups.

26. CEOs should:
    a. Position Quality Operations activities clearly within a company.
    b. Directly provide instructions and control of Quality Operations activities.
    c. Manage the promotional systems for Quality Operations activities.

27. The functions of middle managers in Quality Operations activities include:
    a. Keeping CEOs informed of the status of Quality Operations activities.
    b. Allowing as much time as needed for the activities without any oversight.
    c. Not affording opportunities for members to work together.

28. Middle managers take the initiative in:
    a. The introductory period.
    b. The development period.
    c. The period in which spontaneity is established.

29. Setting up a company-wide organisation to promote Quality Operations activities includes
    establishing:
    b. A Quality Operations activities chaired by the CEO secretary within the TQM promotional secretariat.
    c. In-house registration and reporting systems outside of QOC control.

30. The functions of the Quality Operations activities promotional secretary include:
    a. Preparing the environment conducive to Quality Operations development.
    b. Selecting the Quality Operations activities leader.
    c. Publishing Quality Operations activities news and other public relations materials.

31. When contacting Quality experts from other companies it is imperative to:
    a. Respect confidentiality of information.
    b. Divulge all information freely.
    c. Use specific examples from your own company when discussing problems.

32. In setting Quality objectives the top step of the planning pyramid is:
    a. The Quality objectives.
    b. The Quality policy.
    c. The Quality planning.

33. Key to the data processing procedure is:
    a. Randomising all items for consideration as not to prejudice selection.
    b. Combining causes with results.
    c. Converting facts into numerical scores whenever possible.

34. There are _____ principal pathways to introduce Quality Operations activities.
    a. Three
    b. Seven
    c. Ten

35. The number of employees that should participate in Quality Operations activities is:
    a. Half.
    b. Some.
    c. All.

36. There are _____ modes of participation in Quality Operations activities.
    a. Two
    b. Three
    c. Five

37. Participation by the maximum number of employees in Quality Operations activities:
    a. Gives everyone a break from the daily work routine which boosts morale.
    b. Means the company loses money by production stoppage, so it is not possible.
    c. Has a superior morale boosting factor and provides a sense of unity.

38. When selecting Quality leaders:
    a. It does not matter if they have no leadership capabilities.
    b. Select those already with proven quality leadership.
    c. Base the decision upon seniority.

39. Quality Operations activity meetings should:
    a. Be planned in advance.
    b. Occur spontaneously.
    c. Discourage introducing new ideas.

40. Quality Operations activity meetings should be scheduled at least:
    a. Every week.
    b. Twice a month.
    c. Every six months.

41. Benefits from Quality Operations Assemblies are for:
    a. Both presenters and listeners.
    b. Presenters only.
    c. Listeners only.

42. The problem of self-evaluation is:
    a. Most employees are too critical of themselves.
    b. Most employees assess themselves too leniently.
    c. Most employees cannot take the process seriously.

43. A basic requirement for implementation of the Quality System is that senior management:
    a. Lends its authority to the QOC.
    b. Creates the environment that supports active participation.
    c. Carries out the objectives and activities.

44. CEOs and middle management are involved with TQM in:
    a. Leadership roles.
    b. Non-participating roles.
    c. Support roles.

45. It is important to establish _____ system(s) for promoting Quality Operations activities.
    a. One
    b. Three
    c. Five

46. Quality Operations meetings should be held:
    a. In or close to the workplace.
    b. Far away from the workplace for a change of environment.
    c. Always in outdoor locations.

47. When presenting why a particular quality focus was selected, it is important to explain:
    a. Why there was no action plan involved.
    b. How the particular focus financially impacted on the company.
    c. How the initial problem was identified.

48. The QOC should not pay attention to:
    a. The monetary achievements by a company.
    b. The improved performance achievements by a company.
    c. The improved process achievements by a company.

49. The basic principle of TQM is:
    a. Quality is always added on.
    b. Quality comes first.
    c. Quality is last.

50. In the Quality Information Feedback Loop, quality improvements to products and/or services directly affect:
    a. Customers/patients.
    b. Senior management.
    c. Employees.

Chapter 7. Problem solving

Root Cause
Analysis

It is an established truth that there will always be problems in work processes whether it be due to human error or mechanical fault. What is more important is that they are identified immediately, reported to whomever will act upon them, and that emergency action is taken to contain and minimise any damage, while at the same time, finding out what is causing the errors, and preventing them from occurring again. There are established systems to use that will help in recognising and dealing with problems. (Chapter 8 provides detailed guidelines on using statistical methods to help solve problems through analysing and interpreting data.)

7.1 Recognising abnormalities

If managers and employees are to identify problems as soon as they occur, they need to have an awareness of what to look for. This requires not being alert just for problems, but for anything unusual, anything that is considered different from the norm; what we would label as an abnormality. An abnormality is not always a defect or a problem, but it often indicates that there is a hidden problem or the risk of a potential problem. Always treat an abnormality as a danger signal, find out what is causing it, and take measures to prevent it recurring.

There are two types of abnormalities:

a. Those that can be identified through inspections.
b. Those that cannot be identified through inspections.

Typical abnormalities that are found during inspections are:

a. An increased number of errors or defective cells in a product.
b. A new type of defect not encountered before.

These may be found by inspectors/auditors during periodic inspections, or by employees conducting their own inspection. Often they are only seen when examining control charts.

In periodic inspections the current values of the product or process are compared with the standard or target values expected. The term "dispersion" is used to describe how the current values differ from the standard or target values. Some dispersal is anticipated and acceptable within reason, but if dispersion goes beyond that, then you have an abnormality. Wherever possible, convert the values observed on inspection into numerical figures, and present them visually in graphs or charts so that abnormalities can be easily evaluated. A control chart is a typical graph for identifying abnormalities. It shows how the values are dispersed and will immediately display the presence or absence of any abnormalities.

However, dispersion during the production process is not always caused by abnormalities. It may also be caused by random chance, even when standard processes are strictly adhered to. These cannot be eliminated but their impact can possibly be reduced. A control chart will show the dispersion type. When it shows that dispersion results from abnormalities, investigate the causes and take appropriate action.

There are some abnormalities that are not spotted on inspections such as equipment which is producing values, readouts, or results that do not appear abnormal when examined but in reality are false and the true value is outside the acceptance limits. Unless there is reason to suspect that the results are false, it is unlikely that anyone will ever notice them and therefore the equipment will never be reported as being faulty.

Similarly, there are numerous occasions when technicians because of their experience have a suspicion that something is not working properly. They may not know exactly what the problem is, but they have a strong feeling that there is something unusual about the cell products. This intuition is honed from working years within the laboratory and should not be ignored or dismissed. They should immediately call their supervisor who should spontaneously examine the product. If they confirm that there is a suspected difference from the norm, then production should be suspended immediately and a non-conformance report is filed with Quality Assurance regarding their identification of a suspected abnormality. An investigation will then be initiated to assess the accuracy of the reported abnormality.

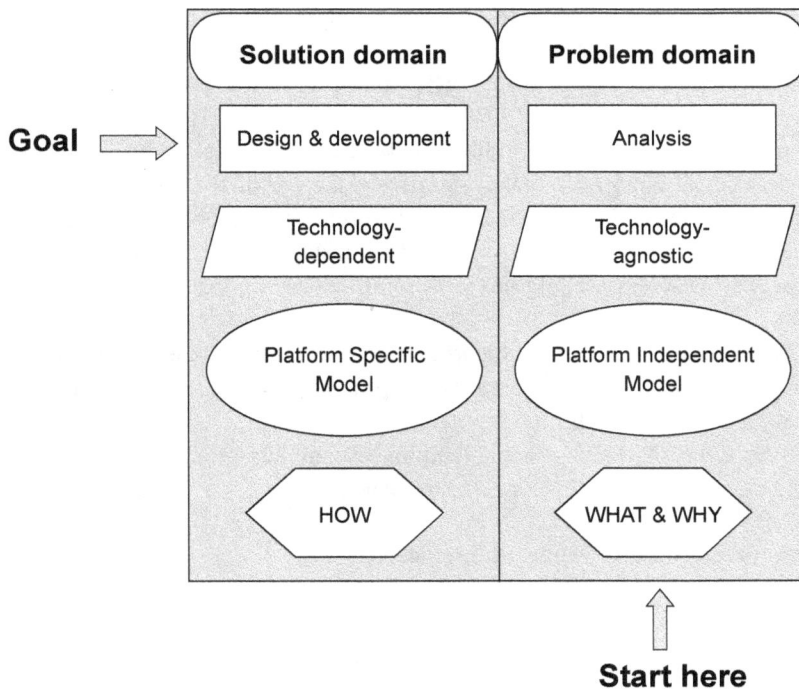

Every problem has a solution.

7.2 Reporting abnormalities

If abnormalities are to be dealt with properly they must first be reported. As simple as this may sound, many abnormalities never get reported and subsequently, never investigated. Good abnormality reports will accurately describe what happened, when it happened and to whom it happened. This information must then be passed on quickly to whomever is in a position to act upon it.

To ensure that employees report abnormalities, you need to:

a. Create a company culture where they are willing to report abnormalities.
b. Draw up rules for making abnormality reports but ensure that they are user friendly.

It is important that employees adopt the responsibility of reporting to their superiors whenever they suspect that something is wrong. However they will often find it a burden to have to write even a short, simple report. That is why it is essential that the rules for reporting are not so cumbersome that employees refuse to cooperate through avoidance.

A company should therefore:

a. Take every opportunity to explain to employees that there will be no improvements possible unless managers know the unpleasant facts; simply, if the managers are not told these facts, the result will be bad news for everyone, especially the employees that will constantly be attempting to deal with the unreported problem and its consequences.
b. Encourage employees to report abnormalities that others may consider not worth reporting. Often what is being overlooked is merely the tip of the iceberg. Only through thorough investigation can the degree of the problem be properly assessed.
c. Have its management establish a trust-based relationship with employees. The response to reporting problems should never be punitive. Two good ways of doing this are:
   i. Demonstrate to employees that action will always be taken on the reports they file and their voice will not be ignored Deal immediately with any outstanding reports that have identified chronic problems.
   ii. Develop a practice of managers and employees solving problems together. Sharing the responsibility builds a much stronger team.

Ensure that all abnormalities are reported, including:

a. Abnormalities that may seem too minor to be reported.
b. Abnormalities that have been recurring for a long time, and are no longer considered merely abnormalities (including operational deficiencies).

The primary rules for abnormality reports are:

a. All abnormalities must be reported to superiors and include those identified during the production process, from acceptance of raw materials to product shipment, as well as those conditions under which new problems have emerged. Reports of abnormalities are important and under GMP/GTP rules they must all be documented and registered in a log.

b. Abnormality reports should have a fixed format. This ensures that abnormalities are properly processed and that recurrence prevention measures are undertaken to prevent the abnormality appearing again. This format should be easy to use and should relate to the actual realities of the work.

c. Report abnormalities at once. Emergency reports must quickly reach those in charge of implementing emergency actions. Some reports may require production suspension, cessation of shipments, and the sorting and isolation of abnormal products – otherwise abnormalities may get worse or damage subsequent processes or even reach customers/patients resulting in adverse reactions. These reports must therefore be passed quickly to those with the authority to decide what to do.

d. Examine the abnormality from several perspectives. What may not be viewed as a problem in one area or department may have a very different reality from another point of view.

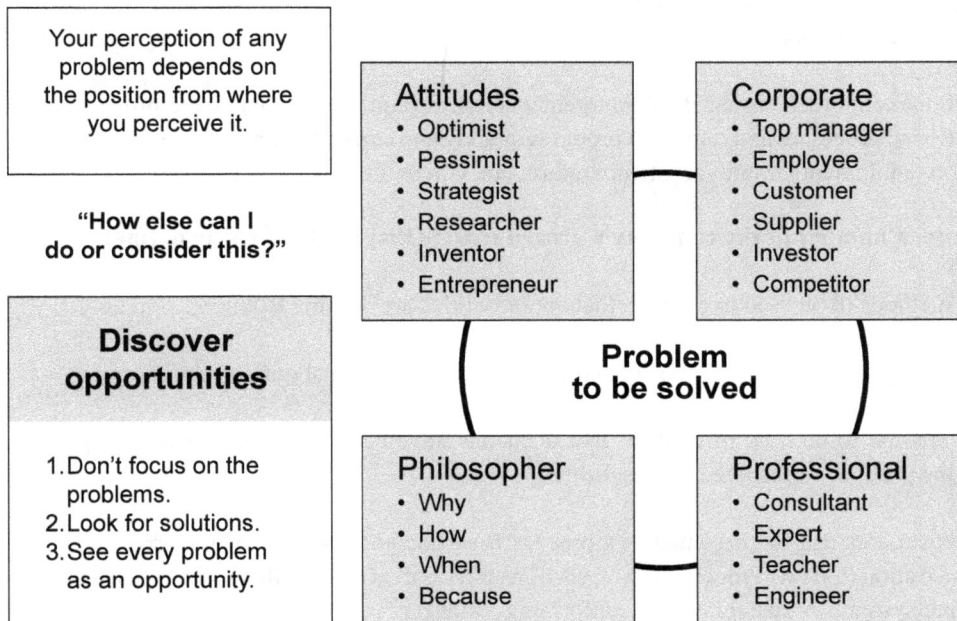

Your perception of any problem depends on the position from where you perceive it.		

"How else can I do or consider this?"

Discover opportunities		
1. Don't focus on the problems.	**Attitudes**	**Corporate**
2. Look for solutions.	• Optimist	• Top manager
3. See every problem as an opportunity.	• Pessimist	• Employee
	• Strategist	• Customer
	• Researcher	• Supplier
	• Inventor	• Investor
	• Entrepreneur	• Competitor

Problem to be solved

Philosopher	**Professional**
• Why	• Consultant
• How	• Expert
• When	• Teacher
• Because	• Engineer

Look at a problem from a different perspective.

7.3 Emergency actions with non-conforming products

Abnormalities include any products that are determined to be not of a suitable quality. Officially, these are known as non-conforming products. The first step is to take emergency actions to stop any further non-conforming products being produced and to deal immediately with those that have been already made and/or released. Afterwards, the investigation for the causes begins in earnest.

Take emergency actions to:

a. Prevent abnormalities getting worse.
b. Prevent abnormalities causing damage in subsequent or downstream processes.
c. Identify non-conforming products and confirm exactly the nature of their non-conformity.

Emergency actions should be accompanied by preliminary inspection reports. These will also help to prevent the number of abnormalities increasing, or adversely affecting subsequent or downstream processes.

To ensure that emergency actions will be as effective as possible, decide in advance:

a. Who will be in charge.
b. Who will be responsible for taking emergency actions.
c. What procedures should be followed.

This will make it much easier, if an emergency action is required, to change the process conditions, suspend processes and segregate products quickly and correctly. (Process conditions are the location, equipment, employees, temperature, etc. where the process is carried out.)

Therefore, a number of the company's procedures (SOPs) need to clearly define:

a. The methods to be used to confirm that abnormalities are starting to emerge and how to begin countermeasures.
b. The management level at which a final decision will be taken about non-conforming products.
c. The method to be used to confirm that decisions have been implemented correctly and the problem is resolved successfully.

These procedures can be presented in a process flow document known as a Corrective and Preventive Action (CAPA) Process. In fact, GMP, as well as the GTPs, calls for construction of the following system for managing non-conforming products:

Control of non-conforming products. This will identify both CAPA.

To prevent non-conforming products from entering the treatment chain:

a. Establish the methods for identifying and separating non-conforming products.
b. Establish the procedures for implementing these methods.
c. Identify who has the responsibility.

In this context, non-conforming products include finished products as well. Therefore it is not simply a case of identifying abnormalities in the company's production chain, but also looking at possible non-conformities both pre and post company's own manufacturing role.

This means having SOPs for purchased materials whose abnormalities were identified during acceptance inspections and the subsequent actions:

a. Products left unfinished because of abnormalities that emerged during production processes.
b. Review and dispose of any non-conforming products.

Clearly define:

a. Responsibility for confirming the contents of non-conforming products.
b. Authority for taking action for dealing with non-conformities.
c. That employees are not allowed to process any non-conforming products in a random, arbitrary manner.

The GMP introduces the following four methods for processing non-conforming products:

a. Reworking.
b. Acceptance with or without repair by mutual agreement.
c. Using them for alternative applications.
d. Rejecting and/or disposal.

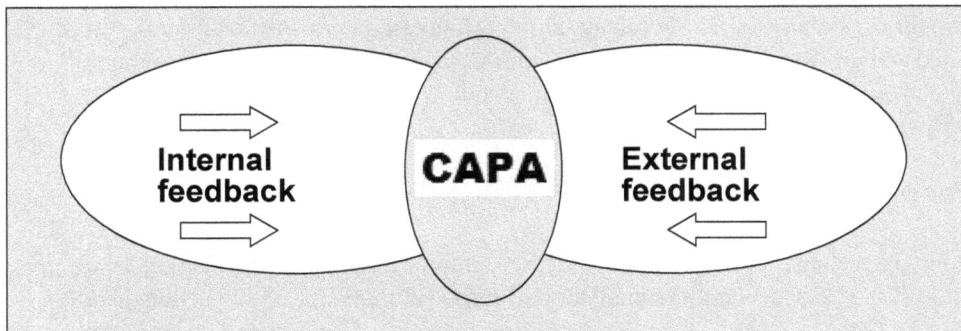

Corrective and Preventive Actions

7.4 Preventing the recurrence of abnormalities

Pay particular notice to any causes of abnormalities that are related to work processes and job procedures, and try to get to the root causes. Once the root causes are identified, take prevention measures to stop these causes recurring or resulting in any further problems. The combination of emergency action, when it is necessary, and systematic recurrence prevention measures, will provide stable production and satisfy the requirements of the CAPA system.

Causes can vary from one abnormality to the next without ever having any definable relationship. It may only be obvious and immediate causes that differ, as an examination of secondary and third level causes of different abnormalities often points investigators to the same root cause. Unless the root cause can be identified, then the problems will most likely reoccur. If the investigation does not lead to the real problem but only superficially identifies related secondary causes, then the therapy and therapeutic product will remain at risk.

Once the fundamental causes have been identified, proper recurrence prevention measures can be undertaken. One definition of recurrence is the re-emergence of abnormalities due to the same cause or causes within a year of supposedly being corrected. It is important to identify whether an abnormality is a new one or simply the same old one that has reoccurred. It should be noted that if the problem is a recurring one, then the previously identified root cause was likely incorrect and therefore problems obviously exist within the CAPA investigative process that has been adopted.

Recurrence prevention measures can be implemented in three levels:

Level 1. Measures aimed only at the specific problem identified.
Level 2. Measures aimed at identical operations conducted by the company.
Level 3. Measures aimed at the overall production system.

Most problems can be dealt with on the three level system as identified problems almost always run deeper that the originally presenting problem. The practices of one area are almost certainly being conducted in other areas and therefore the same problems will exist in similar functional areas as well as may be spreading into other operational areas of the company as well.

It is essential that managers participate in the process of taking appropriate action to prevent abnormalities and are not totally reliant on the QA investigators only performing the investigation and arriving at the proper corrective actions. Since management is in possession of knowledge and experience for the area it is responsible for, it has the ability to provide insights that QA and others would not be aware of.

7.5 Rules for processing abnormalities

It is important to have an established procedure outlining the rules for dealing with any abnormalities. This SOP sets out the methods and responsibilities for all the actions that have to be taken in order to identify, contain, document, investigate, correct and confirm the correction process for abnormalities. These rules are the primary operational basis for the contents in all SOPs within a company that are related to CAPA, non-conformities, deviations, out of specifications, etc.

The general rules are:

a. Take emergency (immediate) actions.
b. Contain the problem. (Safeguard the product.)
c. Investigate the causes of the abnormality. (Identify root cause.)
d. Implement countermeasures. (Continuous evaluation.)
e. Confirm the effects of countermeasures. (Data analysis.)
f. Standardise the successful countermeasures. (Change control.)

The specific rules are:

a. When abnormalities appear, classify them into at least three ranks (A, B and C or Critical, Major and Minor) based on the impact they would have on quality, cost and the volume of production. A fourth classification which is "Not Significant" and therefore no investigation is necessary and therefore needs no further mention.
b. Report abnormalities as soon as they appear according to the prescribed method identified in the SOP.
c. Enter the nature of the abnormalities, their suspected immediate causes and any actions taken with process or the products.
d. Take emergency actions immediately to contain the situation and then prepare an abnormality report. Abnormality reports are designed to record the true causes and the recurrence prevention plans and to confirm the effectiveness of these plans.
e. Use a table, checklist or chart for managing the chain of events in the abnormality processing procedure.

Deal with the three classifications of abnormalities as follows:

a. Primary or A: Report all "Critical" abnormalities to the CEO, and confirm the progress of recurrence-prevention plans at monthly meetings of the Quality Operations group.
b. Secondary or B: Report all "Major" abnormalities to the production managers/ director. Section managers manage the recurrence-prevention plans, and report the progress of recurrence-prevention measures to the Quality Operations group usually within 60 days after the filing of the report of abnormalities.
c. Tertiary or C: Process "Minor" abnormalities by the technicians and employees in the sections or departments where they happened.

By ensuring that all abnormality reports are immediately forwarded to Quality Assurance then it can be assumed that the CAPA investigative process will begin immediately and as a result recurrence-prevention measures will also be implemented successfully. But this assumes that the QA department is a highly functional, experienced group that knows how to deal with corrective actions. Therefore, the ability of a company to deal with recurring problems rests with the actual "quality" of its Quality Assurance department.

When the targets for recurrence prevention measures are not met (i.e. the measures are not confirmed as successful) they are registered as chronic deficiencies. They should be prioritised as themes for improvement, and action taken to resolve them. But more importantly, a company must have the ability to assess why they were not dealt with successfully in the first place and this would indicate another problem that will need to be investigated and corrected. It is not unusual to find that the root cause of problems may lie in the fact that the company lacks the proper personnel to investigate root cause.

Develop the SOP on How to Resolve Non-conformities. Use of a flowchart like the one below aids in understanding the framework involved.

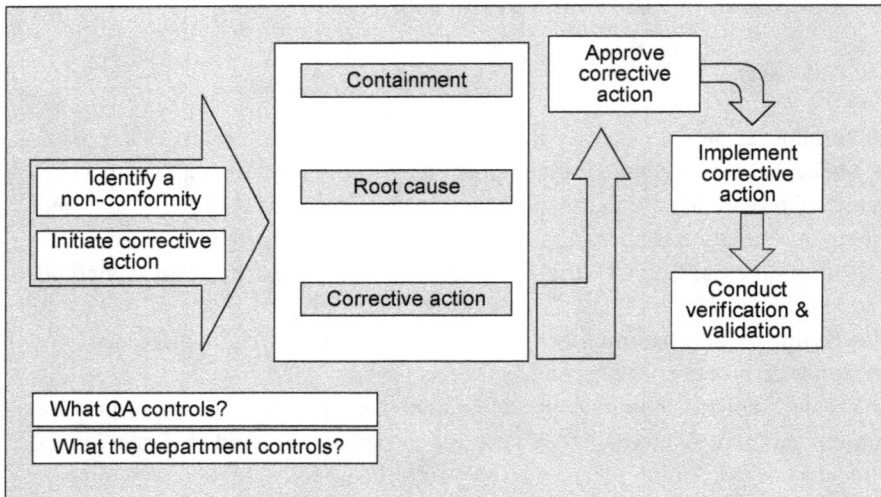

The CAPA flowchart

The Non-conformities Resolution SOP should have the following details:

a. Reporting the non-conformity
b. Ranking the level of non-conformity
c. Writing the improvement plan
d. Charting the progress of non-conformity processing
e. Emergency responses and actions
f. Implementation of the improvement plan
g. Measuring effectiveness of the improvement plan
h. Preventive actions for recurrence

7.6 Base problem solving on facts

Any solutions adopted should be based on facts, rather than on subjective judgments. Even though factually based, there is still a requirement for the invaluable insights gained from experience and intuition. Without experience it is impossible to know what kind of factors to look for, and how to interpret the evidence once they have been identified. Experience, knowledge, intuitiveness, and logic are all key elements. Subjective impressions must be objectified. This means gathering and quantifying them similar to facts if possible.

Actual data, when it has been collected and arranged properly during the investigation, reveals facts that may have not been obvious to subjective observation. Such data provides clear support for hypotheses, and allows positive, effective action to be taken.

To establish the facts take the following steps:

a. Observe firsthand information and examine the site where the problem occurred.
b. Decide on the quality characteristics that will be evaluated during the investigation. When these quality characteristics are determined and recorded in a quantitative data format, they become characteristic values.
c. Formulate clearly the objectives for which the data is being collected.
d. Make certain that the data is collected accurately.
e. Analyse this data using statistical techniques.
f. Obtain accurate information from the analyses upon which to determine actions.

Step	Problem solving	Stage	Decision making
1	Identify the problem	1	Frame the decision
2	Explore alternatives	2	Innovate to address needs and identify alternatives
3	Select an alternative	3	Decide and commit to act
4	Implement the solution	4	Manage consequences
5	Evaluate the situation	4 & 1	Manage consequences and frame the related decisions

Steps of problem solving

7.7 Managing dispersion

Once data has been collected, it needs to be interpreted. Simple mathematical averages are the most common way of interpreting data, but they will fail to give a true perspective as to what the data actually means. Measuring how the data is dispersed provides a more accurate picture. Dispersion refers to how the different items of data are spread or scattered in relation to the expected distribution, i.e. in relation to the standard or target values.

Initially, when the data appears to indicate a problem, clarify as to whether it is the average or the dispersion that suggests there is a problem. Depending on which it is, the approach to correcting the problem will be entirely different. Problems indicated by averages of the data can be solved relatively easily by merely reviewing the processing conditions and any other factors that directly affect the results. Problems indicated by dispersion require far more intensive trending analysis and appreciation as to which factors can impinge on the product and thereby affect the results.

When problems are indicated by dispersion, base countermeasures on whether:

a. The range of dispersion (the distance of the maximum and minimum data points) from the standard or target values (in some cases this is a regulatory standard) is acceptable, but the average is skewed (distorted or biased).
b. The range of dispersion is too wide and outside of acceptable limits.
c. There are outliers (an outlier is an item of data, or a value, that falls well outside the dispersion range of the rest of the data) but generally the data points are within normal range.

Dispersion may be due to chance or to abnormalities. There will always be some dispersion even when the materials and work methods are those prescribed by the standards and this dispersion is usually acceptable as this type of chance dispersion stays within a defined range. The values tend to form a bell curve, with the average in the centre in a pattern known as the normal distribution.

Dispersion created by abnormalities may result from the following factors:

a. Employees not following the operational standards.
b. Change in reagents or raw materials.
c. Loss of experienced employees and new recruits working in the lab.

These are but a few of the more common factors that can skew the average, causing the presence of outliers.

Dispersion in the quality of a product is a direct consequence of dispersion within the manufacturing process. Any unexpected dispersion necessitates finding out the cause of this event as it can be assumed that whatever the cause of the problem, it has strongly influenced the results. By identifying the degree, time and any change that may have occurred simultaneously to the dispersion event, then a correlation can be established which points towards the actual cause.

7.8 Control charts

Control charts are a key tool in interpreting data. They can distinguish between dispersions caused by accidental factors and dispersions caused by abnormal factors, and can show whether the process is in a stable condition or whether a directional trend or pattern is occurring. The typical Shewhart control chart consists of a centre line (CL) and upper and lower control limits (UCL and LCL). UCLs and LCLs are based on calculated values. This was one of the first control charts ever produced and has been in use for almost a century.

When characteristic values that indicate process conditions are plotted as data points on the control chart, and all the points fall within the upper and lower control limits, or there is no bias in the way the points are distributed (i.e. they are not distributed in any particular repetitious manner), the process is said to be "under control". When the plotted points fall outside the control limits or there is a bias in the way the points are distributed, the process is deemed to be "out of control". In other words, an abnormality has emerged in the process causing an irregularity in the pattern of the data points. An investigation of the possible causes of the abnormality should be undertaken immediately and countermeasures ensue. Chapter 8, Statistical Process Control, provides more detailed guidelines on using control charts.

The following criteria indicate when the process is out of control:

a. When one or more plotted points fall outside the control lines.
b. When the points indicate a bias. This can be determined as:
   i. When seven or more points form a chain above or below the centre line.
   ii. When a large number of points are on one side of the centre line, e.g. 10 out of 11 consecutive points.
   iii. When five or more consecutive points form an upward or downward line.
   iv. Other changes which show periodicity.

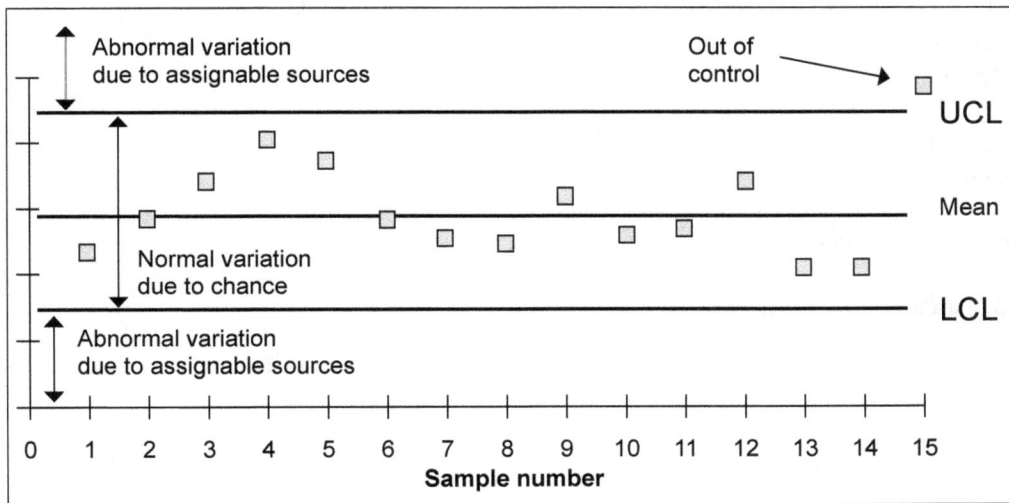

Example of a control chart

There are several kinds of control charts commonly used:

a. x-R control charts (average and range). These are used in the management of variable data. **x** and **R** represent a sub-group average and sub-group range respectively. The **x** control charts are used for monitoring changes in the sub-group average (variation among sub-groups), while the **R** control charts are used for managing dispersions within a sub-group (variation within a sub-group). These two charts are paired for use.
b. **p** control charts and **pn** control charts: These manage processes in which the characteristic values of discrete values are considered. **pn** control charts are used when the number of samples (n) is constant and the number of rejected units (pn) is considered. When the number of samples (n) is not constant, in other words when the ratio of defects (p) is considered, **p** control charts are used.

c. **c** control charts (defects per production unit) or **u** control charts (standard defects per production unit) may be used depending on the characteristics of measured values.

The following directions, steps 1 through 4, are for the implementation of the **x-R** control chart, which is the type most commonly used.

Step 1. Gather data

In principle, more than 100 pieces of data should be collected for this chart to be accurate. This data must be relatively recent, so that it will be likely identical to what any future processes are expected to produce as results.

Step 2. Classify the data

Assemble the data into sub-groups and arrange it by measuring times or lots. The number of data items that one sub-group contains is known as the sub-group size. It is represented by the letter **n**. Usually, **n** is set between 2 to 6 inclusive. Sub-group sizes should be uniform. The letter **k** represents the number of sub-groups made by data classification. Normally, anywhere from 20 to 25 sub-groups are assembled for this type of chart.

As an example, the application of the numbers could appear as follows:
Sub-group size n = 4
Number of sub-groups k = 30
Number of all data items N = n x k = 120

Step 3. Calculate the average for sub-groups

Calculate the average (represented by $\bar{x}$) for respective sub-groups. The value $\bar{x}$ is calculated using the following formula:

$$\bar{x} = (X_1 + X_2 + ... + X_n)/n$$

The value $\bar{x}$ should be calculated to one decimal place more than the measured values.
For example, assume sub-group 1 is calculated as follows.

$$\bar{x} = (0.52 + 0.50 + 0.51 + 0.50)/4 = 0.508$$

The average $\bar{x}$ for all other sub-groups should be calculated in the same manner.

Step 4. Calculate

Calculate **R**, range per sub-group, where
R = [maximal value within the sub-group] – [minimal value within the sub-group]

For example, the value for sub-group 1 is calculated as follows:

$$R = 0.52 – 0.50 = 0.02$$

Range **R** for all sub-groups should be calculated in the same manner.

To calculate the more detailed x R-bar chart, follow steps 5 through 6.

Step 5. Calculate $\bar{\bar{x}}$

Add up the value $\bar{x}$ for all sub-groups and divide this total by **k**, the number of sub-groups.

$$\bar{\bar{x}} = (\bar{x}_1 + \bar{x}_2 + ... + \bar{x}_i)/k$$

Value $\bar{\bar{x}}$ should be calculated to four decimal places (one more than the sub-group averages). Assuming that there were average values for the 30 sub-groups as follows:

Five sub-groups were 0.510, eight sub-groups were 0.508, seven sub-groups were 0.511, four sub-groups were 0.512 and six sub-groups were 0.509, then average of the sub-groups would be:

$$\bar{\bar{x}} = [(5 \times 0.510) + (8 \times 0.508) + (7 \times 0.511) + (4 \times 0.512) + (6 \times 0.509)]/30 = 0.5098$$

Step 6. Calculate $\bar{R}$

Add up the value **R** for all sub-groups and divide this total by **k**, the number of data items in a sub-group. The resulting figure is value $\bar{R}$.

$$\bar{R} = (R_1 + R_2 + ... + R_i)/k$$

Value $\bar{R}$ should be calculated to one more decimal place than the range values.
Assuming that the range values for the 30 sub-groups were as follows:
Ten sub-groups 0.02, eight sub-groups were 0.03, five sub-groups were 0.05 and seven sub-groups were 0.06, then the average range would be:

$$\bar{R} = [(10 \times 0.02) + (8 \times 0.03) + (5 \times 0.05) + (7 \times 0.06)/30] = 0.037$$

An X-bar chart may look something like this:

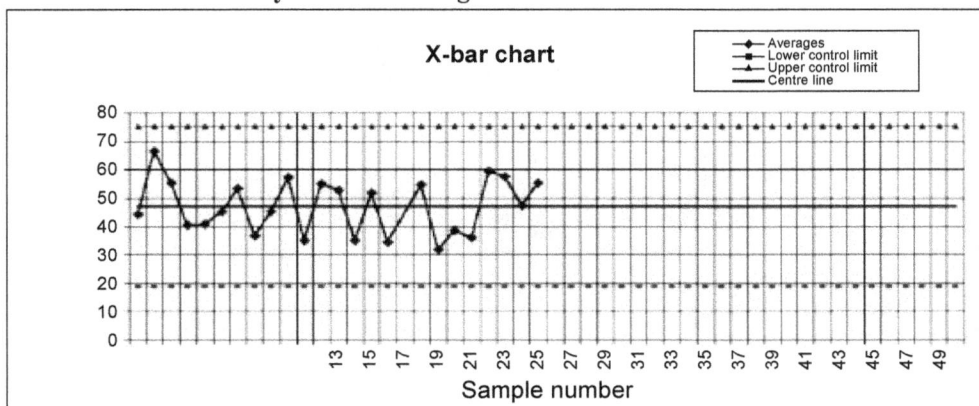

An R-bar chart may look something like this:

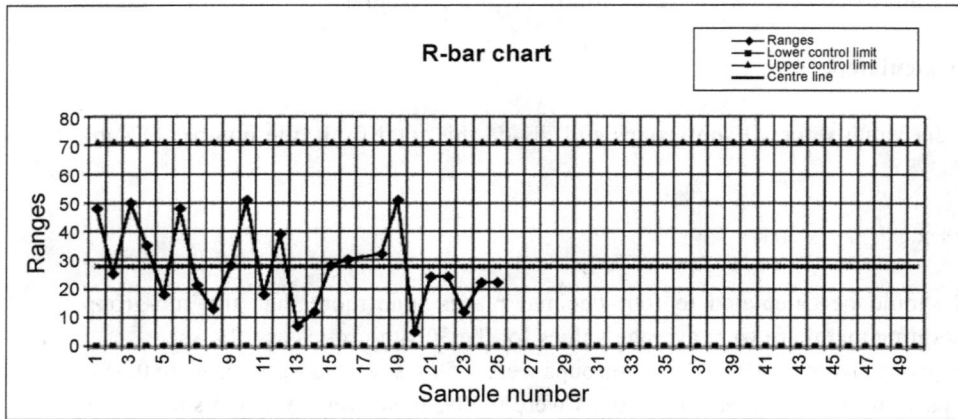

By combining them into overlapping charts a different perspective results:

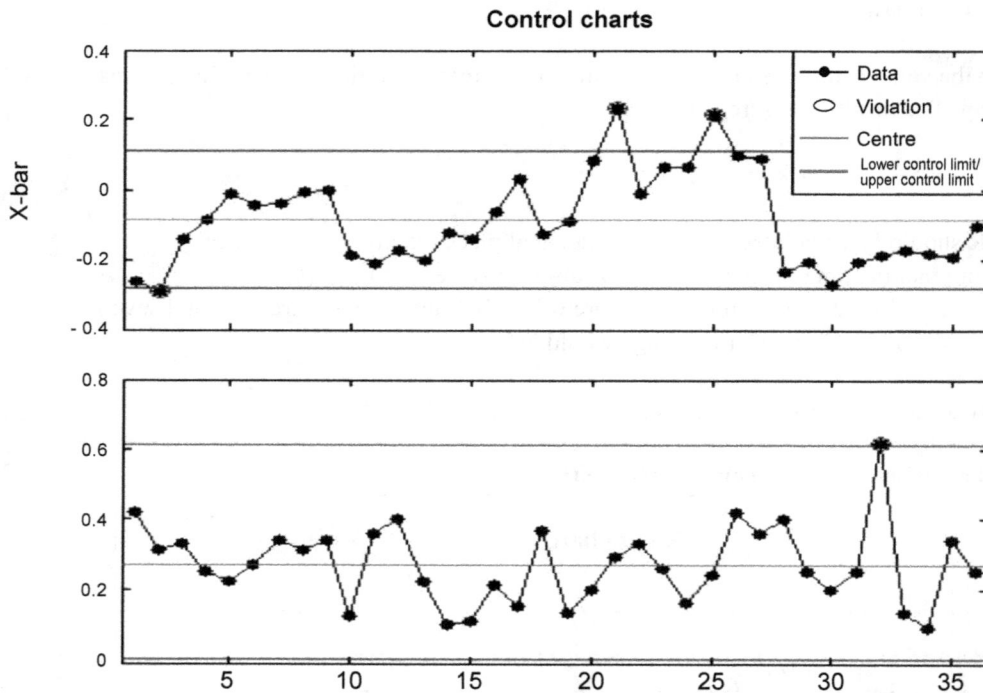

The x R-bar chart above when viewed as a single control reference indicates that overall there were four sample lots that were of a concern.

Sometimes we over-react to problems. We try to solve them immediately without fully assessing the situation. Problems can be handled much more effectively if they are approached systematically. For this it is necessary to have an excellent Problem Solving SOP, and to follow it step by step. A Corrective Action Stepwise Process is a good procedure for solving problems scientifically, logically and rationally.

A Corrective Action (CA) step consists of the following eight components:

a. Select a focus to work on.
b. Clarify the problem and set the targets.
c. Get a clear understanding of the effects that the problem has caused.
d. Analysis: Investigate the causes.
e. Devise and implement recurrence prevention measures.
f. Confirm the effects of these measures.
g. Standardise the new methods.
h. Reflect on the problems left unsolved and consider future countermeasures.

CA Steps provide a neat problem-solving procedure – one that is clear and easy to follow, and keeps investigators on target.

To select a focus, first of all identify all the problems in your workplace, compile them on a chart, evaluate them and select the most suitable as focal points.

To identify which problems need to be looked at:

a. Tasks that constantly present trouble to the technicians.
b. Tasks that are difficult to perform properly.
c. Tasks that have not been completed successfully.
d. Tasks that are hazardous and therefore present safety problems.
e. Tasks that often produce abnormalities or defects in the cell product.
f. Tasks that frequently require reworking of final product due to problems.
g. Tasks that take longer to complete than the standard time frame.
h. Tasks repeatedly resulting in wastage of labour, money, materials, or time.
i. Continuous complaints regarding subsequent production processes, or from other departments, laboratories or customers/patients.
j. Repeated failures in achieving work targets.
k. Failures to implement processes properly by staff.

It may be necessary to work on isolated components of a much larger problem and consider each of those parts as a problem all to itself. When taking this approach, sight of the over-riding, larger problem must still remain as a target when dealing with these smaller component problems.

Selecting the appropriate focus for identifying the problem is not always straight forward. Sometimes the problems are more speculative in nature than factual when trying to determine exactly where the problem lies. One can speculate for example that a lack of preventive maintenance will result in more problems manifesting but that is not true in every case. It can be assumed that there is a likelihood that this is true and by upgrading the preventive maintenance programme problems can be avoided but if there are inherent issues with the cells, or the procedures, or other aspects of the product, then failures will still result despite the upgrades in maintenance. Being able to speculate on what are the more concrete factors that will result in failure or abnormalities and giving them a priority focus is a skill that needs to be developed and honed. This avoids the hit-or-miss approach that many companies implement when trying to deal with their corrective action policies.

Once the problem to be dealt with has been identified, then the next step is to more specifically define exactly what the problem is. Therefore it is more important to obtain a clear and objective view of the actual problem rather than immediately begin thinking of solutions to what might not be a problem at all.

Clarifying the problem:

a. Step one:
    i.   Has the problem actually manifested or is it latent?
    ii. Is it to do with maintenance or with an interruption of the production process?
    iii. Is it a problem of Quality (Q), Cost (C), Productivity (P), Safety (S), or Personnel (M)?
    iv. Is it restricted to one department or does it also affect others?
    v.   How does it relate to corporate, departmental, or laboratory policies?
b. Step two:
    i.   Suggestions as to immediate corrections and advanced determination of the assessment of the results of improvement will be made objective.
    ii. Identify the critical points, criteria to be examined in order to identify abnormalities and for judging potential problems,
         a) the inspection items,
         b) acceptable limits and tolerance for each inspection item, and
         c) the control limits (UCL and LCL) of any control charts used.

Note: An allowable tolerance is the acceptable range given in technical standards and action limits for target values. The action limit is the limit represented by the highest or lowest value in a quality control chart. If the actual values fall outside these limits, a correction in the process is required. Target values are the desired outcomes from a process. (Accuracy)

For abnormalities where there are no standardised critical points or inspection and checking items determined, then establish the criteria on a case-by-case basis, quantifying the data wherever and whenever possible.

Set the target values and target dates for the resolution of the problem. Target dates are essential in order to measure progress of the problem solving activities.

An awareness of the anticipated results of solving the problem by both tangible and intangible effects so that progress can be measured when performing problem solving activities. Tangible effects are easy to identify but an intangible effect such as improved working environment may prove a little more difficult.

7.10 Next level in CA Steps procedure

CA Step strategy continues as follows:

a. Develop a clear understanding of the effects of the problem.
b. Investigate the causes through thorough analysis.

It is necessary to fully appreciate the negative impact that the problem has caused or will cause, before any attempt is made to analyse it. Sometimes it may prove difficult to distinguish between the causes and the effects of the problem, as the line between them blurs in a chicken vs. egg consideration. In order to get a clear perspective of what are the actual effects, it is first necessary to make an objective evaluation of the dispersions in the data related to the problem. Not every dispersed data point will be the effect of the same causal event.

Evaluate according to the following four points:

a. **Time**: Are the effects corresponding to a particular time of day, the day of the week, or any other time factors. For example:
   i. Disorders in certain equipment are concentrated during the evening shift.
   ii. There are many patient complaints about infusion of cell products performed on either Mondays or Fridays.
b. **Place**: Check for the specific places where abnormalities occur.
   For example:
   i. Accidental contamination during storage is most intense in product stored on the top shelf of the south wing of the warehouse.
   ii. Damage to cellpack bags for a particular cell product are concentrated in the upper left corner of the bags.
   iii. Adverse reactions occur in Hospital A but do not occur in Hospital B for exactly the same treatment.
c. **Symptoms:** Examine the characteristics of the abnormal conditions and the circumstances in which they first appeared.
   Example:
   Burst cell bags are the result of corner leakage or split seams. The burst bag is the effect, the symptom is how the leakage is occurring at either the corner or the seams. What causes the leakage at either location has to be determined, especially if one condition occurs more commonly than the other.
d. **System:** Check if the effects from the use of different machines or personnel performing the same work has a different effect.
   Example:
   The product is being manufactured using three different lines known as A, B and C but the abnormality ratio for system B twice as high as the other two lines.

Use sound research methods:

a. Work with data that is based on facts.
b. Pay attention to positive dispersions: dispersions are not always negative.

Exceptionally good conditions can also be evidence of an abnormality. (ie. If it is too good to be true, then it probably is not.)

c. Go to the source and inspect actual details of the problem and the process on site.

Investigate the "real cause" by using two steps:

a. Firstly, identify as many probable causes as possible.
b. Secondly, decide which probable causes are the actual cause.

Identify the many probable causes:

a. Investigate as thoroughly as possible whether the abnormalities that have been identified are the results of any changes that were instituted.
   i.   Have the technicians changed?
   ii. Have the facilities, equipment or machines changed?
   iii. Have the base materials, reagents or other materials changed?
   iv. Have the procedures changed?
   v.   Has the working environment changed?
b. Look at the concrete facts and ask the most important question "Why?". This takes the investigation beyond superficial causes to identifying likely root causes.
c. Use cause and effect diagrams and other tools to examine the probable causes that emerge in steps a and b. This may actually lead to even more probable causes, as one event may be precipitated by another. Organise all the probable causes both sequentially and symptomatically in the process of tracing back to the original root cause.

From the probable cause list, identify the root cause:

a. Establish which of the probable causes actually had an impact and effected the reported problems. Use cause and effect diagrams (ie. fishbone, dichotomous key, branching diagrams, etc.) to narrow the list and construct a hypothesis on how the abnormality may have initially developed.
b. Check this shortened list of possible causes against the known facts in order to identify which of them are genuine. Carry out reproducing experiments to show the relationship between these possible causes and the effects if possible.

7.11 The final CA Steps

The final four stages of the CA Steps process are as follows:

a. Develop and implement recurrence prevention measures.
b. Test and analyse the effectiveness of these countermeasures.
c. Once proven effective, then standardise the new methods.
d. Reflect and check if there are still any problems left unresolved.

It is important to distinguish between emergency response actions and recurrence preventive measures. Emergency actions are undertaken as containment responses to eliminate the phenomena (the immediate, visible problems, e.g. stop producing more non-conforming products) while recurrence preventive measures are taken to eliminate not only the phenomena but the cause of similar future problems. Investigation and preventive measures should be taken even if symptoms appear to disappear after emergency responses are taken.

In devising recurrence preventive measures:

a. Devise a series of measures against each confirmed cause.
b. Brainstorm as many corrective measures as possible, even if it is assumed that the correct measure has been identified.
c. Assess all the proposed measures and eliminate the least useful based on the following
    points. an the anticipated effects be achieved?
    ii.   Will the costs of the measure be acceptable?
    iii. Is the corrective measure plan technically feasible?
    iv. Can the tasks be performed adequately?
    v.   Are there any safety problems?

Prepare implementation plans for recurrence preventive measures:

a. Identify those responsible for performing the testing.
b. Determine the testing schedule.
c. Outline exactly which tests will be performed.
d. Hypothesise any possible negative outcomes or side effects.
e. Determine the criteria for deciding upon successful implementation.

Confirmation of successful implementation of recurrence preventive measures:

a. Target values were achieved following implementation.
b. Successfully met the criteria and investigative points that first indicated there was a problem.
c. Countermeasures followed post implementation to ensure no other negative effects occur nor any unanticipated side effects.

When the recurrence preventive measures prove effective standardise them as new operational methods by SOP revision:

a. Submit the appropriate change control document request to Quality Assurance prior to making changes.
b. Establish revised standards and delete those procedures which are no longer appropriate.
c. Investigate the impact of the changes on any other operations, and revise those standards as well if the procedures are affected.
d. Communicate to all related departments and laboratories the changes made and

distribute copies of the new standards for review and training.

e. Clearly specify in any changes the who, what, where, when and how.

f. Specify the timeframe in which implementation of the new procedure will take place.

g. Commence training programmes for employees using the new standards.

h. Quality Assurance signs off when change control has been performed successfully.

i. Follow up on the results: Use control charts and control graphs to monitor the continuation of the positive effects of the change.

Problems are rarely solved completely on the first attempt. Always make clear which components of the problems have been resolved and which have not. Then make plans to handle the unsolved components in the future.

Review the methodology used to resolve the problems for future direction:

a. Compare the actual data obtained following implementation of the corrective measures against the hypothesised expectations. Account for any differences.

b. Reflect upon the methods used for deciding action plans, setting targets, and implementation, in order to identify areas that could be improved in the future.

c. Examine the next area to focus upon within the company for future corrective analysis.

7.12 Preventing problems from arising

The best way to avoid problems is to anticipate them, and eliminate them before they actually occur. This is called preventive action. All employees should be on the look-out for any issues or potential factors that may lead to problems. The company should be ready to respond and take appropriate action when they are identified.

An identified potential factor should be examined thoroughly in workplace meetings, assessed, and one or more of the following types of countermeasure taken following the assessment.

For example, two passageways cross in the production area, hence a potential source of cross contamination is identified as a potential factor:

a. Eliminate the factor in question (e.g. replace it by separating the corridors through building reconstruction).

b. Remove the possibility of recurrence (e.g. separate use of the passageways by time so that the two passageways are not in use at the same time).

c. Reduce the possibility of recurrence (e.g. install an automatic alarm device, flashing light, etc. when someone is using one of the passageways).

d. Increase the level of caution (e.g. post a "Caution" sign and instructions on avoiding contact).

The ability to predict problems arising is directly proportional to the level of experience that personnel have in dealing with such problems. This ability can be enhanced by the company implementing a culture of creative ideas and methods that aid in predicting problems by the use of tools designed for this purpose.

The FMEA Analysis Table:

FMEA (failure mode and effects analysis) uses an approach to predict problems that could arise with product components. FMEA is widely employed as a design method for increasing reliability. It attempts to predict problems by focusing on specific units within the manufacturing process. Simply stated, it is a probability list of the various problems that might occur within a restricted range, which then predicts the occasions in which they could emerge, and then assesses the chances of them occurring and their possible impact. The probability score, which is the multiplication of these factors together provides an insight into potential risk and that problem which has the highest score, and hence the greatest risk, is the one to be tackled first.

Recommended reading

1. Quality System Inspections Reengineering Team, U.S. Food and Drug Administration. Guide to inspections of quality systems, 1999.

2. Trautman, Kimberly A. Quality System Regulation 21 CFR 820 Basic Introduction. Centre for Devices and Radiological Health, U.S. Food and Drug Administration.

3. U.S. Food and Drug Administration. 21 CFR 820.100: Corrective and preventive action.

   (Note: CFR stands for Code of Federal Regulations.)

Self testing multiple choice questions

1. Abnormalities can be defined as:
   a. Deviations from abnormal conditions.
   b. Defects.
   c. Deviations from normal conditions.

2. Problems occur when abnormalities are:
   a. Resolved.
   b. Noticed and reported.
   c. Ignored.

3. Abnormalities are not always:
   a. Discovered upon a thorough inspection process.
   b. Negative in their effect.
   c. Indicated by control charts.

4. In inspections the obtained values should not be examined against:
   a. Target values.
   b. Standard values.
   c. Historical values.

5. Dispersion may be caused by:
   a. The CAPA process.
   b. Control charts.
   c. Chance.

6. When an employee thinks that a machine sounds different he/she should:
   a. Keep quiet about it.
   b. Report it.
   c. Wait for a day or two to see if it recurs.

7. To encourage employees to develop a habit of reporting abnormalities managers should:
   a. Explain that if managers do not know all the unpleasant facts the employees themselves will suffer.
   b. Financially punish the employees every time they overlook a problem.
   c. Give concrete instructions that even minor abnormalities should be reported.

8. Very minor abnormalities should be reported:
   a. Always.
   b. Sometimes.
   c. Never.

9. Abnormality reports should be delivered in writing:
   a.   Always.
   b.   Often.
   c.   Never.

10. The three-point procedure to carry out before reporting an abnormality is:
   a.   Confirm the problem, select the report format, fill it in precisely.
   b.   Get a fellow employee to confirm the problem, think about it, forget about it.
   c.   Confirm the problem on the spot, try to fix it yourself first, fill in the report.

11. Reports that are aimed at preventing abnormalities causing further problems should result in an immediate:
   a.   Suspension of product.
   b.   Cessation of all laboratory functions.
   c.   Containment of any abnormal products.

12. When abnormalities produce non-conforming products, the first priority is to:
   a.   Find out what is causing the problem.
   b.   Stop more non-conforming products being made.
   c.   Remove the abnormalities.

13. To prevent abnormalities escalating a company should have:
   a.   SOPs dealing with Corrective and Preventive Actions.
   b.   Grace periods that overlook non-conforming products while corrections are made.
   c.   Prepared statements to deal with potential complaints.

14. To prevent non-conforming products being used, make clear:
   a.   The methods for labelling these products.
   b.   The methods for segregating and disposing of these products.
   c.   The procedures for preventing them recurring.

15. The options for dealing with non-conforming products after investigation do not include:
   a.   Reworking the products.
   b.   Using them for an alternative application.
   c.   Using them under risk but without patient consent.

16. An examination of secondary and third level causes of different abnormalities lead:
   a.   Often to a variety of root causes.
   b.   Often to the same root cause.
   c.   Always to a variety of root causes.

17. In the definition suggested in the chapter, an abnormality is not regarded as a recurrence if it appears again:
   a.   After six months.

b. After one year.

c. After three years.

18. The second stage of implementing recurrence preventive measures involves measures aimed at:

a. Operations in which problems have been identified.

b. Identical operations.

c. The overall system.

19. When abnormalities appear they should be classified into _____ ranks depending on the negative impact they have.

a. Three

b. Four

c. Five

20. Enter abnormalities as soon as they appear in a Prompt Report of Abnormality Situations, giving the details of:

a. The nature of the abnormalities.

b. The investigated root causes.

c. The disposal of all involved products.

21. Major category abnormalities should be reported to:

a. The CEO.

b. The production director.

c. The technicians only.

22. Abnormalities are registered as chronic deficiencies when:

a. They cause non-conforming products.

b. Targets for measures to prevent their recurrence are not met.

c. The non-conforming products they cause have to be disposed of.

23. Subjective impressions:

a. Should be ignored.

b. Should be objectified.

c. Are not involved.

24. The steps to be taken to establish the facts about abnormalities do not include:

a. Examining the actual site where the problem occurred.

b. Clarifying the objectives for which data is being collected.

c. Using statistical techniques to analyse the data.

25. Averages are a way of:

a. Assessing problems resulting from indirect factors.

b. Interpreting the data.

c. Measuring dispersion.

26. Countermeasures need to be taken if:
    a. The range of dispersion is within the upper and lower limits.
    b. There are outliers.
    c. There is very little dispersion from the accepted average.

27. If the materials and work methods are those prescribed in the standards:
    a. There will be no dispersion.
    b. There will never be an error.
    c. There will always be dispersion.

28. Dispersion may result from:
    a. Changes to the operational standards.
    b. Always using the same materials.
    c. Employees becoming too experienced.

29. Shewhart control charts have:
    a. One line.
    b. Two lines.
    c. Three lines.

30. When all the characteristic values plotted on the chart lie between the lower control limit and the centre line:
    a. The process is out of control.
    b. The process is under control.
    c. The process needs to be investigated.

31. x-R control charts respectively represent:
    a. Sub-group average and sub-group range.
    b. Sub-group range and sub-group average.
    c. Defect variation and defect ratio.

32. **p** control charts are used:
    a. When the ratio of defects is considered.
    b. When the number of samples is constant.
    c. When the number of samples is not constant.

33. A process is out of control when:
    a. Seven or more points form a chain above or below the centre line.
    b. Three or more points form a consecutive line above or below the centre line.
    c. Five out of 11 consecutive points are on one side of the centre line.

34. The CA story involves eight consecutive steps. Which of the following is correct?
    a. The second step is to clarify the problem and set targets.
    b. The fourth step is to devise and implement solutions.
    c. The eighth step is to standardise the new methods.

35. The CA story's final step is about:
    a.   Confirming the effectiveness of the solutions.
    b.   Rewriting the SOPs to include the corrective measures.
    c.   Reflect on the corrective measures and consider a new focus.

36. If you identify parts of a large problem as component problem you should:
    a.   Specify the overall aims of the larger problem in the component problems.
    b.   Specify new aims for the component problems and focus only on these.
    c.   Do not specify anything new.

37. To clarify the problem you should first ask this question:
    a.   Has the problem actually appeared or is it latent?
    b.   Was it a process interruption which does not require an investigation?
    c.   Has anyone been harmed as only then would it be an emergency?

38. Before you attempt to analyse the data you should have a clear picture of:
    a.   The root cause.
    b.   The likely causes of the problem.
    c.   Possible containment measures.

39. Abnormalities are not caused by differences in:
    a.   Employees.
    b.   Strategies.
    c.   The working environment.

40. If it is not possible to use experiments to confirm the relationship between possible causes and the results:
    a.   Use hypothetical cause and effect diagrams.
    b.   Evaluate the effectiveness of the correction based on subsequent production.
    c.   Then it is impossible to correct the problem.

41. To devise recurrence preventive measures:
    a.   Devise measures against each confirmed cause.
    b.   Be careful not to devise too many possible measures.
    c.   Implement corrective measures only against the important causes.

42. To confirm tangible effects of the corrective actions taken:
    a.   Confirm whether the values now achieved all lie on the centre line of the chart.
    b.   Use different criteria that were used to get a grasp of the initial problem.
    c.   Confirm the effects of each countermeasure.

43. To standardise the new methods:
    a.   Ensure that related departments are fully informed of them.
    b.   Ensure that changes are made to existing SOPs and related departments are informed.
    c.   Use control charts to monitor the continuation of the good effects only.

44. When factors that can lead to abnormalities have been assessed the actions that cannot be taken is:
    a.   Eliminating the dispersion completely.
    b.   Reducing the possibility of recurrence.
    c.   Removing the possibility of recurrence.

45. FMEA attempts to predict problems by:
    a.   Focusing on a broad range of units in the manufacturing process.
    b.   Focusing on specific units in the manufacturing process.
    c.   Increasing the reliability of specific components.

46. A process is out of control when a minimum of ____ plotted point falls outside the control lines.
    a.   One
    b.   Three
    c.   Five

47. C and U control charts are two types of charts that measure:
    a.   Success rates.
    b.   Technical errors.
    c.   Product defects.

48. To clarify a problem, identify if it is an issue of:
    a.   Quantity, Expenses, Productivity, Safety, and/or Mechanical.
    b.   Quality, Cost, Productivity, Safety and/or Personnel.
    c.   Quality, Cost, Proficiency, Security and/or Morale.

49. Tangible effects of the corrective actions are those that can be:
    a.   Qualified.
    b.   Quantified.
    c.   Both qualified and quantified.

50. An "Allowable Tolerance" is an acceptable range:
    a.   Given in the technical standards.
    b.   Established independently by a company.
    c.   That permits 99% of product to pass.

Chapter 8. Statistical Process Control

Basic concepts

There are many issues and problems that cannot be solved simply by examining the equipment, watching staff, and counting cells. In most cases, data has to be collected over a period of time, and then analysed and interpreted in order to identify the real problem. Data provides information by transferring objective facts into numerical values that can be manipulated. When data has been collected properly, then the statistical methods and tools in this lesson provide various means to analyse and interpret that data so that it can provide answers to the existing problems. Being aware that problems can arise at almost any point during a process is fundamental in establishing Quality Control over the entire process.

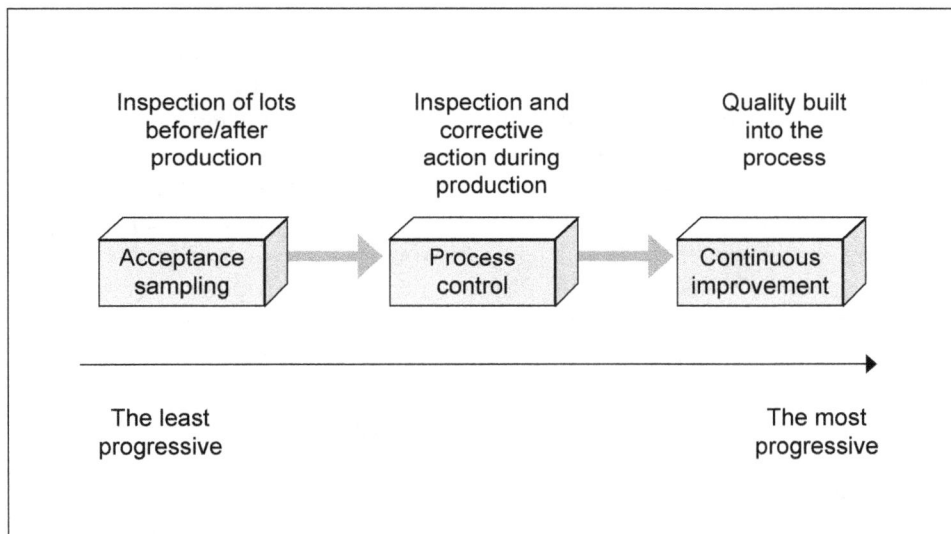

Phases of quality control

8.1 Data comprehension

Quality in production deals with two categories of data:

a. Data that demonstrates the quality conditions of a product: This data is drawn from the quality characteristics of the product, and is used to check that these characteristics conform to the standards or specifications, or from the other perspective, just how far they deviate or are dispersed from the regulatory targets provides clues as to the problem.

b. Data demonstrates conditions of the process that produces the immunotherapy: This data indicates which factors in the process are effecting the quality characteristics of the final product. It indicates both the correctness of the processes and the conditions in which the processes have been set and maintained. Either of these can influence and cause major problems.

Data for quality characteristics includes data from inspection records and supportive test data. Inspection records include records of acceptance inspections, in-process quality inspections (self-inspections), between-process inspections, finished product inspections, and delivery inspections. All these contain records of quality characteristics. In-process quality records also contain data for process conditions.

a. The acceptance inspection record confirms that delivered raw materials, cell lots and reagents can be accepted and that they conform to desired specifications necessary for the expansion procedure. During this inspection, data and the status of acceptance (yes or no) are recorded. The test report attached to the delivered product is also checked, and then filed as attachments to the acceptance inspection data.

b. In-process quality record (self-inspection) contains in-series-event quality control conditions and the resulting quality characteristics that help to ensure product quality through a stable process. It contains data for quality characteristics and the process conditions, which is then used to verify smooth process functioning. The ability to maintain and sustain the process under controlled conditions provides the first judgment whether the process is actually under control or not.

c. Between-process inspections determine whether partially-completed products are actually ready to be sent to the next step of the process. This inspection is carried out at specific predetermined points and any product that does not pass will not be moved on to the next phase. The records of this inspection are used to improve Process Control and reduce variance of quality in the manufacturing of therapeutic products. These records are particularly important in evaluating quality characteristics that may no longer be apparent during the next stage of the process.

d. The finished product inspection record verifies the quality of finished therapeutic products. It contains data for quality characteristics examined according to company and international standards and forms the basis of subsequent test reports submitted to the regulators whenever required. Products are inspected according to standards in order to verify that they meet the requirements of the particular customer/patient and the quality requirements specified for immunotherapies.

e. The delivery inspection record is a quality record checked immediately prior to the infusion process into the patient to make certain that packing conditions, markings, and all documentation met required specifications.

f. Supportive test records in biotechnology would include efficacy tests, accelerated tests, environment tests, and potency tests which are used to evaluate the stability over time of those product traits that tend to deteriorate. Such tests play a vital role in the product development and design stages. Routine safety tests with accurate data are crucial.

Other data includes but is not limited to:

a. Record of abnormalities, defects and complaints: An abnormality report is issued when defects are found in any of these inspections.
b. Record of disposal of defective product: A non-conformity (defect) report is issued, together with data for quality characteristics, when a product is found to be rejected in relation to the specifications or the company standards in the inspections.
c. Record of reworked products: Reworking is the term for the actions taken to make rejected products meet the specifications. Records for reworking also provide data for product quality characteristics.
d. Record of complaint handling: Keep a record of quality problems that customers/ patients have complained about, the conditions in which the product was used, the causes that were investigated, the measures adopted, the preventive measures taken, and the results of surveys on similar therapies, as well as the actions taken based on these results.

> To obtain accurate data for tracking processes, record the process conditions according to company standards. Keep a record of all data for process conditions in the different steps of the process with the intention of improving the process by clarifying the relationship of this data to the quality characteristics through careful analysis.

Documentation is required for the following reasons:

a. Data of the actual process conditions provides invaluable information in determination of root cause if problems occur downstream.
b. Records of quality levels are required by the customers/patients whom do not have any oversight into the processes. Therefore in-process quality records provide a level of reassurance that they require.
c. It is crucial to base management decisions on indisputable facts, rather than to rely solely on concepts, experience, or intuition. This requires accurate data for process and quality conditions.
d. Archiving must be treated as a specific process with written procedures in order to ensure the quality of records and data. To achieve this, the data must be recorded in a specified format on a specified medium (whether paper, computer, etc.) and checked and approved by a proper sign-off authority. Without proper documentation then the archiving cannot be performed and maintaining records for 7 to 10 years is a requirement.
e. Quality records include not only the data for each product or manufacturing processes but also includes graphs, control charts, histograms, or check sheets generated by statistical techniques.

Therefore documentation must be viewed as far more than simply paperwork for control of processes but are in fact viable documents serving a variety of purposes for different regulatory and management groups.

8.2 Checking data integrity

Managers should confirm that process data is reliable and correct and has been recorded accurately according to the company standards as outlined in the company SOP on recording data. Managers should ensure that work is carried out in a consistent manner and that control and improvement activities are promoted effectively. A key task in achieving this is to inspect and verify the data daily, rather than several days after the event, and maintaining daily management records.

The SOP on proper record handling should include:

a. Record all necessary data on formalised and accepted recording sheets.
b. Avoid inaccurate descriptions, omissions, illegible writing, improper descriptions, or improper corrections. An SOP on proper filling of forms and how to correct errors on forms must exist.
c. Submit record sheets to the manager responsible for the test without delay, according to the prescribed procedures.
d. Record sheets should provide spaces for the technician responsible for the test to confirm the results, insert comments or observations, enter the date of the record and provide a signature or initials for identification.
e. In the event of improper, incorrect, or ambiguous descriptions or of work records that do not meet corporate standards, the manager responsible for the test must obtain an explanation from the technician responsible for the record, confirm the contents, and issue written instructions for any corrective actions that need to be undertaken.
f. After confirming that the record is correct, the manager responsible for the test must sign it or mark it with a seal in the specified column.

Various preformatted check sheets:

Check sheet for recording each abnormal product:

Cell production lots that have a tendency towards abnormalities are entered in advance on the recording sheet, and this is checked whenever a defect occurs. This check sheet shows the frequency of defective cells in each lot and identifies the most likely cause and probable impact. Expressing the record as a diagram makes it easy to see trends in defective therapy lots.

Check sheet for recording the cause of each defect:

This check sheet compiles data for different equipment, work classifications, operators, and work times, and is useful in finding the mechanical causes for a problem.

Check sheet for recording the distribution of data characteristics:

This check sheet records measurements for cell type, cell volumes, cell numbers, and other characteristics and sorts them into relevant data to clarify the profile and dispersion of distributed data. It makes it easier to record individual values and to process data because it arranges the data in frequency charts.

Check sheet for recording defect positions:

This check sheet marks defect locations on a product sketch, such as an "x" marking the spot. It is useful in determining defect positioning, clustering, defect categories, and rationale for their concentration. For example, failures of sterilisation in an autoclave may be identified to one particular location in the autoclave where temperatures do not reach the required minimum for the correct duration.

Be sure to record the following data items for process conditions and quality characteristics:

a. Title, number and version number of the applied standard.
b. Title and identification number of the product.
c. Date and time.
d. Title of process.
e. Title of equipment.
f. Title of work.
g. Name of technician.
h. Name of recorder.
i. Lot number of material.
j. Data for process conditions (temperature, speed, pressure, concentration, contents, time, and other information).
k. Name and number of equipment used.
l. Name and number of measuring instrument.
m. Conditions of the sampling test and frequency of checks.
n. Data for quality characteristics, whether accepted or rejected.
o. Detailed descriptions, report and actions taken as a result of defects.
p. Essentially, record all data with reference to the correct descriptions of the measuring instruments, testers, inspection equipment, and test methods used in the test. It is imperative that these measuring instruments are checked and properly calibrated as per the calibration SOP.

Data should be precise and accurate, and reliably expressed in significant digits, to allow proper processing of the data.

8.3 Understanding data diversity

Data collection terminology:

Use standardised methods for collecting data so that judgments can be made quickly and accurately based on the data and subsequently appropriate improvement can be performed.

Population: A group of entities whose characteristics are to be investigated or studied or a group of entities from which samples are to be taken.

Sample: Part of a population selected to establish its characteristics.

Variable: A quality characteristic value that can be measured as a continuous quantity, but not counted as separate items, e.g. length, colour. Data derived from such values is referred to as variable data.

Discrete value: A quality characteristic value that can be counted, e.g. the number of defects or rejected products. Data derived from such values is also referred to as discontinuous data.

The purpose of data collection is to gain an understanding of user needs, the quality characteristics of a product, the quality conditions of a process, and the quality of raw materials. Once data is collected, then it is possible to investigate or inspect a population.

Data to which statistical techniques are to be applied is obtained by measuring items sampled from processes, lots, or populations. The data thus obtained has sampling and measuring errors in addition to the dispersion that results from changes in production elements. To give an accurate estimate of a population's characteristics by applying statistical techniques to the data, the unit entities and unit measurement values that compose the population should be sampled at the same probability and a correct measuring method used to obtain the values.

Random sampling may give the impression that any results are governed only by chance but in reality, because the random sampling is repeated many times, this results in increasing degrees of statistical regularity.

Data rule 1: Therapeutic products should not be sampled by the workers responsible for manufacturing them (conflict of personal interest – nobody wants to make themselves look bad at their job).

Data rule 2: Sampling should be witnessed by a person with authority (sampling cannot be checked retrospectively). Recording of sample results is also time sensitive and should not be performed retrospectively.

Data rule 3: Those taking the population sample should be aware in advance of the significance, importance and purpose of sampling.

Data rule 4: Random sampling is based on a table of random numbers. These tables can be run from an Excel spreadsheet.

The following are important points in collecting data:

a. Data collection must have a purpose. Clarify the purpose of collecting data, and be sure that the data collected is significant and pertinent for that purpose and expresses objective facts.

b. Before using data recorded in the past, confirm the purpose for which the data was collected is similar to your intended purpose, and examine the context, history, and constraints in its collection. Do not use data collected for different purposes, from different populations, or in different measuring conditions. Do not use improperly sampled data or data measured using different equipment. Historical data therefore must be highly compatible.

c. Record any data collected for process conditions in a form that allows immediate stratification for different elements. This will permit analysis of the cause and effect relation between process conditions and quality characteristics to be determined more easily.

d. The sampling method used must be appropriate to the purpose of sampling. In other words, do not collect the data in a prejudicial, biased, or predetermined manner.

e. In using data, be attentive to objectivity and reliability. Technicians must be trained in methodology of maintaining instrument precision, measuring methods, and methods of rounding and recording values. Check the conditions by which the instruments are calibrated and maintained in order to eliminate any introduced errors by technicians.

f. Confirm that correct data has been recorded free of errors. If any calculations are involved, check that they have been calculated properly.

g. Ensure that no part of the data has been overlooked or omitted.

Using diagrams to analyse data:

Before commencing the analysis, ensure that the links exist between cause and effect. Characteristic diagrams are useful for doing this. Characteristics are effects that result from the various processes.

The list of characteristics include:

a. Quality (Q): Appearance, volume, weight, purity, strength, evenness of distribution, number of rejected products, number of complaints, ratio of rejected products, etc.

b. Cost (C): Cost of raw materials, reagents, equipment, services, labour, waste and rejects, marketing, loss due to complaints, etc.

c. Safety (S): Contamination ratios, number of technician accidents, number of near miss incidents.

The list of characteristics is dependent on those values deemed significant due to impact on the final product and therefore will not necessarily be the same in all cases and for each project.

It is usually the case that most factors affecting quality can be lumped into specific causal classes. This is true in every case, and routinely expressed as the 5Ms because they all can be referenced by a single word beginning with the letter M.

The 5Ms:

a. Manpower (technicians, employees).
b. Materials (includes reagents).
c. Machines (equipment and facilities).
d. Methodologies (processes).
e. Measurements (includes sampling techniques).

To extract the most information from a characteristic diagram, it is necessary to identify only those features that seem important and are influential (causes). Therefore, it is necessary to determine whether the factors being examined are real causes and that will necessitate studying them in the environment in which they would likely have an effect, such as the production site. Doing so is known as verification. Use the following procedures to extract and verify factors:

Extraction of factors:

a. Examine the production site in order to have a more realistic appreciation of the process and environment in which the problem occurs.
b. Identify as many potential factors for cause as possible.
c. Identify possible factors by searching for reasonable possibilities rather than complicated and convoluted possibilities. Think simple.
d. Investigate the relevance of factors by associating them with "how" they could cause the problem. "How" they arise in the first place. "How" they can be eliminated, mitigated or mollified.

Statistical techniques:

Use statistical techniques on collected data to estimate the mean and dispersion of the population. For this purpose, the company should train a person or persons specifically in the role of statistician. It is the expectation of the regulatory authorities that companies will have someone employed in this capacity full time. Employees may be sent to external seminars to learn statistical techniques and how to apply them in their production processes. These techniques will enable them to increase their control over these processes and make improvements to them. However, such training is useless unless full TQM training is scheduled and performed continuously for employees at each organisational level because the statisticians will continually identify problems otherwise which is counterproductive.

Any in-company training in statistical techniques is to provide all employees with a general working knowledge of statistical Process Control techniques. Separate training should be provided for each organisational level as different levels will have different requirements of understanding. In a later chapter on Education and Training, there are guidelines on providing training in statistical techniques in the context of TQM training.

Functions and values calculated from data samples express their characteristics. Statistics are employed to estimate population parameters and to verify hypotheses. There are certain basic statistical calculations that are fundamental to any data management programme.

Fundamental statistics:

a. Mean value.
b. Standard deviation.
c. Range.

Mean:

The mean is the average of a set of numbers determined by the sum of the values (x) in the data set divided by the number of values (n) in the data set and represented by $\bar{x}$.

$$\bar{x} = \Sigma \frac{x}{n}$$

From these values we can estimate the general features of a population distribution.

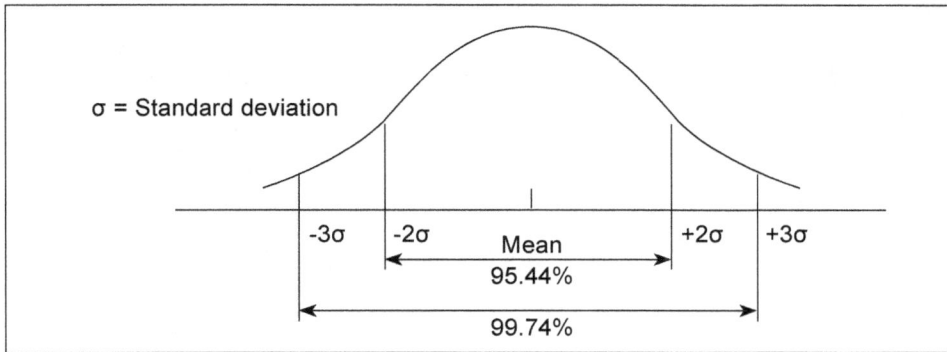

σ = Standard deviation

-3σ -2σ Mean +2σ +3σ

95.44%

99.74%

A standard deviation chart

Random variables:

These are variables that take values according to the rules of probability. Random variables are either discrete or continuous, depending on the nature of their values.

Expected value:

When values of a random variable x are observed repeatedly, the ultimate mean value is called the expected value of **x**.

Median:

When measurements are arranged in size, the value at the centre of the sequence is called the median.

a. Use the measurement at the centre for an odd number of measurements.
   Example 1: 5.7, 6.0, 6.1, 6.2 and 6.3 → Median = 6.1 (mm)
b. Use the mean value of the two measurements at the centre for an even number of measurements.
   Example 2: 5.7, 6.0, 6.1, 6.2 → Median = (6.0 + 6.1)/2 = 6.05 (mm)

Although a median value is less precise than the mean value, you can obtain it directly from the numbers listed without performing a calculation for an odd number of measurements.

Range (R):

The difference between the maximum value (UL = upper limit) and the minimum value (LL = lower limit) of a data set is called the range and is denoted by **R**.

R = UL – LL or in words, range equals upper limit minus lower limit.

The range therefore can never be a negative value.

$R \geq 0$ always!

The range is used for 10 measurements or less, ideally for five or six measurements. Example: Find the range of measurements 6.7, 5.2, 6.1, 6.6 and 6.8.

$R = 6.8 - 5.2 = 1.6$

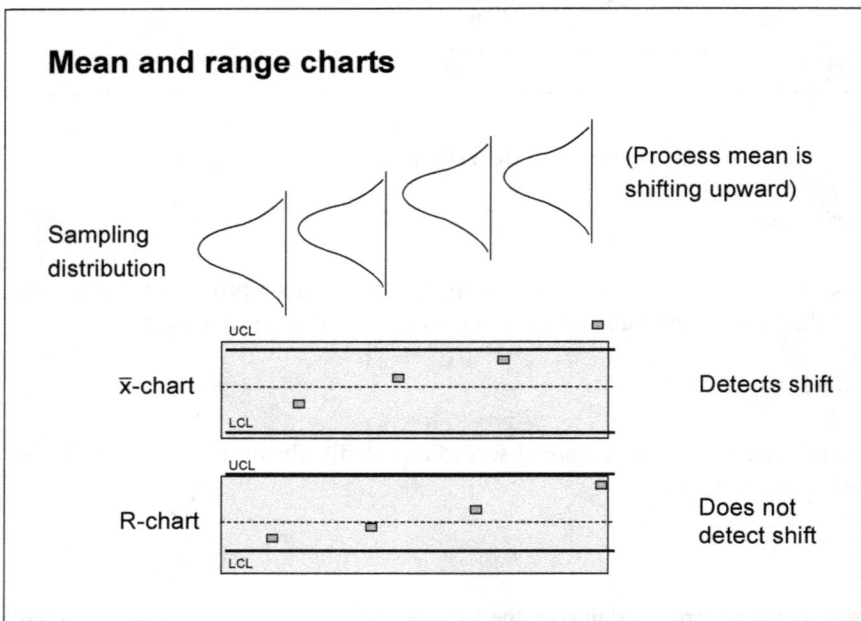

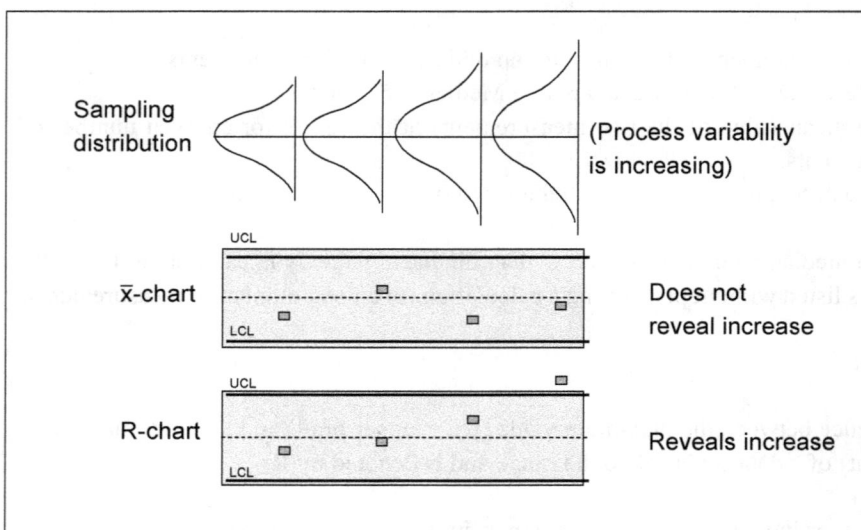

Sum of the squares (S):

The sum of the squared differences between individual measurements and the mean value is called the sum of the squares and is denoted by **S**.

$$S = (x1 - \overline{x})^2 + (x2 - \overline{x})^2 + ... + (xn - \overline{x})^2$$
$$= \Sigma(xi - \overline{x})^2 \text{ ... Formula for definition}$$
$$= \Sigma xi2 - (\Sigma xi)^2/n \text{ ... Formula for calculation}$$

where the term $(\Sigma xi)^2/n$ is called a correction term (CT).

Variance (V):

The sum of the squares divided by (n-1) is called the variance or (unbiased variance) and is denoted by **V**.

$$V = S/(n-1)$$

The sum of the squares is larger when the number of measurements is larger. However, the variance gives a dispersion unrelated to the number of measurements.

The value (n-1) is called the degrees of freedom.

Standard deviation (s):

The square root of the variance is called the standard deviation and is denoted by *s* or σ $s = (\sqrt{V})$.

Since the sum of the squares and variance are related to squared measurements, values based on the measurement unit cannot be directly compared. The standard deviation does not have this drawback.

When measurements of a population are normally distributed, having a population mean value μ and a standard deviation σ, the mean value $\overline{x}$ of **n** measurements sampled at random from this population is also normally distributed, with a mean value μ and a standard deviation $\sigma/(\sqrt{n})$.

Using graphs to analyse data:

It is important to know how to apply graphs to process control and improvement. This involves creating well defined data by expressing raw data as numeric values and then plotting these numeric values on a graph. What will appear on the graph is a figure rather than simply numbers. This figure will show at once the relations of measurements with lengths, areas and angles. You can use different types of graphs to compare the magnitudes of different measurements and see at a glance the measurement changes over time. This allows you to quickly appreciate the conditions of process control and the effects of improvement and to communicate this information to others.

The advantages of using graphs are:

a. Anyone can draw them.
b. The status and conditions of a process can be judged at a glance.
c. They can present a large amount of information simply.

There are different types of graphs: bar, polygonal line, ribbon, pie, radar chart, Z-plots, Box-plots and violin graphs. Selection of the appropriate graph is based on purpose, design expressions and ease of understanding. A simple dot plot for counting runs reveals a lot.

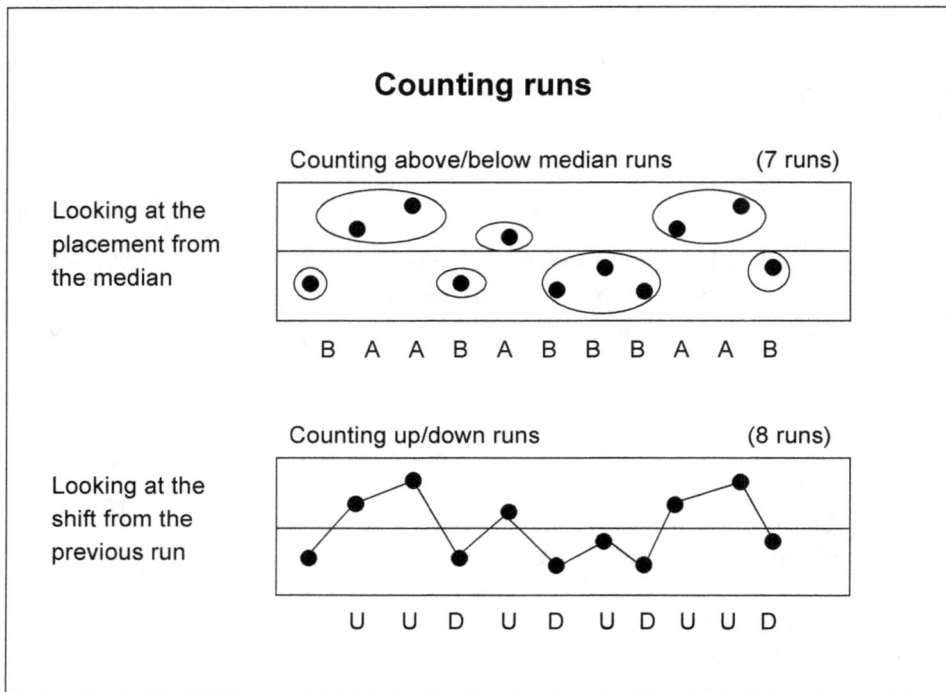

Counting runs

Counting above/below median runs   (7 runs)

Looking at the placement from the median

B   A   A   B   A   B   B   B   A   A   B

Counting up/down runs   (8 runs)

Looking at the shift from the previous run

U   U   D   U   D   U   D   U   U   D

Bar graphs:

Bar graphs express data using the length of a bar to facilitate comparison between the magnitudes of different data, such as inventories by product or plant.

Polygonal line graphs:

Polygonal line graphs plot time on the X-axis against a characteristic on the Y-axis. Adjacent data points are connected by line segments to indicate trends or changes in the characteristic over time, such as the quantity of manufactured products, number of rejected products, or number of re-workings per unit of time.

Pie graphs:

Pie graphs are circular graphs, in which the circle is sectioned along the circumference in proportion to the ratio of each component. Expressing component magnitudes by the central angle and radially spread sections makes it possible to view at a glance the relative sizes of the components.

Band graphs:

Band graphs appear rectangular, with the longer side sectioned according to the ratio of the components. Band graphs are used to compare or note changes in the ratios of components over time.

Radar charts:

Radar charts feature lines radiating from the centre of a circle that section the circle into portions of the same size according to the number of items. These lines are scaled. An item value is plotted on each line, and data points on adjacent lines are connected with line segments. Radar charts are useful in representing and comparing a large number of items, and in tracking progress towards targets.

Scatter diagrams:

Scatter diagrams are used to determine whether a relation exists between two characteristics by plotting pairs of data on an X-Y coordinate, or by plotting one characteristic on the Y-axis and another on the X-axis. When a relation exists between the two characteristics, they are said to be correlated. Where one characteristic increases as the other also increases, they are said to be positively correlated. When one becomes smaller as another becomes larger, they are said to be negatively correlated. When two characteristics are positively correlated, data points scatter within an ellipse tilted to its right. When they are negatively correlated, data points scatter within an ellipse tilted to the left. When no relation exists between the two characteristics, they are said to be not correlated, and their data points are found to be scattered within a circle.

A coefficient of correlation represents the degree of correlation between two characteristics and takes a value in the range from -1 to +1. A value close to -1 indicates a strong negative correlation. A value close to +1 indicates a strong positive correlation. A value close to 0 indicates that the correlation between the two characteristic is weak.

An equation for regression represents the relation between two characteristics **y** and **x**, where **y** is called an objective variable and **x** an explanatory variable. The equation for regression takes different forms depending on which variables are used, and the numbers of variables.

The following equation expresses **y** as a linear equation of **x**:

$$y = a + bx$$

where **a** is a constant or intercept and **b** is called a regression coefficient.

8.4 The QC tools

Pareto charts:

Pareto charts use bars to express phenomena and causes, grouped by item, such as rejected lots, reworked lots, discarded lots, claims, accidents and failures. Polygonal lines are added to show cumulative frequencies.

When examining a Pareto chart with defect items on the X-axis by frequency and the number of defects or amount of losses and their cumulative amounts on the Y-axis, it can be seen as an effective device to pick out a small number of vital items than a larger number of unimportant items. The Pareto chart is widely used to choose problems and subjects for study and discussion at the planning stage for Quality Operations activities and to confirm the results of an action once the action is performed.

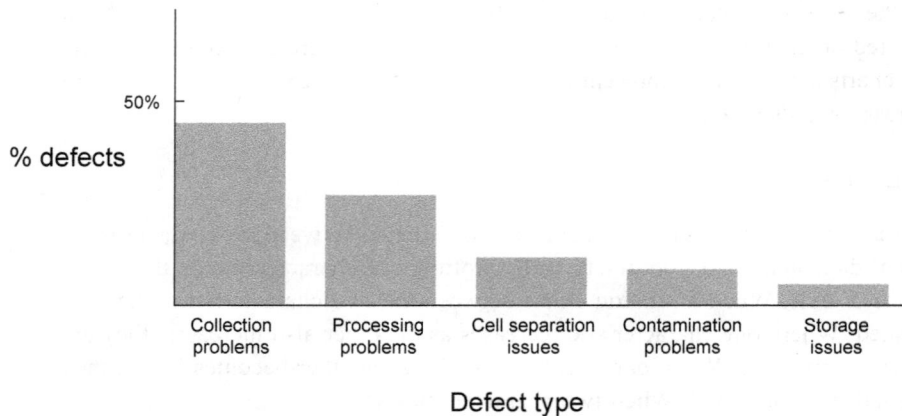

Pareto chart of autologous cell therapy

To draw a Pareto chart collect data stratified by cause (factor) or effect (characteristic, phenomenon). Do not use data for items from different classifications, levels, or causes (factors). Be sure to enter totals for cases, amounts and times, and the period of observation.

a. Adopt the amount of loss along with the number of defects, defect ratios, and the number of claims on the Y-axis.
b. Determine a period of observation appropriate to the purpose.
c. When a Pareto chart indicates little difference between different items (strata), change the method of classification or the characteristic on the Y-axis to bring out the important items.
d. After determining the most significant item, draw a secondary Pareto chart for that item alone.

Histograms:

A histogram for the distribution of numeric statistics is used to enable users to grasp data at a glance. It uses columns to express frequencies of data from different categories. The range of distribution is divided into several sections. Histograms deal with variables such as length, weight, temperature, and volume, all of which can be obtained by measurement.

Histograms clarify the following data features:

a. The profile of distribution of data.
b. The centre of distribution of data.
c. The dispersion of data.
d. The relation of data to standards.

Data points:

a. You should have at least 50 data (n) points, though 100 is better. A small number of data points produces a highly variable distribution profile. As a general rule the square root of the number of data points gives the appropriate number of sections.
b. Determine the scales of the Y and X axes so that the diagram is roughly square.

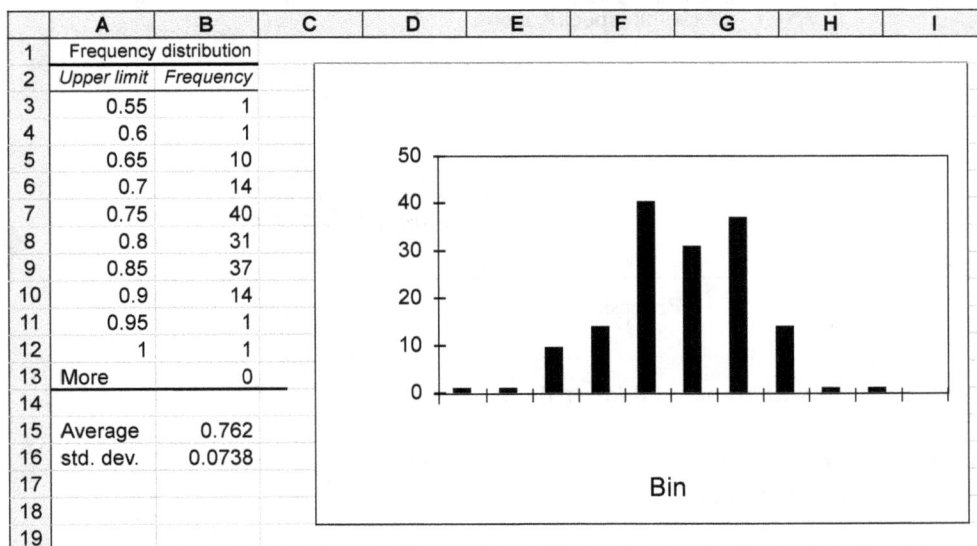

	A	B	C	D	E	F	G	H	I
1	Frequency distribution								
2	Upper limit	Frequency							
3	0.55	1							
4	0.6	1							
5	0.65	10							
6	0.7	14							
7	0.75	40							
8	0.8	31							
9	0.85	37							
10	0.9	14							
11	0.95	1							
12	1	1							
13	More	0							
14									
15	Average	0.762							
16	std. dev.	0.0738							
17									
18									
19									

Histogram

Process Capability Indices:

This text presents the methods of calculating the Process Capability Indices C_p and C_{pk}, using the frequency table of a histogram and effective use of the indices. Process Capability is a qualitative capability for a process, a scale used to evaluate the distribution of important product characteristics obtained through the process by comparison with specifications or standard values.

Process Capability is determined by the relationship between the dispersion of product characteristics and standard values. It is usually expressed by the Process Capability Index (below). The Process Capability should be evaluated when the process is stable.

The Process Capability Index (C_p or C_{pk}) compares a histogram with the standards to evaluate whether a process has a capability that satisfies the standards. The methods below are used to calculate the Process Capability Index for two-sided standards with upper and lower limits, and one-sided standards.

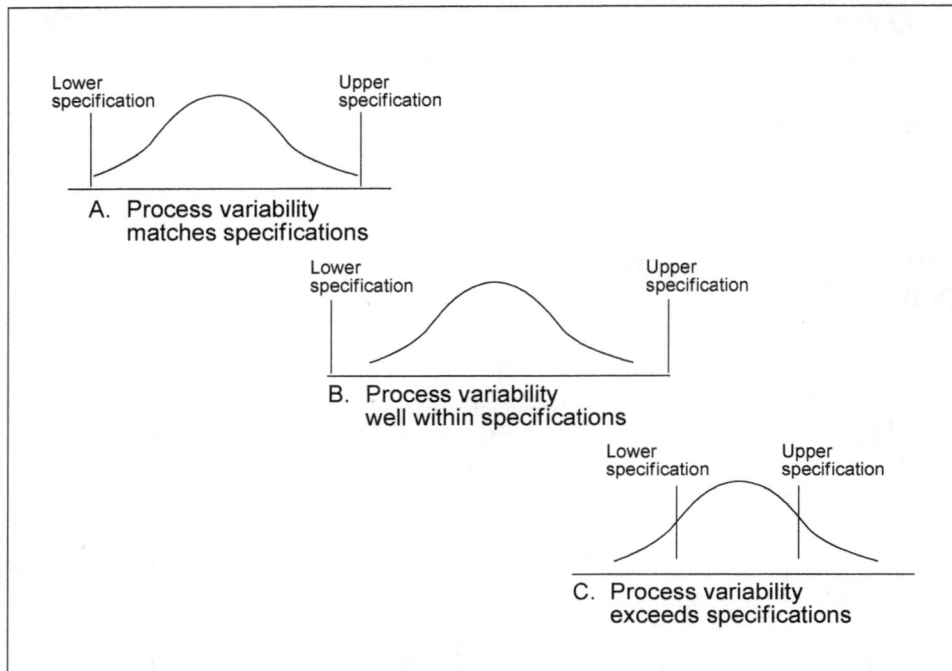

Process Capability

Two-sided standards with upper and lower limits:

These standards set both the upper and lower limits of standard values. Process Capability is calculated with the following formula:

$$C_p = \frac{UL - LL}{6s} = \frac{\text{upper limit} - \text{lower limit}}{6 \times \text{standard deviation}}$$

Even when the Process Capability Index C_p is sufficiently large, characteristics do not satisfy a standard if the difference between the centre (M) of the standard and the mean value **x** of the characteristic is too large, or if the distribution of the characteristic is biased from the standard.

Stratifying data:

This text describes the methods of applying stratification to process control and improvement. Stratification refers to the division of a population into different strata (groups). This should be done when individuals composing sub-populations are similar to others within the same sub-population, but differ markedly from individuals in other sub-populations. In other words, compare apples with apples and oranges with oranges.

By stratification, items in the data that has been collected are divided into groups according to either cause or effect. Elements of a group will therefore share characteristics that differ significantly from the characteristics seen in a different group. Data for rejected products can be divided into groups (strata) by causal characteristics such as technician, equipment, process method, etc. Each group will have a distinctive common feature.

Procedure 1: Clarify the items to be solved or the contents to understand.

Procedure 2: Determine the items to be stratified.

Procedure 3: Collect data for stratification.

Procedure 4: Compare stratified items using an appropriate method. If a difference is found, investigate the causes. If no difference is found, adopt other items for stratification and start again from procedure 2.

a. By result (characteristic): This examines differences between factors by stratifying data by result (characteristic).

b. Stratification by factor (cause): This examines differences between characteristics by stratifying data by factor (cause). Draw a characteristic diagram to pick out a factor that seems to significantly affect results and stratify data by that factor. If no difference is found, select another factor. For example, if no differences are found between defect ratios classified by technicians, try classifying data by piece of equipment used.

Comparison of stratified data is a valuable way of checking data. This will clarify the reasons for defects and the factors that affect characteristics.

It is important to stratify qualitative information as well, and not just data expressed through numeric values.

An effective way to analyse a process is to stratify data according to various factors chosen on the basis of past experience.

Present the data stratified for different levels of factors in check sheets, histograms, graphs, characteristic diagrams, scatter diagrams, and control charts to clarify the differences between different strata. Once a difference is found between groups, investigate its cause. This should help to solve the quality problem.

8.5 The concept of dispersion

Control charts for each process:

A control chart is a polygonal diagram for plotting the mean value of a characteristic, the ratio of defects, or the number of defects. It shows changes in characteristics (control characteristics) and can therefore be used to check for process abnormalities. It is used to confirm that a process is stable.

A control chart has three lines:

a. A centre line (CL) which represents the average value for all data collected.
b. Two lines that indicate control limits, the upper control limit (UCL) line and the lower control limit (LCL) line.

Points representing quality or process conditions are plotted on the chart. When these points fall between the control limit lines and do not show any particular trend, the process is stable. When the plotted points fall outside the control limit lines or show a trend, something unusual is causing this. When this happens find and eliminate the cause.

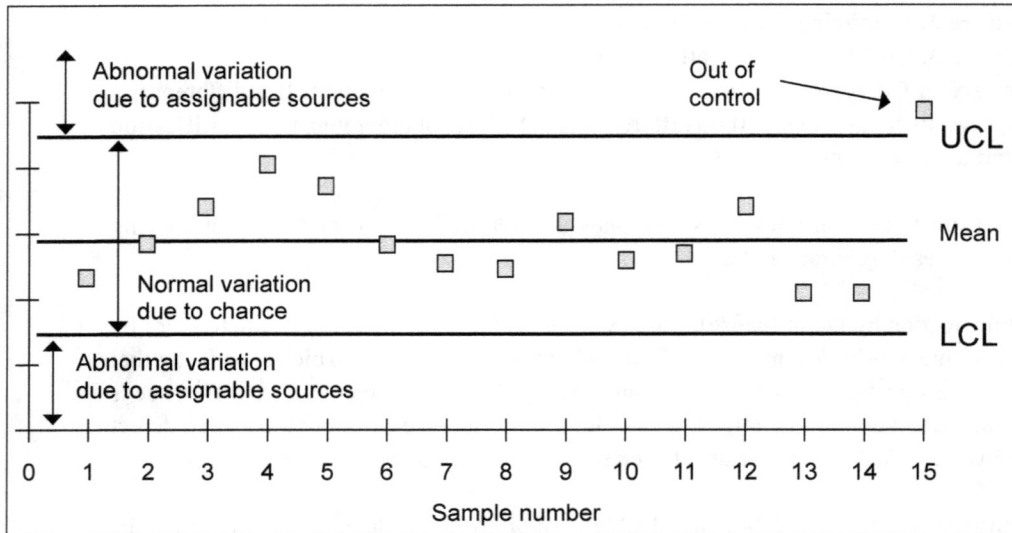

Example of a control chart

Control characteristics represent process results and provide the control status of the process. Variables or continuous data include figures for length, mass, time, strength, content, yield, and purity, and such data that can be measured in the usual way.

Discrete values or enumerated data include figures for the number of rejected products, the number of defects, the ratio of rejected products, the average number of defects, and other such countable values. A percentage indicating a ratio to a total is a variable if the numerator is a variable, and a discrete value if the numerator is a discrete value.

Sub-groups are sets of measurements divided into sections when differences appear in terms of time, product, or material while checking whether they are stable. The term used for dividing measurements into groups is called sub-grouping. The number of measurements included in a sub-group is called the group size. Ideally, group size is 2 to 6 for x-R control charts and 100 to 1,000 for **p** control charts.

Two types of dispersion are found in data collected from a process. The first, dispersion by chance, is inevitable even in well-controlled processes. The second, dispersion due to abnormal causes, can be avoided through adequate process control. Calculate the standard deviation due to dispersion by chance and set control limit lines at a distance three times the standard deviation from the centre line (the mean value of the distribution). If the process is stable, data for the control characteristics will disperse between the two control lines.

If the control lines are drawn in this way, only three out of 1,000 measurements should fall outside these lines, given unchanged process conditions or environmental conditions, machine conditions and material specifications.

This is a concept derived from the probability theory related to normal distributions. Processes are regarded as stable when the data points plotted between the control lines do not exhibit a run, trend or periodicity, because the dispersion is within tolerance or an allowable range under the set conditions. This is called dispersion by chance.

Controlling a process with a control chart calls for periodic observations. In order to confirm that the process is normal, you must plot data on the chart and check that data points do not fall outside the control limit lines or indicate a trend.

There are two types of control charts. One is used for process analysis, and the other is used for process control.

Control chart for process analysis:

This chart is formed by drawing control lines based on the data already recorded, provided that process conditions at the time of data collection are clear. If no data is recorded, record the process conditions precisely and collect data for the control chart. (Initially a control chart can be used to identify specific assignable causes of variation. These assignable causes must then be eliminated to achieve a state of control).

Control chart for process control:

This chart is used to determine the presence of abnormalities in a process. It is formed by the daily plotting of data and makes use of the control lines of a control chart for process analysis. (Once the process is confirmed as being under control with the control chart for process analysis, then it is possible to extend the chart's control lines and use them as a control chart for process control.) The control lines of a control chart for process analysis are drawn as broken lines (----), and those for process control as lines composed of dashes and dots (- . -). The centre line is a solid line in both charts. (After using the control chart for process analysis, it can be used to maintain processes in a stable condition).

Control charts for continuous variables:

This text describes how to draft a control chart for continuous variables and to apply it to control and improve processes. These charts are used to assess quality conditions with a small number of samples. There are three sub-types of control charts for continuous variables:

$\bar{x}$-R control chart:

The $\bar{x}$-R control chart is composed of an $\bar{x}$ control chart, used to check changes in the mean value, and an R control chart, used to check changes in dispersion. It represents the largest volume of information among the different control charts.

$\bar{\bar{x}}$-R control chart:

The $\bar{\bar{x}}$-R control chart uses the median $\bar{\bar{x}}$ in place of $\bar{x}$ for the groups in $\bar{x}$-R control charts, eliminating the calculation of $\bar{x}$. Although not as efficient in detecting abnormalities as control charts, its applications and method of use are the same as those for control charts.

X control chart (X-Rs control chart):

The X control chart uses individual measurements **X** in the following cases:

a. When only one measurement is available (e.g. cell numbers per day).
b. When the process is almost the same, and a single data point is sufficient for representing the process (e.g. the concentration of a component).
c. When obtaining certain measurements is time consuming and costly (e.g. some chemical analysis).

Control charts for discrete values are:

a. **pn** control chart
b. **p** control chart
c. **c** control chart
d. **u** control chart

The **pn** control charts are a special case of **p** control charts in which **n** is constant. There are no fundamental differences between these two control chart types.

Unlike the $\bar{x}$-R control chart which combines two kinds of control charts, control charts for discrete values use only one chart. The method of determining control lines differs from that for continuous variables, but the basic concept is the same.

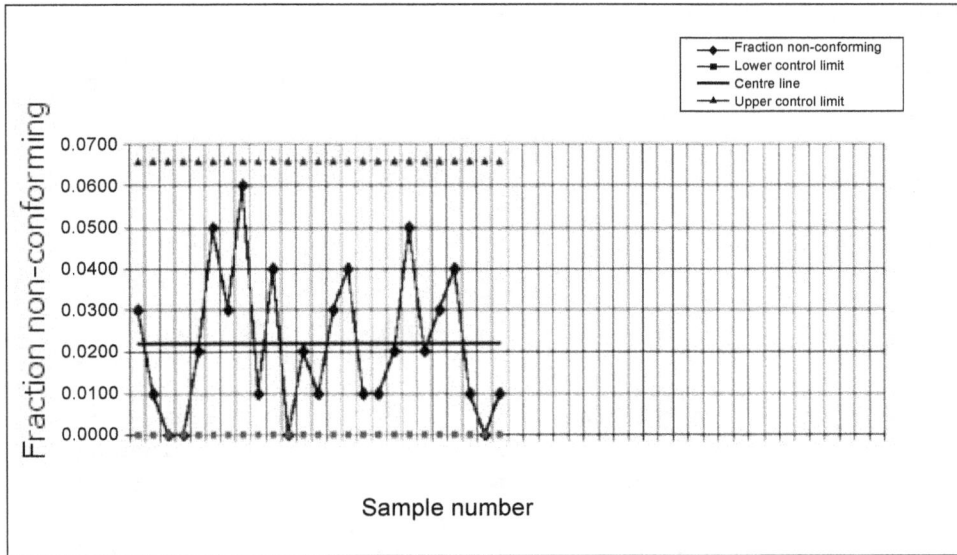

Attribute p chart

Key points in drawing a control chart:

To use a control chart for discrete values to analyse and evaluate a process effectively, you must collect data for at least 20 groups. The average ratio of rejected products (P) used for **p** control charts is not the average ratio of rejected products in each group, but is calculated using the following formula:

$$\bar{p} = \frac{\text{Total number of rejected products}}{\text{Total number of inspected products}}$$

This provides an average weighted by sample size (n), which differs from group to group. Although the values of control limits differ for different sample sizes (n) in **p** control charts, they can be made constant in simplified **p** control charts that use the average of **n** which differs from group to group. The average of **n** ($\bar{n}$) is obtained using the following formula:

$$\bar{n} = \frac{\text{Total of the values of } \mathbf{n}}{\text{Number of groups } \mathbf{k}}$$

When this method is used, the values of **n** should not differ significantly from group to group, but should fall in the following range:

$\bar{n}/2$ to 2n

The charts are relatively easy to use, since control limit lines do not fluctuate. However, to evaluate data points close to a control line, the control limit lines must be determined exactly, using the correct value of **n** for the group in question.

8.6 Interpreting control charts

Control charts are characterised by a centre line, and upper and lower control limit lines that are used as a scale to check data dispersion around the centre line.

Circumstances for which a data point falls outside the control limit lines are referred to as "out-of-control" states.

Data points fall outside limit lines:

a. By chance.
b. Due to a process abnormality.

Falling outside by chance has a probability of about 0.27% in the Shewhart chart (three sigma chart), or three times per 1,000 plots. Because of this low probability, the fundamental principle of using a control chart is to interpret an out of control incident as being due to a process abnormality and to find the cause of the abnormality.

The criteria that are used to judge whether a process is controlled or stable are presented below. Ideally, the status of a process should be evaluated with 25 or more measurements.

A process can be considered to be controlled in the following cases:

a. Data points do not fall on or outside the control limit lines, nor do they mark a trend.
b. 25 or more successive data points fall within the control limit lines.
c. Among 35 successive data points, only one data point for which an abnormality is not detected falls outside the control limit lines.
d. Among 100 successive data points, only one or two data points for which an abnormality is not detected fall outside the control limit lines.

A process is considered abnormal in the following cases:

a. Data points are on and outside the control limit lines.
b. Though all data points fall within the control lines, a run of 7 data points falls on one side of the centre line (the median line).
c. The process can be deemed abnormal when data points frequently appear close to the control limit lines. On either side of the centre line, divide the range between the centre line and the control limit line with another new line at a position two thirds of the way from the centre line. If data points fall between the new lines and the control limit line on either side in the following manner, the process is abnormal:
   i.   2 of 3 successive data points fall in the range.
   ii. 3 of 7 successive data points fall in the range.
   iii. 4 of 10 successive data points fall in the range.

8.7 Using control charts

Control charts are particularly effective in controlling processes. When an abnormality is seen in a process, immediately identify the cause and correct it. Systematise procedures for this purpose so that quick action can be taken when an abnormality is confirmed. Use a Non-conformance Report to report abnormalities promptly to the Quality Assurance Unit responsible for taking the first course of action.

Process analysis determines the factors that affect the characteristics of a process, and how they affect it, and sets out the actions needed to improve the process.

To use control charts effectively:

a. Draw stratified control charts.
   The cause of problems can be located by drawing control charts stratified by time, technician, or equipment, comparing mean values and dispersions, and looking for differences in defect ratios.
b. Devise an efficient method for sub-grouping.
   This requires data for each lot, classified by raw material, equipment, technician, day and time of work, working condition, and other causal factors.
c. Devise an efficient sampling method.

The method of sampling determines whether process conditions are reflected in $\bar{x}$ control charts. Select a sampling method that gathers data points in the $\bar{x}$ chart representative of each group. Clarify changes in process conditions from a technological viewpoint.

8.8 Applying various statistical techniques

Modern quality control makes wide use of statistical techniques. It is essential to recognise the value of these techniques in controlling and improving processes in a variety of company activities, to be able to select those that are appropriate, and to apply them effectively.

Effective QC tools:

a. Test and estimation.
b. Correlation and regression analysis.
c. Design of experiments.
d. Multivariate analysis.
e. Statistical sampling.
f. Sensory test.
g. Reliability test.

Statistical techniques can be applied to:

a. Cell product design.
b. Dependability specification and estimation of stability and longevity.
c. Process Control and research of process capability.
d. Data analysis, performance evaluation, and non-conformity analysis.
e. Process improvement.
f. Safety evaluation and risk analysis.

To eliminate product non-conformances:

Apply the QC tools consistently to stabilise processes and keep them under control.

To improve quality and achieve zero defects:

a. Pursue optimum conditions to attain the highest quality characteristics through experiment design and multivariate analysis.
b. Develop new products and new technologies.
c. Apply statistical techniques to develop and introduce new technologies and to raise product quality.

Keep in mind their interdependence:

a. Select an objective.
b. Set targets.
c. Assess the present situation.
d. Analysis: Investigate and analyse the causes.
e. Devise and implement recurrence preventive measures.
f. Confirm the effects of these measures.
g. Standardise, maintain and control the new methods.
h. Reflect on any problems left unsolved, and consider future countermeasures.

8.9 Reporting a process abnormality

This text describes how to improve quality and productivity by establishing a system that enables managers to recognise process abnormalities when they occur and to ensure and verify that action is taken.

Use a process abnormality report sheet to:

a. Report process abnormalities promptly.
b. Verify that proper actions are taken.
c. Analyse and correct abnormalities to prevent their repeated occurrence.
d. File abnormalities, organise research on countermeasures, and provide a reference to determine priorities for laboratories and facilities.

The report should contain:

a. A unique identification number.
b. Process conditions: Control chart number, name of process, product, control characteristics, lot number, lot conditions, technician, and other information.
c. Contents of abnormality: Date of occurrence, time, conditions of abnormality, person detecting abnormality.
d. Cause: The cause, if identified, and the views of the opinions and impressions of the person in charge.

e. Actions: Temporary measures, immediate actions taken against cause and process, date of correction and adjustment, notification of other departments, and related information.

f. Preventive action: Evaluation of proposed measures to prevent repeated occurrence.

g. Measures to prevent repeated occurrence: Countermeasures implemented to prevent repeated occurrence, future outlook, date that all countermeasures will be fully implemented, and results to date.

h. Confirmation of countermeasures and future control methods once fully implemented.

i. Responsible personnel, person reporting, those receiving the circulated report, filing departments, and related information.

Adopt a standard for the report, including the following written into an SOP:

a. What data should be entered? Who completes the report and when? How many copies should be issued?

b. Method of circulation.

c. Determine the system to be maintained until the final solution of the abnormality, especially before implementing measures to prevent repeated occurrence and confirming the results.

Recommended reading

1. Committee for Proprietary Medicinal Products (CPMP). Points to consider on switching between superiority and non-inferiority (CPMP/EWP/482/99), 2000.

2. European Medicines Agency. Guidelines on the choice of the non-inferiority margin (EMEA/CPMP/EWP/2158/99), 2005.

3. European Medicines Agency. ICH Q1E: Evaluation of stability data, 2003.

4. European Medicines Agency. ICH topic Q5E, Step 5: Note for guidance on biotechnological/biological products subject to changes in their manufacturing process (CPMP/ICH/5721/03), 2005.

5. Goldenthal, A.E. Setting alert and action limits for vaccine titres in the biologics industry, 2012. https://www.academia.edu/5308442/, accessed 29 August 2016.

6. International Organization for Standardization. ISO/IEC 13528:2015, Statistical methods for use in proficiency testing by interlaboratory comparison. https://www.iso.org/obp/ui/#iso:std:iso:13528:ed-1:v1:en, accessed 19 September 2016.

7. U.S. Food and Drug Administration. ICH Q1A(R2): Guidance for industry Q1A(R2) stability testing of new drug substances and products, 2003.

8. U.S. Food and Drug Administration. Statistical guidance for clinical trials of non diagnostic medical devices, 1996.

Self testing multiple choice questions

1. Data for quality characteristics does not show:
   a.   The conditions of the process.
   b.   The accuracy of the protocol.
   c.   The specifications of the product.

2. Inspection records do not include records of:
   a.   Acceptance inspections.
   b.   Between-process inspections.
   c.   Patient pre-infusion examinations.

3. The between-process inspections are used to determine whether:
   a.   Products are ready to be shipped.
   b.   Partially completed products are ready to be sent on to the next process.
   c.   Final product conditions are satisfactory.

4. Principle 1 of process control states that:
   a.   To ensure integrity, the regular technician should also take the sample.
   b.   The person taking the sample should be different from the regular technician.
   c.   It does not matter who takes the sample.

5. To carry out efficient sampling you do not need to:
   a.   Blind yourself to what you want to control.
   b.   Determine a sampling interval.
   c.   Determine a sampling method.

6. A key management task in ensuring that control and improvement tasks are promoted effectively is to inspect data:
   a.   Daily.
   b.   Weekly.
   c.   Monthly.

7. Maintaining daily management work records includes:
   a.   Submit record sheets to the employee responsible for the test at the end of each week to confirm they did the work.
   b.   Record sheets should be marked for confirmation and comments by the supervisor on general conditions during the test.
   c.   When records are incorrectly completed the test manager must get an explanation from the employee who completed the record.

8. The check sheet for recording distribution of characteristics makes it easier to process data because it presents the data in:
   a.   Root cause diagrams.

b. Frequency charts.

c. Highly technical language.

9. The purpose of in-company training in statistical techniques is to provide all employees with:

a. An overview of QC and other statistical techniques.

b. A working knowledge of QC and other statistical techniques.

c. An in-depth knowledge of QC and other statistical techniques.

10. A population is:

a. A group of people who live in the same area.

b. A group of entities whose characteristics are to be studied.

c. A selection of items from a group of items.

11. Variable data is composed of:

a. Discrete values.

b. Discontinuous data.

c. Continuous values.

12. Random sampling gives a result:

a. Governed only by chance.

b. Of statistical regularity.

c. That is error free.

13. Products should _____ be sampled by the workers responsible for manufacturing them.

a. Sometimes

b. Always

c. Never

14. Before data recorded in the past is used:

a. Confirm the purpose for which it was collected.

b. Remove any constraints on its collection.

c. Do not talk to the employees who collected it as this will cause bias.

15. If data points exhibit periodicity, the process is:

a. Normal.

b. Abnormal.

c. Rational.

16. A characteristic that cannot be included in characteristic diagram is:

a. Cost.

b. Safety.

c. Morale.

17. The 5Ms do not include:
    a. Material.
    b. Method.
    c. Money.

18. The method for extracting factors that affect results includes:
    a. Examine products at the work site.
    b. Express factors only numerically.
    c. Express factors only in single words or short sentences.

19. Fundamental statistics include the following:
    a. Range only.
    b. Range and mean only.
    c. Range, mean and standard deviation.

20. Random variables are:
    a. Discrete.
    b. Variable.
    c. Discrete or continuous.

21. The value at the centre of a sequence of measurements arrange in size is the:
    a. Mean.
    b. Median.
    c. Range.

22. The sum of **x** measurements divided by **n** is called the:
    a. Mean value.
    b. Median value.
    c. Range.

23. The difference between the maximum value and the minimum value of a data set is called the:
    a. Mean.
    b. Median.
    c. Range.

24. The median value:
    a. Is more precise than the mean value.
    b. Is less precise than the mean value.
    c. Has the same precision as the mean value.

25. When a relation exists between two variables they are said to be:
    a. Positively correlated.
    b. Negatively correlated.
    c. Correlated.

26. When a series of seven data points fall above the centre line the trend is:
    a. Normal.
    b. Abnormal.
    c. Cresting.

27. The advantage with graphs is that:
    a. Very few people can draw them quite easily.
    b. They allow us to judge the process status and conditions at a glance.
    c. They cannot present information clearly.

28. Statistical techniques are incapable of solving problems without the aid of:
    a. Design technology.
    b. Engineering technology.
    c. A trained statistician.

29. To compare or note changes in the ratios of components over time use:
    a. A radar chart.
    b. A bar graph.
    c. A band graph.

30. The value of Pareto charts is that they help us to:
    a. Recognise phenomena and causes.
    b. Recognise the important problems to deal with.
    c. Recognise the full range of problems to deal with.

31. Control lines should be revised in which of the following cases?
    a. When workers, methods of work, materials, or machines change.
    b. When a control chart indicates a change in the process.
    c. When a certain length of time passes.

32. A histogram uses columns to express frequencies of data:
    a. From similar categories.
    b. From different categories.
    c. With similar dispersion.

33. To draw a histogram you should have at least:
    a. 30 data points.
    b. 50 data points.
    c. 100 data points.

34. Process capability is determined by the relationship between:
    a. Different characteristics in the same product.
    b. The dispersion of product characteristics and standard values.
    c. Characteristics in different products.

35. Stratification should be used when individuals within a sub-population:
    a. Are similar to others in the same sub-population, but are different from those in subpopulations.
    b. Are different to each other, but similar to those in other sub-populations.
    c. Are similar to others in the same sub-population, and similar to those in other sub-populations.

36. The procedures for stratifying data do not include:
    a. Clarifying the purpose of stratifying the data.
    b. Incorporating all items in the stratification.
    c. Using an appropriate QC method to compare the stratified data.

37. Use a Gantt chart to:
    a. Confirm causes.
    b. Assess the present situation.
    c. Set a target.

38. A control chart shows that a process is stable when points representing quality:
    a. Fall outside the control lines.
    b. Fall inside the control lines.
    c. Fall inside the control lines and do not show a trend.

39. Sets of measurements divided into sections when differences appear in terms of time, product or material are called:
    a. Sub-sections.
    b. Sub-groups.
    c. Subsets.

40. When control lines are properly drawn and process, environment, equipment and material specifications remain unchanged, it is acceptable to have _____ points outside the line.
    a. 3 out of 100
    b. 3 out of 1000
    c. 5 out of 1000

41. In control charts a run of _____ is regarded as an out-of-control state.
    a. Five
    b. Seven
    c. Nine

42. Controlling a process with a control chart requires _____ observations.
    a. Daily
    b. Weekly
    c. Periodic

43. The chart used to determine the presence of abnormalities in a process is a control chart for:
    a. Process analysis.
    b. Process control.
    c. Abnormality control.

44. The control chart used to control a process with the number of defects, accidents, or failures in a certain unit or during a certain period is a:
    a. **p** control chart.
    b. **c** control chart.
    c. **u** control chart.

45. **pn** control charts are _____ **p** control charts.
    a. Identical to
    b. Not very different from
    c. Very different from

46. In order to use a control for discrete values to evaluate a process, you must collect data:
    a. For at least 10 groups.
    b. For at least 20 groups.
    c. For at least 30 groups.

47. The **pn** control chart is a special case of the **p** control chart when **n** is:
    a. Variable.
    b. Constant.
    c. Unknown.

48. **u** control charts are used to control a process using the number of standard defects per:
    a. Gross production.
    b. Sub-group.
    c. Production unit.

49. A process cannot be considered to be controlled when:
    a. 15 successive data points fall within the control limit lines.
    b. Among 35 successive data points, only one data point for which an abnormality is not detected falls outside the control limit lines.
    c. Among 100 successive data points, only one or two data points for which an abnormality is not detected fall outside the control limit lines.

50. Statistical techniques can be applied to:
    a. Process improvement.
    b. Risk analysis.
    c. Both a and b.

Chapter 9: Inspection

Auditing
product

Internal inspections are a Good Manufacturing Practice (GMP) requirement and therefore subsequently, a requirement for Good Tissue Practices as well. A company must implement an internal audit system to ensure that the products have the specific quality features that the customers/patients require. A management tool for monitoring and verifying the effective implementation of an organisation's Quality Management System is a basic requirement. This monitoring must be capable of identifying areas of conformity and non-conformity against customer/patient requirements, applicable statutory and regulatory requirements, and established planned arrangements in the QMS.

These inspections also provide information upon which the organisation acts in order to improve its overall performance, primarily through identifying opportunities for continual improvement. It is essential that these inspections base all the findings via unbiased means and factual information, focusing on quality performance, and determining that the quality system is effective in maintaining control by checking the prescribed quality objectives are being achieved and the resultant products and services meet specified customer and regulatory requirements.

9.1 Decide which features of a product to inspect

Inspectors/auditors do not look at every feature of a product. This would be impractical as it would take too much time and would be an extremely costly exercise. Instead, they inspect certain quality characteristics which they have identified in a pre-audit plan prior to commencing the audit. The most important quality characteristics are those that have to be proven to be uncompromised and present to provide customer/patient satisfaction.

To prepare a product for inspection, decide on the quality characteristics which the product should possess and therefore must be present. Then for each characteristic, choose one or more inspection items that can confirm that particular quality characteristic. Present all of these, the quality characteristics and the inspection items, in a quality table or checklist. By testing the inspection item the inspector can assess whether the quality characteristics are present and at the right level (within acceptable limits). In some cases the quality characteristics and the inspection items may be identical.

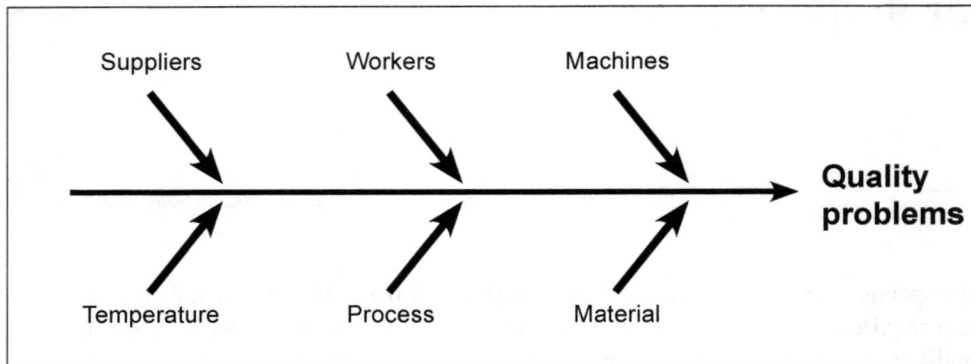

Factors that may cause quality problems

9.2 Establish inspection standards

There are four common inspection points in the production process that must be audited.

These are:
a. Acceptance of raw materials and reagents, etc. received by the company.
b. Intermediate or in-process inspections during that actual production process.
c. Finished product inspections.
d. Distribution/delivery inspections to the customer/patient.

It is necessary to have established operating standards for these four phases in order to ensure that they are performed the same way every time.

Acceptance inspections as mentioned previously, are conducted when accepting materials, equipment parts and reagents into the laboratory or workplace.

Standards for these should always specify the following points:

a. Procedures for selecting a sampling method.
b. Inspection items.
c. Acceptance values: what level of quality is acceptable.
d. Methods for inspections without testing.

Intermediate inspections, also known as in-process inspections, are conducted between processes.

Standards for these inspections specify:

a. Procedures for selecting a sampling method.
b. Inspection items.
c. Inspection methods.
d. Acceptance values.

e. What to do with rejected items.

Final inspections are conducted after all the processes are completed.

Standards for these, like those for intermediate inspections, specify:

a. Procedures for selecting a sampling method.
b. Inspection items.
c. Inspection methods.
d. Acceptance values.
e. What to do with rejected items.

Inspections/audits may be held in different places, over different periods of time, with different instruments, used by different operators and overall, be performed by different inspectors/auditors. There is therefore always the risk of differences both in the criteria on which judgment is based, and in the results. To minimise these differences it is especially important that the standards for each type of product clearly specify the following points in addition to those given previously:

a. The inspection sites.
b. The inspection periods.
c. The sampling method.
d. Sample inspection methods.

Audits need to be performed to meet not only the national but also the international standards, and therefore the adoption of official inspection methods is imperative to satisfy all the regulatory bodies. By using these official methods, quoting of the inspection reference numbers of the standards immediately informs all third parties of the level to which the audits were performed.

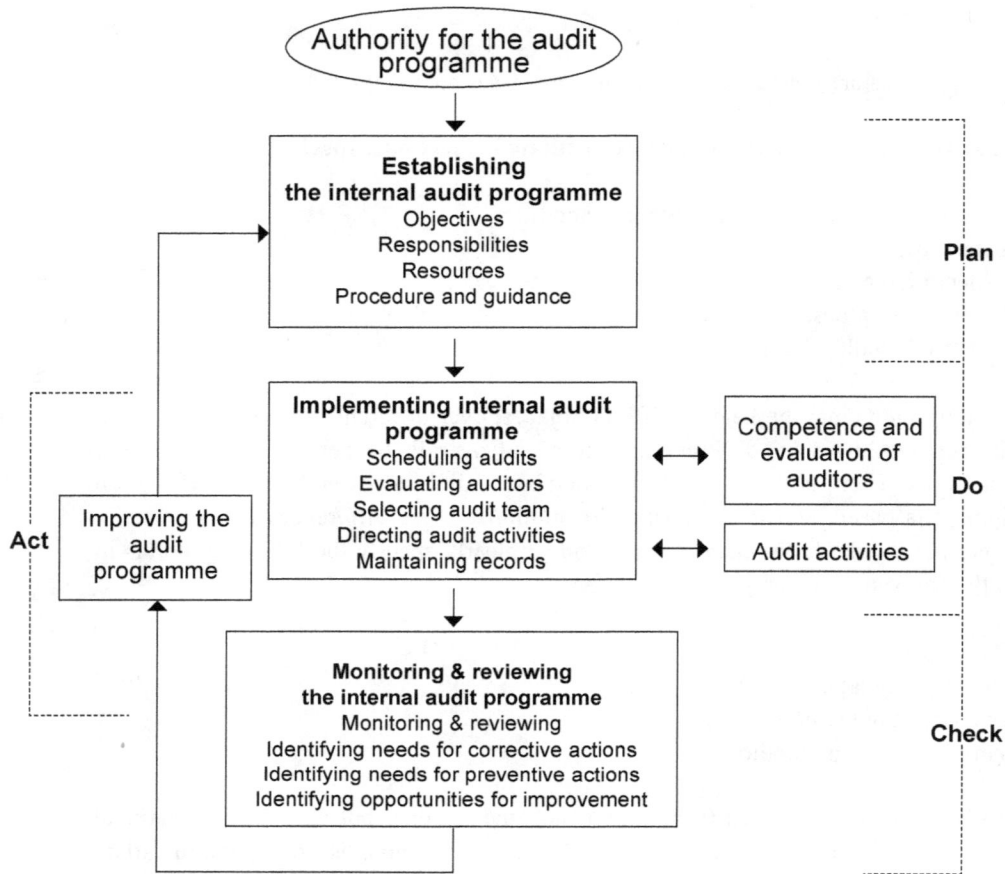

Utilising the PDCA cycle ensures that all facets of the audit are completed.

9.3 Establish specific inspection standards

It is necessary to establish audit specific standards:

a. Standard procedures for preparing inspection standards.
b. Standard procedures for inspecting specific products and product types.
c. Standard procedures for inspecting new and modified products.
d. Standard procedures for writing the inspection report.

A company must also establish standards that describe how to prepare, revise and review the inspection standards.

The standards for inspection preparation should specify:

a. Date, month and year of preparation and revision of the inspection standards.
b. The department or section in charge of issuing the inspection standards.

c. Forms for keeping records.
d. The scope of the standard: What does it cover?
e. Meanings of terms.
f. Other standards referred to in the standard.
g. Inspection items – the features in a product that are to be examined or inspected.
h. Inspection methodologies.
i. Standard values – what are the acceptance ranges.
j. Methods for selecting samples, sampling size, the number of samples necessary for judging acceptability, and judgment criteria. (In principle, sampling is performed randomly.)
k. Audit periods – when should regular internal audits be carried out?

It is necessary to be strict in establishing inspection standards for new and modified products, since there is little past quality-related information to refer to. The best way to do this is to increase the number of inspection items or increase the number of samples.

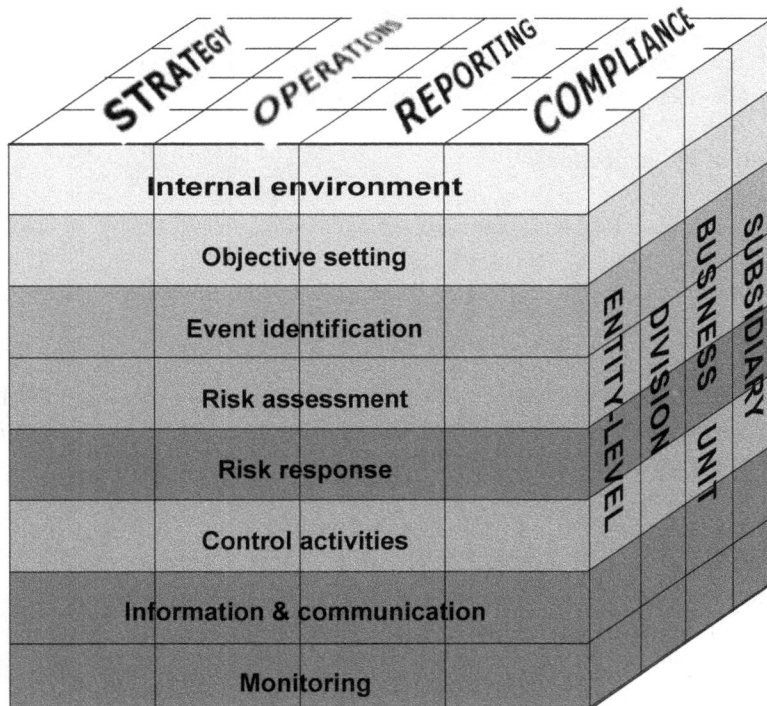

The Rubik's Cube of inspection and auditing criteria

9.4 Select, train and monitor inspectors/auditors

To ensure that inspections are carried out to the highest degree of accuracy:

a. Inspectors/auditors are trained to follow the domestic and international standards exactly as prescribed.
b. Inspections/audits are best performed as a team so that every member of the unit's performance is checked by other members ensuring accuracy and completeness.
c. Inspectors/auditors undergo continuous education and training as the standards are constantly improving in a dynamic environment.

It is imperative that inspectors/auditors follow the standards precisely, otherwise there will be variances even when inspections are carried out by the same inspectors/auditors and certainly when they are carried out by different inspectors/auditors.

To ensure that this is so:

a. Set up an official company system that allows only properly trained and qualified people to be appointed as inspectors/auditors.
b. Usual qualifications for inspectors/auditors:
    i. Have completed a science or engineering course at college or university level.
    ii. Have attained a recognised quality certification.

After completion of all inspections:

a. Confirm whether inspections have been carried out as prescribed.
b. Give a seal of authorisation to acceptable reports.
c. Give immediate instructions on any corrections that are required, and confirm the results of these corrections.

Set up a programme to educate and train new inspectors/auditors in the specific skills and knowledge they will require, including the ability to make impartial judgments, and establish standards for their certification (including working experience).

A curriculum for new inspectors/auditors should cover:

a. Basic immunology.
b. Methods for operating measuring devices.
c. Data processing methods.
d. Inspection methods.
e. Methods for judging if units can be accepted.

Provide periodic education and training for inspectors/auditors who are already performing as such in a company so that they can update their skills and technical knowledge, and remain at the cutting edge of their profession.

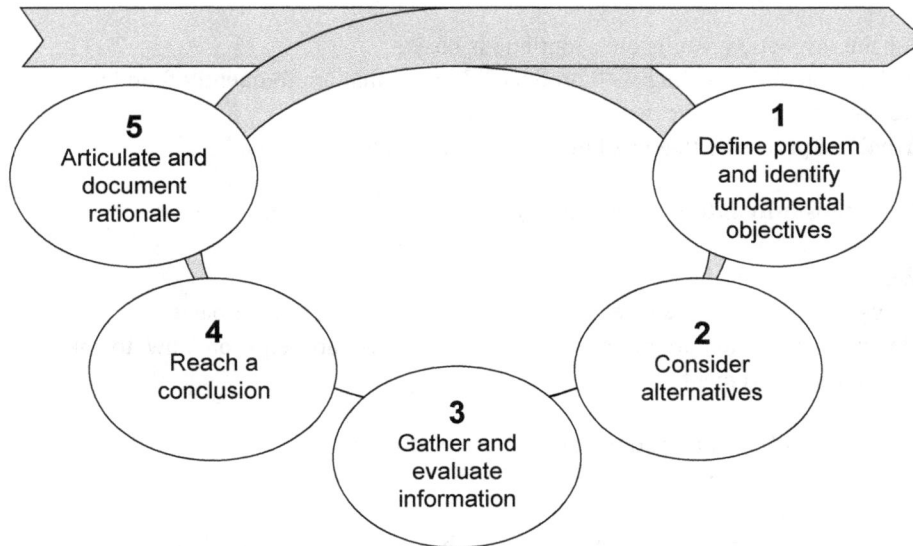

A knowledge base is essential for implementation of the five steps of auditing.

9.5 Deal with rejected products

Rejected products provide a set of urgent tasks:

a. Deal with rejected products: Ensure that they are not comingled with accepted products.
b. Any product that has to be reworked because it was initially rejected must be re-examined after the reworking.
c. Once rejected products are identified, then step one is to reduce the rejection rate if it cannot be immediately eliminated.
d. Use the data found in inspections to improve design and production processes.

Standardise the following procedures:

a. With cell preparations that have not been accepted, remove the rejected units and re-inspect the other bags again as if for the first time.
b. Label rejected cell products that can be reworked and send them for re-purification of the desired cell type.
c. Dispose of those products that cannot be reworked after recording and labelling them appropriately.
d. Inspect all the products again after those that have been corrected through re-purification.

To reduce the rejection rate, take the following actions:

a. Analyse the inspection results on a continuous basis.
b. Establish a system for dealing with product cell types that are frequently found to be rejected, and make effective use of this system.
c. Improve those processes that most often produce rejected products.

To put these steps fully into practice, the QA department should:

a. Analyse inspection results.
b. Publish lists of frequently rejected products (cell types) on a regular basis.
c. Provide instructions to laboratories and production departments on how to take measures for improvement.

The laboratories and manufacturing departments should:

a. Form improvement teams.
b. Set targets for reducing the number of defects that appear repeatedly in inspections, based on instructions from the Quality Assurance department.
c. Establish a system for achieving targets within fixed periods of time.
d. Report the results of improvements to the managers of the Quality Assurance and laboratory departments, and have them sign off.

If the improvement targets are not achieved within the designated periods, the manufacturing and Quality Assurance departments should re-analyse the situation and decide on the next step.

QA must re-analyse the initial set of corrective actions:

a. The inspectors/auditors must establish a system for reporting their results speedily to the Quality Assurance department, using prescribed forms.
b. The QA department analyses these results and presents its conclusions to the management and production departments on any further changes.
c. Management and Production set up systems to use this information to improve product design and production processes as part of the corrective process.
d. The results following correction are reported to the QA department, which confirms the effectiveness of the improvements that were made.

9.6 Keep inspection records

Inspection records provide a clear picture of the quality of processes and finished products, and will indicate where improvements are required.

a. Keep inspection records for each process.
b. Keep records of the measures taken with rejected products.
c. Store the inspection records carefully.

Record inspection results at regular intervals according to inspection types (acceptance, intermediate, final and delivery) and manufacturing processes. These records will also make it easy to trace the history of specific products. They should be maintained for the regulatory designated periods of time and should show:

a. The products that are inspected.
b. The time, date, and location of the inspections.
c. The inspection results.
d. The nature of the defects in the cell products and the circumstances at the time of their occurrence.
e. The cell numbers produced versus the expected values.
f. The names of the inspectors/auditors in charge.
g. Clear descriptions of any significant or related matters.

When this data is arranged in a time series chart (that shows how things change over a period of time) it can reveal a great deal of information. The data can also be converted into control charts and used for statistical analysis. Keep records of the measures taken with any rejected products in prescribed forms and for designated periods of time.

These records should include:

a. The names of those who prepared them or are responsible for them.
b. The names of the particular cell products.
c. Specific information about the cause for rejection.
d. The reasons for taking countermeasures based on impact assessment.
e. The suspected cause of the problem.
f. The conditions prior to the implementation of countermeasures.
g. The places and names of processes where countermeasures were taken.
h. The times and dates of countermeasures.
i. What the countermeasures consisted of.
j. The confirmed effects of countermeasures.
k. Any necessary revision of the Standard Operating Procedure.

This reference data will show the actions already taken to improve quality, and will serve as an important guide for dealing with future defects. The records can also be used for teaching employees. It is essential to analyse all the factors that contribute to cell defects and take appropriate countermeasures.

To store your inspection records properly:

a. Keep them in secure storage, environmentally protected, vermin proofed, and safe against theft and loss.
b. Use storage systems that allow easy data search, such as a computerised archival system.
c. Storage will usually be between seven and 10 years at the behest of most regulatory

agencies but some may even request longer. Product liability law may require the retention of documents for the lifetime of product.

9.7 Deal with after-treatment product related problems

In spite of all the inspections that are carried out, adverse reactions may still result from immunocellular products that have been used to treat customers/patients. To deal with these a company should set up a system to receive early feedback on after-treatment problems, especially from patient complaints, and to route this information to the relevant departments.

Introduce the following procedures:

a. Set up a system to receive early reports of customer's/patient's complaints and other forms of after-treatment feedback.
b. This information is reported to the Quality Assurance department.
c. The Quality Assurance department analyses the feedback and passes its findings to the appropriate production departments.
d. The production departments take appropriate countermeasures.
e. The Quality Assurance department confirms the effectiveness of these countermeasures.

Such a system must be in operation at all times, and should be specified in the company regulations.

Most companies routinely assign the processing of customer complaints to their Quality Assurance department.

A company should carry out systematic post-treatment research, even when no issues of adverse events are reported, to find out how well the therapeutic products have met the needs and satisfaction of the customers/patients.

Introduce the following research procedures:

a. Choose sampling methods.
b. Choose methods for analysing the data that is collected.
c. Set up procedures to gather the data periodically, and to report it accurately to management.
d. Prepare forms for reporting the data, and establish reporting routes. These forms should have plenty of space for recording a broad range of information without any restrictions.

The Quality Assurance department should be pro-active in proposing revisions in the inspection procedures for the products that are to be improved – revisions in inspection items, in methods and in standard values.

The company needs to set up a system that allows quick revisions as soon as further market feedback is obtained. Any system must include all the steps of the Revolving Audit Cycle.

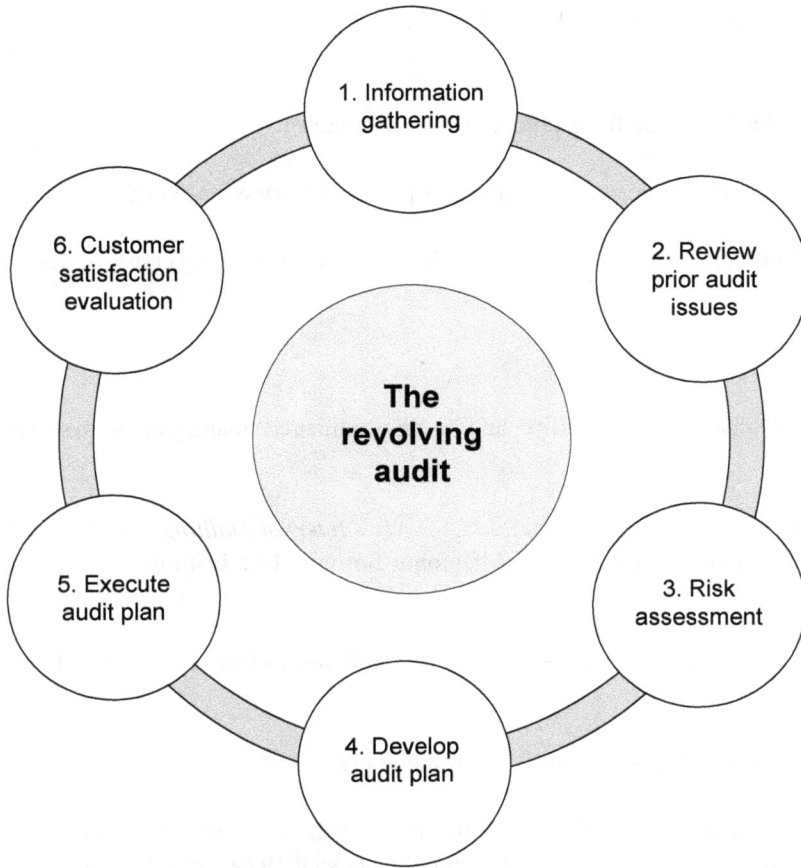

The revolving audit

Recommended reading

1. AuditNet®, The Global Resource for Auditors, www.auditnet.org.

2. Information Systems Audit and Control Association® (ISACA), www.isaca.org.

3. Internal Quality Audit Scheme and Internal Audit Procedure, City University of Hong Kong.

4. ISO 9001:2002.

5. ISO 19011:2002, Guidelines for quality and/or environmental management systems auditing.

6. Sawyer, L.B., Dittenhofer, M.A., & Scheiner, J.H. *Sawyer's Internal Auditing: The Practice of Modern Internal Auditing*, 5th edition. Altamonte Springs, FL: Institute of Internal Auditors, 2003.

7. Stimson, W.A. *Internal Quality Auditing: Meeting the Challenge of ISO 9000:2000*. Chico, CA: Paton Press, 2001.

8. The Institute of Internal Auditors (The IIA), www.theiia.org.

9. WHO – EDM. Basic principles of GMP: Self-inspection, 2012. http://test.futurebeacon.co/wp-content/uploads/2012/10/M07-Self-inspection.pdf, accessed 19 September 2016.

Self testing multiple choice questions

1. The department that has a central role in ensuring that inspection procedures are in line with what customers/patients want is:
   a. The marketing department.
   b. The Quality Assurance department.
   c. The design department.

2. Internal audits are a GMP:
   a. Requirement.
   b. Suggestion.
   c. Recommendation.

3. Successful monitoring must identify product _____ against customer/patient requirements.
   a. Conformity
   b. Non-conformity
   c. Both conformity and non-conformity

4. Inspectors and auditors must look at:
   a. Every feature of a product.
   b. Certain pre-identified quality characteristics.
   c. Only those quality characteristics that are a problem.

5. In some cases the quality characteristics and inspection items may be:
   a. Identical.
   b. Unrecognisable.
   c. In conflict with one another.

6. There are _____ common inspection points in the production process requiring auditing.
   a. Three
   b. Four
   c. Five

7. Acceptance inspections are performed:
   a. When materials, equipment parts and reagents are received from a new vendor.
   b. When materials, equipment parts and reagents are ready to go into production.
   c. When receiving materials, equipment parts and reagents.

8. Intermediate inspections are also known as:
   a. Acceptance inspections.
   b. Process inspections.
   c. Final inspections.

9. Final inspections are performed:
    a. Just prior to completing all the processes.
    b. When all the processes are completed.
    c. At the time of delivery of the final product.

10. When applying the PDCA cycle to audits, the "Act" refers to:
    a. Improving the audit programme.
    b. Establishing the audit programme.
    c. Implementing the audit programme.

11. In order to ensure inspections are carried out to the highest degree of accuracy it is essential
    to:
    a. Hire only trained and certified auditors.
    b. Continually train and educate auditors.
    c. Subcontract a qualified auditing service.

12. By following the standards of auditing precisely it is possible to eliminate:
    a. Any errors in production.
    b. Any variance within the product.
    c. Variance by the same auditor.

13. The usual qualifications for an inspector/auditor do not include:
    a. Quality certification.
    b. University project management backgrounds.
    c. University science or engineering backgrounds.

14. After completion of the inspection, it is necessary to:
    a. Write the report and then distribute to the inspected party within a few weeks.
    b. Report back to senior management and let them resolve any problems.
    c. Give immediate instructions for any corrections.

15. Any product that has been reworked after being initially rejected must be:
    a. Automatically approved.
    b. Re-examined for suitability.
    c. Kept separate from product that was approved initially.

16. Products that cannot be reworked should be:
    a. Disposed of.
    b. Kept as an example for future use.
    c. Stored in a safe place separate from all other products.

17. Inspection records provide a clear picture of:
    a. The quality processes and the customer/patient needs.
    b. The quality processes and the improvements necessary.
    c. The production history and the customer/patient needs.

18. Once completed, inspection records are:
    a.   Disposed of once corrective actions are completed.
    b.   Stored carefully.
    c.   Applied for the product but not for the process.

19. Storage of inspection records is usually for:
    a.   The lifetime of the company.
    b.   The lifetime of the product.
    c.   10 years.

20. If proper inspections are carried out then:
    a.   Adverse reactions will not occur.
    b.   Adverse reactions can still occur.
    c.   Adverse reactions are unrelated to the audit process.

21. Customer complaints and post-treatment feedback is initially dealt with by:
    a.   Senior management.
    b.   Quality control.
    c.   Quality assurance.

22. The effectiveness of countermeasures is confirmed by the:
    a.   Production department.
    b.   Quality Assurance department.
    c.   Senior management.

23. A company should carry out systematic post-treatment research:
    a.   Only when adverse reactions are reported.
    b.   Routinely, even if no issues are reported.
    c.   Only when there are issues reported.

24. The QA department should be _____ in proposing revisions in the inspection process.
    a.   Proactive
    b.   Re-active
    c.   Not involved

25. The Revolving Audit Cycle consists of _____ steps.
    a.   Five
    b.   Six
    c.   Seven

26. The first step in the Revolving Audit Cycle is to:
    a.   Execute the audit plan.
    b.   Review prior audits.
    c.   Gather the information.

27. The fishbone diagram for identifying inspection items suggests there are _____ areas that result in quality problems.
    a. Four
    b. Six
    c. Eight

28. In order to satisfy international regulatory agencies, it is suggested to quote:
    a. The reference numbers of the standards inspected.
    b. All statements made by the operators in regards to their performance.
    c. Any domestic regulations if they do not meet the international regulations.

29. Prior to conducting the audit, it is necessary to:
    a. Confirm that personnel are aware of the SOPs.
    b. Confirm that all the necessary materials and reagents are available.
    c. Confirm and evaluate the competence of the auditor.

30. SOPs for conducting an audit should establish the procedures for:
    a. Inspecting specific products and product types.
    b. Inspecting all products and product types generically.
    c. Inspecting only those products that conform to market standards.

31. The forms used to report market feedback should include:
    a. Inspection results.
    b. Reports on processed complaints.
    c. Proposed new product designs.

32. When auditing a new process or methodology, inspected items and samples should:
    a. Be increased in number.
    b. Be decreased in number.
    c. Remain at the same number as usual.

33. The curriculum of training for new inspectors should cover:
    a. Methods for operating measuring devices.
    b. Data processing methods.
    c. Both a and b.

34. By analysing the inspection results on a continuous basis, a company can:
    a. Reduce the rejection rate.
    b. Eliminate rejections.
    c. Increase the rejection rate.

35. It is imperative that the SOP for auditing stipulates in regards to samples that they are taken:
    a. At the minimum number possible.
    b. At the maximum number possible.
    c. Randomly.

36. The purpose of the audit report is to identify and record:
    a.   Where there are problems in the process and the product.
    b.   Correct performances for both the process and product.
    c.   Both problems and correct performance in process and product.

37. The main purpose of an internal audit is to:
    a.   Demonstrate that a company is in control of its quality systems.
    b.   Ensure everything is acceptable when the regulatory audit occurs.
    c.   Find problems and correct them.

38. One way audit reports can provide a great deal of information is by using:
    a.   A time-series chart to demonstrate decline or improvement in performance.
    b.   The names of the auditors to demonstrate competence.
    c.   Dates of the audit to demonstrate frequency.

39. In the Revolving Audit Cycle, step 5 is to:
    a.   Evaluate customer/patient satisfaction.
    b.   Develop the audit plan.
    c.   Execute the audit.

40. The reason an audit cannot inspect every feature of a product is because:
    a.   It would result in audit bias.
    b.   It would cost too much time and money.
    c.   It would interfere with day to day operations.

41. The fourth inspection point is:
    a.   Finished product inspection.
    b.   Distribution to the patient inspection.
    c.   Post-treatment inspection.

42. When accepting materials it is important that the SOP identifies:
    a.   How to approve it without requiring sampling.
    b.   How to eliminate the need for acceptance values.
    c.   How to select a proper sampling number.

43. Inspection audits may need to use different criteria when involving:
    a.   Different places, at different times, with different people and equipment.
    b.   The same place, at the same time with the same people and equipment.
    c.   It does not matter if there are changes as the criteria must always remain the same.

44. Internal auditors should be selected on the basis that :
    a.   They have complete knowledge of the processes and product.
    b.   They have a general knowledge of processes, products, diagnostics and testing.
    c.   They have minimal knowledge of the company and therefore cannot be influenced.

45. The best way to perform an internal audit is to:
    a. Have an individual perform all the inspections.
    b. Have a broad based auditing team that can investigate all areas.
    c. Have a team selected from management as they know the company best.

46. Internal audits should be performed:
    a. Whenever a problem is reported and on a preset interval such as twice yearly.
    b. According to a schedule of audits only.
    c. Only when a problem occurs.

47. One of the main reasons for contracting auditing services to perform inspections is that they can usually do the inspection _____ than in-house personnel.
    a. More quickly
    b. More cheaply
    c. With less bias

48. Which of the following is not a function of the internal auditor:
    a. Provide a quality assessment of the production process.
    b. Detect any shortcomings in the implementation of the quality systems.
    c. Recommend necessary preventive and corrective actions.

49. The internal audit report is to be considered:
    a. A company document that can be used any way management considers.
    b. An unofficial document but still requiring document protection under GMP.
    c. An official document and therefore cannot be altered or falsified.

50. The primary failure of the internal auditing system in biomedical companies is:
    a. Too many critical findings by the inspector during the audits.
    b. The inspector is compromised in their role by management directives.
    c. Failure of personnel to perform according to the SOPs.

Chapter 10. Education and training

Employee development

The quality of the education and training that the company provides to its staff will ultimately determine the quality of the products and services that a company offers to its customers/patients. Ultimately, this level of training will determine the success of the company's long term strategy and business. The company should therefore approach training and development systematically, implement it thoughtfully, and continuously evaluate and improve it for the mutual benefit of all parties involved. The failure of many biotechnology companies is that they fail to train personnel properly in order to build a team of quality employees for the long term and instead acquire what resembles a revolving door with staff coming and going so frequently that laboratories are constantly in turmoil. In order to maintain staff, build a quality team that is sustainable, it is necessary to provide goals and levels of achievement to keep staff interested and motivated. The first people to be trained should be those responsible for managing and implementing the TQM. The second group should be those developing, delivering, and evaluating education (e.g. curriculum developers, facilitators, and instructors).

Continuous improvement will, therefore, require a cross-functional team effort in which all members have the same basic education. Individual learning styles and instructional methods are not a case of one size fits all. In platform development, several issues should be considered, such as adapting materials to different learning styles and testing and evaluating prototype courses. These prototype courses serve as the basis for evaluating a generalised programme to be used throughout a company.

Once the model programmes have been developed, they must be tested. Evaluation should be based on predetermined criteria and a feedback process. Based on the results of the evaluation, programmes may be modified. Once the courses are completed, they should be scheduled for training of all personnel.

Instructors in the TQM programme represent one of the major stumbling blocks to implementation. Though finding instructors who know a little about various aspects of TQM is not that difficult, finding a single instructor who knows all areas of TQM or is quite knowledgeable in most areas is virtually impossible. One approach to this problem would be to develop a training programme where some instructors may teach only a few courses, requiring in-depth knowledge of a limited number of subject areas, while others may teach a range of courses requiring more extensive training.

Instructors for the on-site courses should be selected from personnel who are respected leaders in their professions, as evidenced by peers and superiors alike. They should have distinguished performance records, have good communication skills, and a strong desire to teach. Preferably, they should have some experience in the theory and application of TQM principles.

Education in TQM should include technical information as well as group development skills such as team building, effective communication, and group problem solving. It is important that the instructors integrate courses concerned with team skills with information about TQM philosophy and practices.

10.1 Establish an employee development policy

The first step in training a company's staff is to have an employee development policy and to communicate this policy to all of the staff so that they appreciate there is an expectation for continuous learning but at the same time they can expect some form of reward for making the effort to progress through the system. Rewards are not always monetary as some find the challenge of advancement, the acquisition of certificates, diplomas, etc. to be a significant reward on their own.

The objectives of the policy should be to:

a. Improve employees' competence in their sphere of operation.
b. Improve employees' understanding of their responsibilities.
c. Improve employees' ability to make good judgments.
d. Improve employees' problem solving capacity.
e. Improve employees' level of knowledge and scientific understanding.
f. Enhance employees' reputation and recognition by their peers.

It is a regulatory requirement to keep and maintain training records:

a. Keep records of the date of training, the instructor's name, participants' names, and the teaching materials and any handouts.
b. Have the person in charge, or the supervisor:
   i. Verify and maintain records for all staff members showing the level of learning they have achieved in the use, operation, and handling of the material, machinery or equipment and/or their level of understanding of any concepts.
   ii. Verify and record whether or not stable product quality is being efficiently and continuously achieved.
c. Use these records to plan and carry out future training.
d. Use these records for progress review and determination of achievement and promotional levels.

Regular follow-up for employees to refresh the skills and knowledge is a requirement of continuous training. Technical staff should be trained for quality function deployment, experiment design, multivariate analysis, and other statistical techniques useful in research and development, and in the application of statistics to their day-to-day tasks and performance.

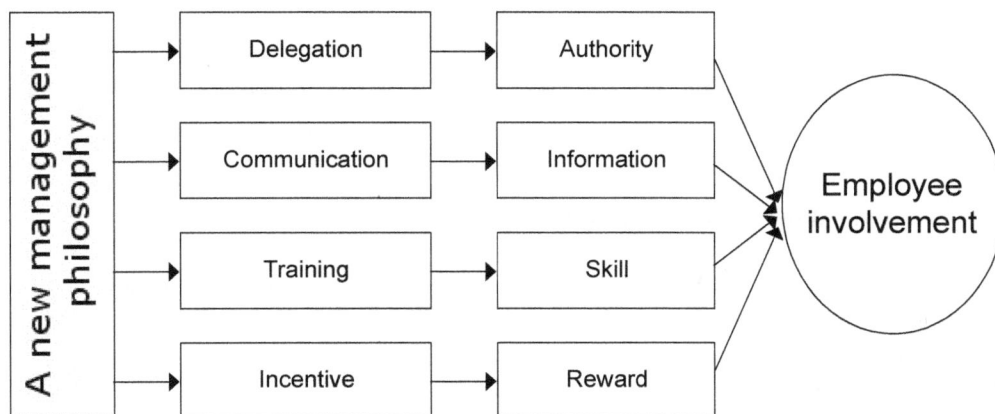

A new management philosophy → Delegation → Authority
A new management philosophy → Communication → Information
A new management philosophy → Training → Skill
A new management philosophy → Incentive → Reward
Authority, Information, Skill, Reward → Employee involvement

Employee involvement at different levels is essential.

10.2 Train employees in using equipment, materials and devices

In immunocellular therapeutic laboratories, the routine equipment used may differ from that found in standard laboratories. In order for technicians to become familiar with a particular piece of equipment it is necessary that they undergo a training programme that is both on- and off-the-job-training.

This training will deal in particular with:

a. How to handle, operate, and maintain materials and equipment.
b. How to avoid danger, promote safety, and prevent damage.
c. How to assess equipment faults and errors before or while they are happening.
d. How to maximise the potential of the equipment output.

Both forms of training must have systems for:

a. Confirming that employees have reached a verifiable level of knowledge and skills through the training before proceeding to the next level.
b. Maintaining a register of those employees qualified to operate and use specific pieces of equipment within the company and to conduct certain specific procedures.

There must, of course, be agreement on how the operations should be carried out, and what exactly the procedures are, before any training can be carried out. These agreed ways of carrying out a company's business should be written down as Standard Operating Procedures. Some companies may use work documents instead of SOPs when they deal with operation of equipment. These work documents may take the form of operation manuals but should have a unique number so that the fulfilment of the operational requirements can be recorded in the employee's training log.

Off-the-job training

In off-the-job training, the operation manuals for equipment as defined in the job standards are used. Instructions should cover function and inspection of any equipment, measuring devices, constituent components, tools, etc.

It should be provided to all relevant personnel, and include:

a. The purpose of the item of equipment, its range of use, the environment in which it should be used, and the conditions for using it.
b. An explanation of the equipment's basic functions.
c. The equipment's structure, principles, capabilities, accuracy, and reliability.
d. The equipment's constituent components (i.e. its permanently assembled parts), attached components, and replacement parts.
e. Any warning notices to ensure operator safety and avoid accidents.
f. Any indications regarding improper handling and how it will affect product quality.
g. Any prohibited activities with the equipment or near to it, in order to prevent damage, malfunction or breakage.
h. Instructions regarding the transport, storage, and handling of the equipment.
i. Instructions for checking and handling the constituent components.
j. All the procedures for handling, operating, inspecting, adjusting and verifying the equipment.
k. Instructions for processing data related to the equipment and for the pre-operation, start of operation, mid-operation, and end-of-operation stages. These all should be in the form of SOPs, Manuals and WDs.
l. Post-operation activities including inspection, verification, cleaning, adjustments, recording, etc.
m. Methods for reporting information on products, and the documents, records, etc. that are used.
n. The handling and management of waste materials after post-activities.
o. Issues that have to be reported to a superior, issues to be recorded, and the filling-in, format, handling and archiving of records.
p. What to do when abnormalities or damages are discovered, the countermeasures to take, and how to report and record such incidents.

On-the-job training

On-the-job training is provided within the laboratory. The qualified technologist or supervisor demonstrates and explains the correct procedures using the actual equipment, measuring devices, raw materials, components, and reagents to be used.

The followings are monitored closely:

a. Ability of handling equipment, reagents and associated materials carefully to prevent damage and loss.

b. Ability to follow and perform operational procedures and methods according to the SOPs, WDs or Manuals.

c. Ability to institute countermeasures should a non-conformance be identified and to follow the appropriate procedures on the reporting of any abnormalities through the Non-conformance, CAPA systems.

d. Proper testing and inspection of equipment in order to maintain their accuracy.

Point of caution:

In the course of training, it is important not to make any changes to equipment and measuring devices and their specifications unless to rectify a problem and in doing so, all proper, designated procedures of change control have been administered.

10.3 Provide formal TQM education and training

Education and training in TQM is particularly important. Setting up a well-structured TQM curriculum for each level of management, for each sub-department, and for each job is vital to success within a company. Typical groupings are top management, middle managers, TQM promotional staff, supervisors and group leaders, technical staff, and general employees. Each of these groups will have different needs, interests and constraints and therefore the training of TQM must be approached differently for each group.

A TQM education and training office as part of Quality Operations would be a valuable resource within a company. Its primary function will be to promote TQM on a company-wide basis, and to facilitate an exchange of information on the evaluation and improvement of training.

The TQM education and training office should:

a. Come under the HR department although its conduct, agenda and training platform will be the responsibility of Quality Operations.

b. Plan TQM training at each organisational level according to basic corporate policies, including seminars on QC (quality control), and reliability and statistical techniques.

c. Obtain senior management approval for a training budget; select the TQM seminars; obtain management approval to register employees in the seminars; and carry out registration.

d. Work with those responsible for training in each department, and hold regular meetings in order to coordinate efforts with them. Supervisors (section managers and department managers) to whom trainees report need to monitor QC, reliability and statistical methods training, until trainees are able to apply these techniques to routine tasks.

e. Verify the status of the TQM education and training promotion schedule, and provide appropriate advice and instructions.

10.4 Maintain a resource of good training materials

Good training materials are an essential training resource. They will provide appropriate content, will meet the needs of the participants, will fulfil the purpose of the course and the overall curriculum, and will allow the annual and monthly training plans to be followed. Review them regularly, and file them systematically. Having a library or at least a reading area where staff can review this material in a supportive and comfortable atmosphere is an asset. Materials should be kept up to date with the latest innovative ideas, research and development projects, and international perspectives to continually motivate employees to adopt new ideas and promote continual upgrading.

10.5 Evaluate and improve company training

Training should be focused on continuous improvement as a primary goal. The biotechnological field is constantly changing and a company must remain open to new and changing technologies if it is to remain relevant.

Quality Operations should:

a. Evaluate what the employees gain from each course.
b. Regularly evaluate the overall training programme, and seek ways to improve it in both content and methodology.

The value of training is best seen in the results that are achieved. If improvements do not occur as a direct result of training then the benefits are not being realised either because the training is deficient or because the personnel involved are not appropriate to the task that they are performing. There are a number of ways to evaluate this: checklists, job observation, reports that participants submit, and instructor evaluation.

a. Use a check sheet to verify that all participants understand what they have been taught, and have acquired the skills to put it into practice. Enter on the check sheet in advance what exactly they are expected to learn.
b. Check if participants can actually carry out the jobs they are trained to do: evaluate whether they can complete all the steps, and with appropriate speed and accuracy; evaluate the results of their work, and their ability to repeat a task precisely and with consistency. Use this as a basis for giving them a grade in on-the-job training.
c. Evaluate participants' submission of reports: Have them report on the knowledge and skills they have acquired, and on the positive impact this will have on the performance of their jobs. Treat this as the basis for giving a grade in off-the-job training.
d. Have supervisors evaluate whether the participants actually achieved their training targets.
e. Get participants to report on actions where they applied the knowledge and skills they have acquired. This is an effective way of evaluating management training in Quality Control.

The head of each department is ultimately responsible for establishing an intra-departmental unit to promote education and ensure training within the department takes place. By coordinating the activities of this unit with those of the HR and Quality Operations departments, and running it effectively with regard to the following provisions:

The unit should:

a. Be aware of the history of formal education, actual job duties, education and training, and official licences held by all the members of the department.
b. Create an annual plan for education and training, a plan for the earning of new licences, and a budget for these activities.
c. Complete a curriculum for education and training for each level in the management by hierarchy, by each sub-department, and by each function.
d. Develop and organise all necessary training materials.
e. Select and register instructors.
f. Be aware of the past education and training programmes and confirm their effectiveness.
g. Maintain records of education and training programmes and all licences earned.

Heads of departments along with HR must evaluate after examining the results of training whether or not the individual has the capacity to perform according to the new requirements. Not everyone has the ability to perform each and every task. Departmental heads must not be afraid to make the decision of shifting responsibilities and removing personnel from certain tasks.

10.6 Encourage employees to pursue self-development

Best results will be achieved by those employees who are motivated to pursue their own development. A company should encourage and facilitate such employees to undertake training in the skills and knowledge, and for the licences or registrations, that each is interested in, and which are necessary both in their own work area and company-wide. This is particularly valuable when employees need to be able to respond effectively to changes in customer/patient needs.

Check where improvements may be needed, and consider how these should be implemented:

a. Think pro-actively about other improvements that could be introduced in the future by the company.
b. For self-development to be effective both for the employee and for the company, the employee should:
   i. Be aware of the company's core business activities.
   ii. Select the skills, licences, and knowledge that they are interested in.
   iii. Talk to their superior about their development interests, and decide how realistic they are.
   iv. Make an effort to achieve them, and keep their superior informed of the outcome.
c. A company should:
   i. Suggest to the employee which specific skills, licences, and knowledge will be most useful.
   ii. Provide access to the necessary training.

iii. Provide the necessary financial support to undertake the training.

iv. Organise an internal system to make it easy to obtain skills, licences and knowledge in-house.

10.7 Set up a system for employees to acquire licences

Establish a licence qualification system that will certify that employees have achieved the level of knowledge and skills required for certain jobs that is domestically and internationally recognised. This system should specify the work that a licence is required for, and the method by which qualifications will be awarded. Organise a merit system to recognise and reward employees.

To set up a licence qualification system, take the following steps:

a. Clarify those jobs and work that require a licence.
   i. Inspecting internal processing.
   ii. Inspecting finished goods.
   iii. Carrying out an internal quality audit.
   iv. Carrying out an external quality audit – a quality audit of subcontractor goods.
b. Define the conditions needed to qualify for a licence:
   i. Employees receive the training designated for a specific job and process, get actual experience on the job within an appropriate period, and acquire an appropriate level of knowledge.
   ii. Employees receive an internal or external licence certification to conduct work for which this certification is judged especially important.
c. Designate a person to be responsible for selecting and approving employees to perform work for which an internal licence is required. Follow established laws and regulations for essential licences.
d. Create a ledger and register the names of qualified employees. Keep the ledger up to date.

10.8 Provide training in new products and new technology

Product therapies and technology are always changing. It is imperative that employees be given training in the latest technologies to ensure that the company maintains a leadership role within the industry. Different professions will require different training focuses as well as different types and levels of expert tools. For example, engineers will need training in advanced statistical methods, while those involved in producing documents may be able to improve office procedures using basic graphic tools. Levels and categories of proficiency will need to be determined. Organising training in this way could assist professional development and training specialists in determining requirements and needed resources for long-term planning.

Such training will be required for:

a. Engineers in the R&D department; they will require both training in the new technology, and the basic knowledge that will enable them to maintain the more advanced technological equipment and systems.
b. QC employees in the workplace, who will have to become familiar with new raw materials, and with production and inspection methods for the new components.
c. The marketing representatives who will have to introduce new therapies and procedures to the relevant hospital and medical venues .
d. Technologists and technicians will need to perform the latest cell differentiation and expansion platforms.

NOTE: Keep training manuals that clearly present the details of new products and new technology, and always keep records of the training given.

Who needs to receive Quality training?

Subject matter \ Who	Top management	Quality managers	Other middle managers	Specialists	Facilitators	Workforce
Quality awareness						
Basic concepts						
Strategic Quality Management						
Personal roles						
Quality processes						
Problem solving methods						
Basic statistics						
Advanced statistics						
Quality in functional areas						
Motivation for quality						

10.9 Provide training for new employees

Training of new people must also be addressed. Employees in transition and new employees will need to be indoctrinated into the TQM system. Since these people will enter their new jobs one at a time, organisations need to have an orientation and follow-on training for these people until they have caught up with their colleagues that have been through the system for much longer periods. Training for new employees involves all the following areas but to varying degrees depending on their positions.

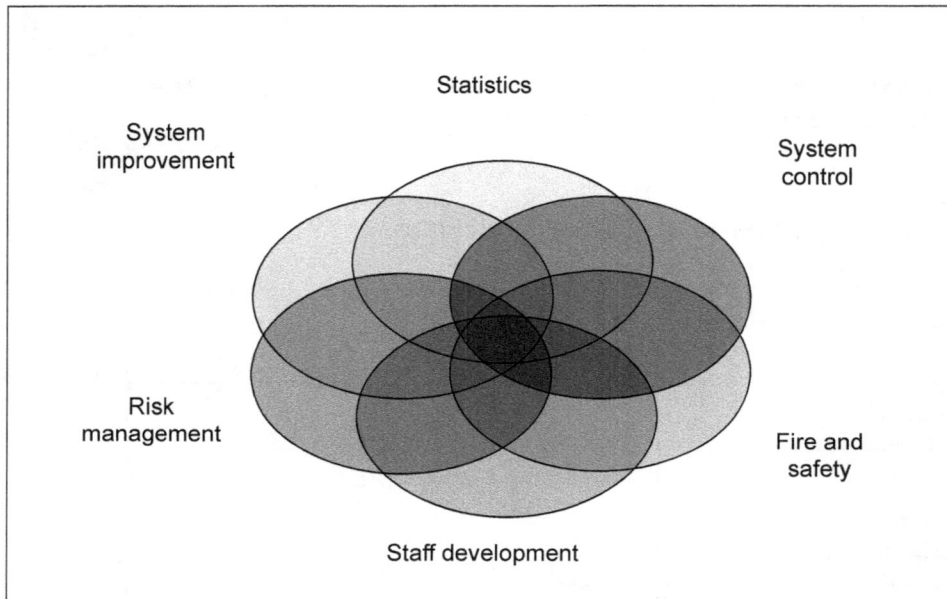

Training for new employees

10.10 Funding

No educational and training system can exist without funding. Therefore, considerations of costs should be done prior to implementation of the training programme.

Several issues related to short and long term funding must be addressed:

a. Investigation, design, and development and implementation costs (personnel, materials, time).
b. Delivery costs (instructors, facilities, materials, time).
c. Sources of funding (which budget will these come from).
d. Distribution and allocation of funds.
e. Accounting and evaluation systems.

10.11 Assessment and evaluation of employees

There are several steps in performing the evaluation:

a. Define skills, knowledge, and abilities required for TQM for that position.
b. Determine differences between the employee's current skills, knowledge, and abilities and the desired ones.
c. Know the overall objectives, what is to be accomplished as a result of the instruction, taking into account that some education will be required.
d. Implement instruction followed by assessment or examination to determine that the employee has learned what was intended.
e. Use the evaluation data as feedback for determining the next phase of continuous improvement.

Recommended reading

1. Cocheu, T. Training for quality improvement. *Training and Development Journal* 1989; 43(1):56-62.

2. Deming, W.E. *Out of the Crisis*. Cambridge, MA: Center for Advanced Engineering Study, MIT, 1986, p. 88.

3. Metz, E.J. Managing change: Implementing productivity and quality improvements. *National Productivity Review*, Summer 1984; 3(3):303-14.

4. Scholtes, P.R., & Hacquebord, H. Six strategies for beginning the quality transformation, Part II. *Quality Progress* 1988; 21(8):44-48.

5. Tuckman, B.W. *Evaluating Instructional Programs*. Boston, MA: Allyn & Bacon, 1979.

Self testing multiple choice questions

1. It is not the objectives of a good employee development policy to:
    a. Improve employees' understanding of their responsibilities.
    b. Improve their understanding of a company's business activities.
    c. Improve their capacity to increase their salary.

2. Rules and procedures for education and training should describe how to:
    a. Set up a schedule for education and training.
    b. Evaluate education and training.
    c. Both a and b.

3. Off-the-job training should provide instruction based on:
    a. The product specification, company policies but not the operating procedures.
    b. The operation manuals, SOPs and WDs only.
    c. All company, technological and procedural documentation that applies to the operation.

4. On-the-job training should have a special focus on:
    a. The purpose of the work at hand.
    b. Cutting costs.
    c. Performance considerations.

5. Records should be kept of:
    a. The names of employees who failed only.
    b. The names of all participants.
    c. The names of employees who passed only.

6. A company should have a well-structured TQM curriculum for:
    a. Each level of management, department level, technical level and general level.
    b. General employees only as management does not require training.
    c. The needs of each individual employee.

7. Top managers should:
    a. Only attend outside TQM seminars for managers given by external instructors.
    b. Attend internal TQM lectures given by internal instructors.
    c. Supervise the implementation of TQM in a company but not participate.

8. Middle managers should:
    a. Attend external and internal TQM seminars for middle managers.
    b. Not attend leadership seminars.
    c. Be the ones to determine a company's TQM policy.

9. The TQM education and training office should:
    a. Work with those responsible for training in each department.
    b. Determine a company's TQM policy.
    c. Learn multivariate analysis and quality function deployment.

10. Foremen and group leaders should:
    a. Study the implementation of TQM and Quality Operations activities.
    b. Take tours of all the departments in a company and copy their ideas.
    c. Not need to attend internal or external TQM seminars.

11. General employees should achieve TQM goals as far as their _____ allows.
    a. Education level
    b. Understanding
    c. Number of years in a company

12. The TQM education and training office should:
    a. Be solely responsible to determine a company's education and training policy.
    b. Promote TQM on a company-wide basis.
    c. Provide a secretarial service to HR.

13. The TQM education and training office should:
    a. Report to the CEO.
    b. Come under the HR department.
    c. Be an independent department.

14. To review internal teaching materials it is only necessary to:
    a. Examine the education records.
    b. Get the teacher to give his/her opinion of them.
    c. Read everything related to the successful implementation of training carefully.

15. After an external seminar is over the education department should:
    a. Have copies of any materials or textbooks used.
    b. Consider it not part of company training and therefore not recorded.
    c. Automatically record all participants as having successfully completed.

16. When using a checksheet to check what participants and trainees have achieved, it is important to know the details of what they should have achieved:
    a. In advance.
    b. At the time.
    c. Later.

17. In checking if participants can carry out the jobs they are being trained for:
    a. Evaluate whether they can complete all the steps, and with appropriate speed and accuracy.
    b. Evaluate their ability to repeat a task precisely but not consistency.
    c. It is not necessary to evaluate their ability to recognise non-conformance.

18. Get participants to report on:
    a. The positive impact the training will have on the performance of their jobs.
    b. The performance of their trainers.
    c. Both a and b.

19. To evaluate and improve your training programme, the person in charge of training should:
    a. Examine the test results and make judgments based on this only.
    b. Examine the training records and test results and get instructor feedback.
    c. Get feedback from instructors which replaces the need to examine test results.

20. Employee self-development should be in skills, licences and knowledge that:
    a. The employee is interested in and related to the work they perform.
    b. Are not necessarily related to his/her work area.
    c. Improve the overall look and importance of the employee's resume.

21. A licensing and registration is recommended for:
    a. Inspecting incoming raw materials to meet GMP requirements.
    b. Demonstrating knowledge meeting the domestic and international requirements.
    c. Confirmation of appropriate knowledge before signing a contract.

22. To qualify for a licence an employee should:
    a. Receive training designated for a specific job or process according to guidelines.
    b. Get actual experience on the job as quickly as possible without formal training.
    c. Obtain sufficient knowledge of the specific process without practical experience.

23. The following is not an objective of employee development:
    a. Improve employee's ability to make good judgments.
    b. Enhance employee reputation amongst peers.
    c. Improve employee's corporate standing.

24. The performance audit for a particular procedure is part of:
    a. Off-the-job training.
    b. On-the-job training.
    c. Both off- and on-the-job training.

25. The TQM education and training office should let _____ be responsible for the conduct, agenda and platform of TQM training.
    a. HR
    b. QO
    c. Itself

26. Companies should provide a _____ for training materials.
    a. Website
    b. Subscription
    c. Library

27. Training for new products and new technology is required by:
    a.   Engineers, QC, Marketing, and technologists.
    b.   Engineers, QA, Cleaning Department, and technologists.
    c.   QA, QC, Finance, and Marketing.

28. Individual learning styles and instructional methods mean that:
    a.   One approach to training is suitable for all employees.
    b.   Training has to be structured to the individual employee.
    c.   Everyone can learn in the same manner.

29. Instructors in TQM represent one of the major stumbling blocks in training because:
    a.   Most will only have a little knowledge from all the areas of TQM.
    b.   They know so much about each area that they become confusing.
    c.   Those with advanced knowledge are often unavailable.

30. The ideal instructor in TQM should have:
    a.   A sense of humour and overlook common mistakes.
    b.   A distinguished performance record, good communication skills and a desire to teach.
    c.   A knowledge of the subject matter but teaching ability is not a necessity.

31. When conducting a performance evaluation it must be accepted that:
    a.   Given enough training, anyone can perform the task.
    b.   Given enough time, anyone can perform the task.
    c.   Not everyone has the ability to perform the task.

32. The management group that requires quality training in all areas is:
    a.   Middle managers.
    b.   Quality managers.
    c.   Top level managers.

33. Funding for any training programmes should be considered _____ the training process.
    a.   Prior to
    b.   During
    c.   After

34. The first step in performing the assessment and evaluation of an employee is:
    a.   Determining the difference between the employee's current skill level and what is needed.
    b.   Defining the skills, knowledge and abilities needed for the position.
    c.   Implementing the instruction and training of the employee.

35. The final step in performing the assessment and evaluation of an employee is to:
    a.   Know what is to be accomplished from the training.
    b.   Determine what training is still required to complete the test.
    c.   Use the evaluation data as feedback to determine the next phase of training.

36. Training of new employees must be implemented in such a way that:
    a. They catch up to the level of existing employees as soon as possible.
    b. They will always remain at a lower skill level than existing employees.
    c. They establish their own pace of learning and a lower skill level does not matter.

37. Problem solving methods should be part of the quality training for:
    a. Management only.
    b. Quality positions only.
    c. Everyone.

38. Advanced statistics should be part of the training for:
    a. Supervisors and top management.
    b. Middle management and top management.
    c. Quality management and specialists.

39. When identifying which skills and licences will be most useful:
    a. The employee should instruct a company as to what they want.
    b. The licensing body should liaise directly with the employee.
    c. A company should suggest it to the employee.

40. Employees should be trained to appreciate how:
    a. Proper handling of materials and equipment will have a positive effect on product.
    b. Improper handling of materials and equipment will not have an effect on product.
    c. Proper handling of materials and equipment can equally have a negative effect.

41. In order to keep a sustainable workforce, a company must provide _____ to keep staff interested.
    a. Financial incentives
    b. Goals and levels of achievement
    c. Promotion opportunities

42. The first group of people to be trained should be those responsible for:
    a. Managing and implementing TQM.
    b. Providing the product according to TQM.
    c. Provision of training and evaluation of TQM.

43. The second group of people to be trained should be those responsible for:
    a. Managing and implementing TQM.
    b. Providing the product according to TQM.
    c. Provision of training and evaluation of TQM.

44. Confirmation that employees have reached a verifiable level of knowledge and skills is:
    a. A requirement of on-the-job training.
    b. A requirement of off-the-job training.
    c. Both a and b.

45. Acquiring knowledge of the written procedure for handling, operating, inspecting, adjusting and verifying the operation should be part of:
    a.   On-the-job training.
    b.   Off-the-job training.
    c.   Neither on- nor off-the-job training.

46. Methods for reporting information regarding products should be part of:
    a.   On-the-job training.
    b.   Off-the-job training.
    c.   Neither on- nor off-the-job training.

47. The ability to institute countermeasures should be confirmed while conducting:
    a.   On-the-job training.
    b.   Off-the-job training.
    c.   Neither on- nor off-the-job training.

48. The plan for the education and training programme should be reviewed and renewed:
    a.   Regularly.
    b.   Every six months.
    c.   Every five years.

49. Risk management is one of the key components of the training programme:
    a.   Only required by some companies.
    b.   True.
    c.   False.

50. Funding for the education and training programme is primarily required for:
    a.   Implementation, field trips, seminars and conferences.
    b.   Salaries, rewards, incentives and prizes.
    c.   Implementation, delivery, distribution and evaluation costs.

SECTION THREE
Working in the laboratory

Chapter 11. Standardisation

SOPs and
policies

11.1 Introduce operation standards and work instructions

It is a requirement that the drafting and implementation of standards for all the company's operations and procedures is undertaken as an initial step in establishing a Quality System. This is followed by the preparation of work instructions (WIDs), based on the standards. These SOPs and WIDs are used for the training and validation of employees in the performance of their jobs.

To introduce standards effectively in the company you will need to:

a. Be clear about the purpose of standards.
b. Recognise the range of items that may be included in a standard.
c. Decide on the procedures you will follow to introduce standards in the company.
d. Prepare work instructions from the standards.
e. Not write meaningless standards that serve no real purpose.

An operation standard is a written description of the best way to carry out an operation or a process so that it will always be done in the same way.

A standard has five main purposes that it serves:

a. To ensure that an operation is done in the same way by every employee, and to the highest level of quality each time.
b. To identify the key points to monitor in an operation in order to prevent production problems and non-conforming products (products that are not of the right quality).
c. To facilitate improvements in operations by stating what exactly the operations are.
d. To improve operator efficiency.
e. To keep every level of the organisation up to date with the current operations in use.

An operating standard usually contains the following categories or sections of information, within the document:

a. **Purpose** – Why are you doing this procedure.
b. **Scope** – The part of the operation covered by this standard. What it specifically applies
    to.
c. **Objective** – Usually to meet the national or international regulatory standard.
d. **Materials and reagents** – Used in the operation.

e. **Equipment** – Including tools and measuring devices required to perform the procedure.
f. **Health, Safety** – And risk management identifying the required hygiene and any protective clothing required.
   Actions for any uncontrolled, adverse events.
g. **References** – A listing of all pertinent documents that support the procedure.
h. **Methodology** – Operation procedures and cautionary remarks: pre-operational checks, preparatory operations, main operations, and final operations. This is how to perform the procedure.
   i. The sequence of the manufacturing process.
   ii. Operating conditions (location, safety aspects, temperature, etc.)
   iii. Operator qualifications, who can do the procedure.
i. **Appendix** – Diagrams, sketches or flowcharts are always useful for clarification.
j. **History** – Which identifies any earlier versions and changes of the procedure and reasons for any of those changes.

A company will need to decide:

a. What procedures need to be drafted, established, revised, archived and removed, and to inform employees of the latest standards.
b. Who is responsible for writing, technically reviewing, and signing off these procedures.
c. The basic format and style of the standards.
d. The specific content of each standard, its title and any logos you might use with it.
e. How to classify the standards in groups, and the classification numbers. Classify standards on a company-wide basis to prevent them being duplicated or contradicting each other.

In drafting the standards ensure that:

a. The instructions are such that, when followed, non-conforming products will not be produced.
b. They are easy to implement.
c. They are as simple as possible. Too much documentation lowers productivity.
d. They indicate procedures for practical actions.
e. They also include methods and criteria to evaluate the quality of the products.
f. They make clear the scope of responsibility, and the authority for implementation and revision.
g. They optimise the entire production process, and not just parts of it.

```
                                    ┌─────────────────┐
                                    │ Revise document │◄─────── NO
                                    └─────────────────┘
                                             │
                                             ▼
┌─────────┐                                                      YES   ┌──────────────┐
│  Start  │                          ┌──────────────┐      ◇           │ Create draft │
└─────────┘                          │Solicit feedback│   General       │   copy of    │
     │       ┌──────────────┐        │from affected  │  agreement      │ change order │
     ▼       │Create draft  │        │ individual    │   ◇────────────►│    [CO]      │
┌──────────┐ │copy of       │        │  or group     │                 └──────────────┘
│New document│─►│document    │──────►│              │                        │
│or change to│ └──────────────┘      └──────────────┘                        ▼
│existing    │                                                       ┌──────────────┐
│document    │                                          YES          │  Submit for  │
│needed      │ ┌──────────────┐  ┌──────────────┐   ◇               │formal review │
└──────────┘   │Train affected │  │Process release│ Document          │ and approval │
     ▼         │individuals and│◄─│of document   │◄─approved          └──────────────┘
┌──────────┐   │create training│  └──────────────┘   ◇
│On effective││records       │                        │
│date, update││└──────────────┘                       NO
│master and  │◄───                                     ▼
│controlled  │                               ┌──────────────┐
│procedure   │                               │Revise document│
│copies      │                               │  and CO      │
└──────────┘                                 └──────────────┘
     │
     ▼
┌─────────┐
│   End   │
└─────────┘
```

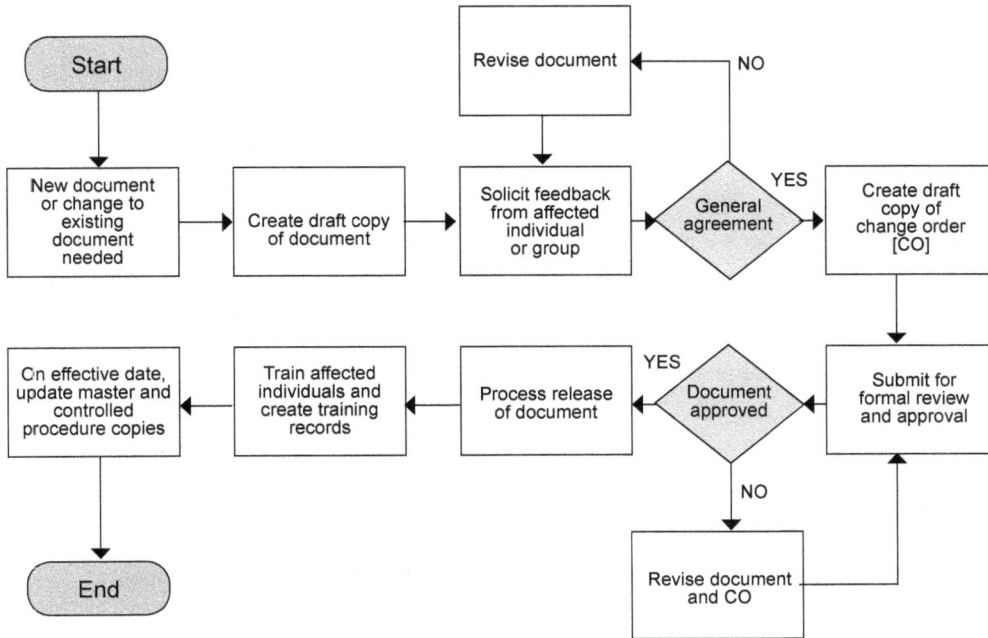

The operating procedure roadmap

A work instruction is a clear, simple written description of the daily production plans and related information, based on the standards, that is given to the employees before they begin operations. It is always a good idea to meet employees before they begin an operation and go over the work instructions with them. In any case, employees should always read them before they begin operations.

A work instruction specifies who should do what by name, when, where, why and how.

It should include:

a. The names of the employees who will do the work.
b. The machinery and material they will use.
c. The method they will use.
d. The environment in which they will do it.

In TQM this is referred to as 4M1E (men, machinery, material, method, and environment). Include columns for the results of operations: the hours worked, the values or quantity of items completed, the quality characteristic values, and any abnormalities or defects. These columns may then be included in operation reports to assess effectiveness of SOP implementation.

Always include the numbers of the operation standards in the work instruction, since the detailed work methods are given in the operation standards, and it may be necessary to refer to the actual SOP.

11.2 Maintain the standards

When standards have been drafted and implemented, there are three important ongoing actions to take:

a. Ensure that employees are following the standards.
b. Keep the standards up-to-date.
c. Use the standards to check that operations have been carried out correctly.

Draw up rules to ensure that the standards are maintained and revised properly.

There are a number of actions to be followed:

a. Carry out periodic reviews to see how the standards have been implemented, and how effective they are.
b. Revise standards whenever:
   i. There are changes in the production process.
   ii. Out-of-control events occur even though operations are carried out according to the standards. (These are events well outside the control limits of what is acceptable.)
   iii. There are changes in outside conditions, such as raw materials, equipment, or target quality.
c. Put a person with practical experience in charge of establishing and revising each standard. Include his name in the standards and delegate the necessary authority to him/her.
d. Include in each standard the dates when it was established and revised.
e. Be sure to remove from the workplace any operation standards that are no longer in use.

Include notes with each standard giving the history of its establishment and revision.

These notes may include:

a. Controversial items found during the review.
b. Explanation of items specified in the standards.
c. The rationale for the standard values, and for the classification of the standard.
d. Relationship with external standards if any.
e. Any design changes that have been made.
f. Issues that have to be decided, including future policies.

11.3 Educate and train employees to follow the standards

Standardisation will greatly improve quality in a company – but only if employees always follow the standards.

There are three reasons why employees may not follow the standards:

a. They are not aware of them: This is a problem of training.
b. They cannot perform the operations correctly: A problem of skill-sets.
c. They refuse to follow them: A problem of motivation.

First of all educate employees to use the newest standards correctly. Then set up a method of checking that they are using them. Do this systematically. Find out in each workplace what percentage of operations are using newly established standards (including old standards that have been revised) and then find out what percentage of these are being followed. HR must make a point of hiring personnel with specific skill qualifications from recognised colleges and universities. This ensures that they have basic procedural knowledge and have demonstrated technical capabilities in a controlled and reviewed environment. If SOPs are not being followed because an employee lacks motivation, find out why, and then take whatever steps are necessary to ensure that they are followed in every operation. Continuous failure to follow the procedure should result in the termination of the person from that position.

Educate employees in the value and use of standards so that they will:

a. Be able to read and understand the standards and work instructions.
b. Recognise the importance of standardisation in maintaining quality, and realise that operations must always be carried out according to the standards.
c. Understand the quality level that is to be reached in the manufacturing process.
d. Understand the methods to be used to confirm that these quality levels are reached, and the measures to be taken when they are not.

Grade technical skill levels in order to monitor employee progress:

a. Define between three to five levels of skills to be achieved.
b. Assess the current skills that operators have and sign off at that level.
c. Set the target level of skills that they should reach with time frames.
d. Make clear to employees the methods that will be used to evaluate the new skills.
e. Begin any training required in order to achieve the next skill level.
f. Retrain operators whenever operation methods are changed.

Training methodology or technical training can be taught in the two aforementioned ways: on-the-job training during operations, and off-the-job training before and after operations. Off-the-job training can be used to provide training in practical skills, technical education, and awareness training and may include workshops and group discussions, as well as the more traditional lectures. Standardise the methods of training, and always use common, everyday terminology. When training is completed, check the level that the employees have achieved. If necessary carry out re-training, and record the results.

Benefits. By providing operators with systematic and continuous training in operations, you will ensure that performance of man and equipment is optimised, that quality is maintained and improved, that productivity is raised, and that safety is improved.

Several key quality concepts:

a. Train employees to recognise the important quality characteristics of their products and job procedures, and to be committed to producing only quality products. They should be aware that rejected products decrease work efficiency, increase costs, and break delivery promises – and ultimately destroy customer/patient confidence.

b. Consciousness of quality is also important to consider in terms of tidiness and cleanliness. The environment and how it is maintained by the employee is a reflection of their overall attitude towards quality.

Developing quality consciousness:

a. The company and employees must be well informed of the present quality level of production: the ratio of defects, process capability, etc.

b. Make good use of Quality Operations activities.

c. Use every opportunity to promote quality consciousness. State "quality first" on corporate policy documents, and encourage employees to develop posters and slogans that can be put on display.

11.4 Organise standards on a company-wide basis

When standards have been established for all the operations in a company, they will most likely number in the hundreds. With so many, duplication or contradiction of other standards can become a problem. To avoid this, organise standards carefully on a company-wide basis; classify them; decide who has the authority to establish, revise, or withdraw them; and draw up rules to govern all this in a Quality Manual.

The usual areas governed having standards and governed by Standard Operating Procedures tend to be universal for most biotechnical production companies.

Typical manufacturing standards:

a. Product standards define the quality requirements that a product must satisfy. They specify a realistic quality level, given the current level of technical skill in the workplace. They are established by senior management.

b. Raw material standards specify the raw materials used to produce therapeutic products will meet the regulatory standards. They should:
   i.   Make clear the quality characteristics of raw materials and reagents to be used.
   ii. Define the purchasing specifications for raw materials.
   iii. Specify the procedures for inspecting incoming raw materials.
   Note: Raw materials include parts and supplementary materials.

c. Inspection standards describe the test categories, inspection methods, evaluation standards, etc., for raw materials, work-in-process (WIP), and finished products.

d. Job standards describe the work procedures for production (including the operation of equipment), inspection, shipping and transportation, working conditions, safety issues, etc.

e. Standards for maintaining equipment describe the methods for inspecting, maintaining, and preserving equipment.

f. Standards for maintaining measuring devices describe the identification, calibration, maintenance of accuracy, and control methods for measuring devices.

g. Packaging standards describe the methods for packaging products and WIP, and give the specifications, and identification methods for packaging materials, etc.

h. Work standards define the broader business rules, procedures, methods and report formats.

i. Supplementary standards are additional standards which include work procedures, operation procedures for equipment, operation procedures for measuring devices, work instructions, safety instructions, and one-point lectures used for basic training.

Rules governing the use of standards will vary depending on the policy implementation by Quality Assurance. As a small company, a company is much easier to control under the QA umbrella but as it grows in size, then the QA department must also grow accordingly and this lends itself to different interpretations, auditing skills, etc. of the QA personnel. There are certain things which should not be influenced by personal bias or skills of QA personnel, no matter how large a company becomes:

a. Basic rules for the general use of standards.

b. Rules for product and manufacturing standards.

c. Common rules for formatting and classifying standards.

The basic rules will include:

a. Rules for basic policies.

b. The roles of top and middle managers.

c. How to set up committees.

d. The role of marketing and promotional departments.

e. The functional roles of all departments.

f. How to promote policy management, education, the use of statistical methods, standardisation, diagnosis and the utilisation of Quality Operations.

The rules for product and manufacturing standards will include:

a. Rules specify quality level and inspection items (the points that are examined in an inspection) for products and raw materials.

b. Standards specify criteria, sequences, procedures and methods in design, manufacturing and inspections.

c. Specifications document the manufacturing and testing methods.

Common rules that the formatting of standards and their classification include:

a. Numbering system, units.
b. Symbols, graphic symbols.
c. Terminology used.

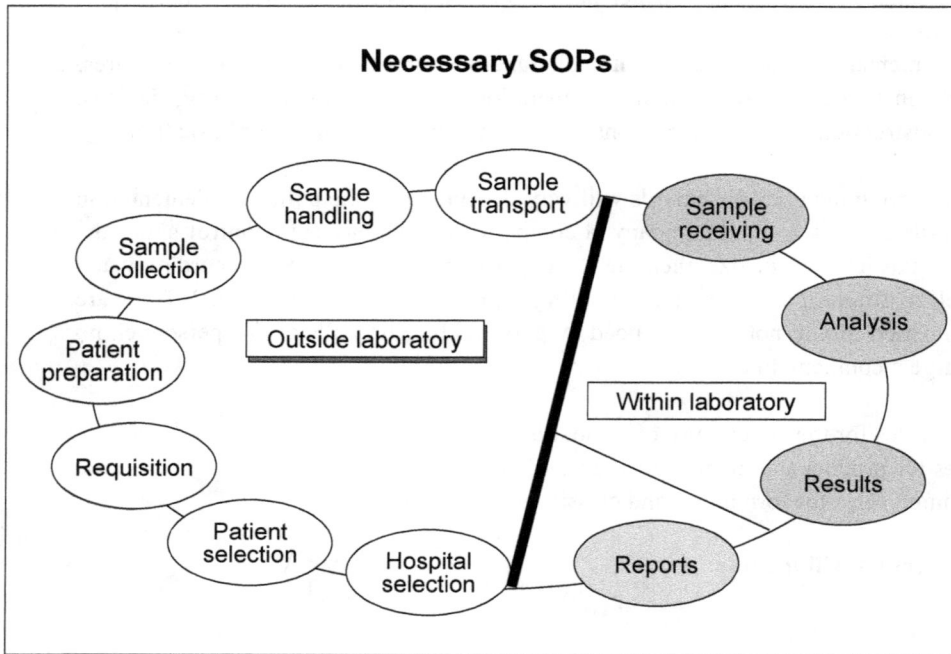

Necessary SOPs

11.5 Assess and promote standardisation

If standards are to make a lasting and effective contribution to quality, three important actions need to be undertaken:

a. Set up a system to regularly assess standardisation throughout the company.
b. Set up the Quality Assurance unit to promote standardisation.
c. Involve employees in continuously improving standardised operations.

Check to see that the standards have been implemented properly.

To do this you will need to have:

a. An SOP on the method for checking operations and confirming that they have been carried out correctly.
b. SOPs that clearly lay out the rules for handling things that do not go according to plan – when the quality characteristics of products are not what the standards say they

should be. These are known as abnormalities or out-of-control events. Therefore SOPs for Non-conformance, Out of Specification and Deviation Handling are required.

There are two ways of checking:

a. Check if operations have been carried out according to the standards. If not, find out why and take whatever action is necessary. This could mean reviewing the standards. To review the standards, compare the quality characteristics specified in the standards with the quality characteristics of the products that have been produced according to the standards.
b. Check if the effects (or results) of the operations, what are considered the quality characteristics of the products, satisfy the standards. When abnormal quality characteristic values appear, find out the cause. If necessary, investigate the operation conditions (e.g. raw materials, equipment or target quality) and revise the standards.

The methods for carrying out these checks should be stated clearly in the operation standard sheet or QC process chart.

The main points to decide are:

a. Who is to check what, how frequently, where, and how.
b. The sampling and measuring devices to be used.
c. The handling methods to use when abnormalities occur.
d. The boundary samples to be used, if these are needed.

When the checks have been completed write up reports of what has been done, and have these confirmed by supervisors and Quality Assurance managers.

Use control charts to provide ongoing monitoring of the process check points. This will help to stabilise the manufacturing process, and show where further improvements may be needed.

The assessment system should be set up by senior management from each department and should be run by a company-wide Quality Operations department. Use the system to periodically check if standardisation has been implemented as planned and is functioning properly.

Carry out a company-wide Quality Audit at least once a year, and departmental assessment at least once every six months. Inform each department in advance of the important items on which its assessment will be based.

These assessment items may include:

a. Standards related to claims, quality and similar matters.
b. Standards related to new or modified products and operations.
c. The percentage of standards that have been established and implemented.

The Quality Operations department and Quality Assurance unit should carry out the following actions:

a. Plan standardisation.
b. Maintain the standards system.
c. Monitor how standardisation is progressing.
d. Plan and implement assessment.
e. Plan and implement education and popularisation.

The ultimate purpose of standardisation is improvement. Standardising the daily work of the entire workplace will not only reduce abnormalities and deficiencies. It will create a system that brings continuous improvement in quality.

Managers must create an environment where employees are motivated to look out for problems in their workplaces, and report them. Managers can then examine these problems, select those whose solution will bring the greatest improvement, and find solutions to them, often with the participation of employees. Standards are improved when problems are recognised and solved.

When simple improvement activities are introduced, revise the operation standards immediately using the Change Control System. Complex improvement activities should be implemented as a project activity or as a Quality Operations activity.

Evaluate the effects of improvements in terms of quality, productivity, quantity, safety, efficiency, the productivity index, and cost-effectiveness. Summarise and record the results, and use these records to motivate employees to seek further improvement.

11.6 Investigate when standards are not being implemented

It can never be repeated enough that standardisation will greatly improve quality in a company but only if employees are consistent in following the standards.

There are three main reasons why technicians are not consistent:

a. They have not bothered to undergo the training or read the standards.
b. They cannot do the operations correctly: this is not always just a problem of skill but a physical incapability to perform as required. This can be a reason why rejected products occur even when the operator follows the standard. Not everyone has the ability to perform a specified task, the same way not everyone will become a great chef merely by following a recipe.
c. They refuse to follow them: this is a problem far more serious than merely motivation. There are individuals within any company that intentionally and purposely refuse to cooperate for reasons of their own. These individuals must be identified and removed from positions of responsibility that impact on the product.

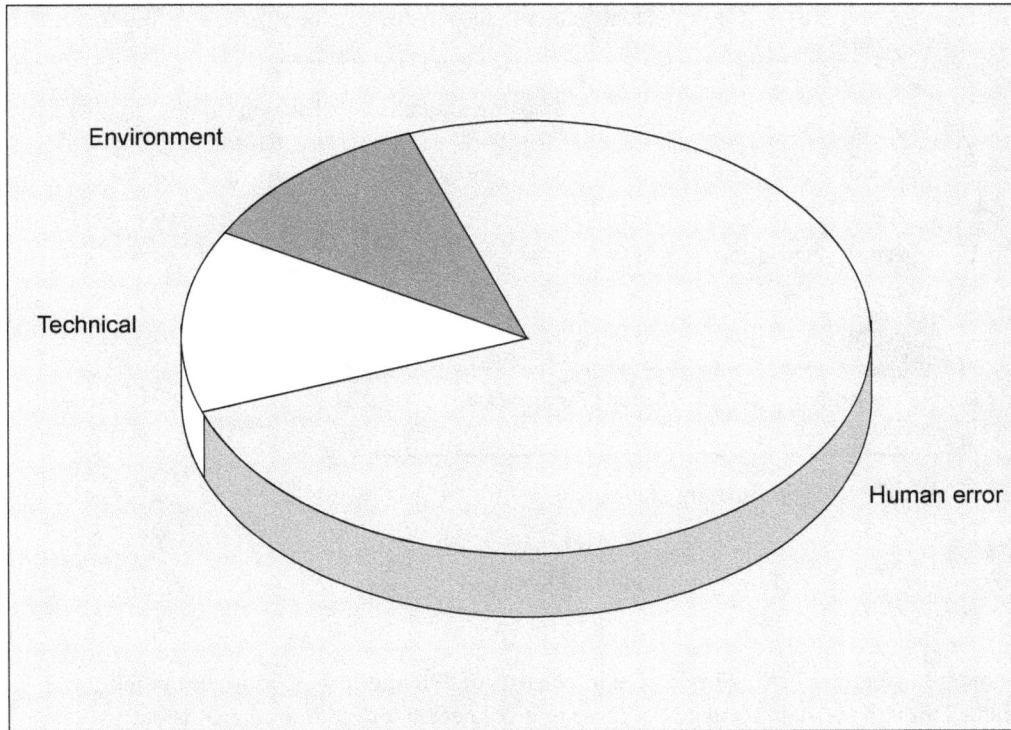

Probable cause of SOP deviations

To ensure that your employees overcome their reluctance or procrastination in undergoing training and reading the standards, you must:

a. Educate them in the value and use of standards while identifying the punitive action should they not adhere to them.

b. Provide the technical training needed to carry out the operations with emphasis on the fact they can only hold the position if they successfully pass a testing/trial regimen.

c. Provide training in quality consciousness.

d. Audit their operations periodically afterwards in order to ensure they do not backslide.

e. The individual that continue to refuse to cooperate will be terminated.

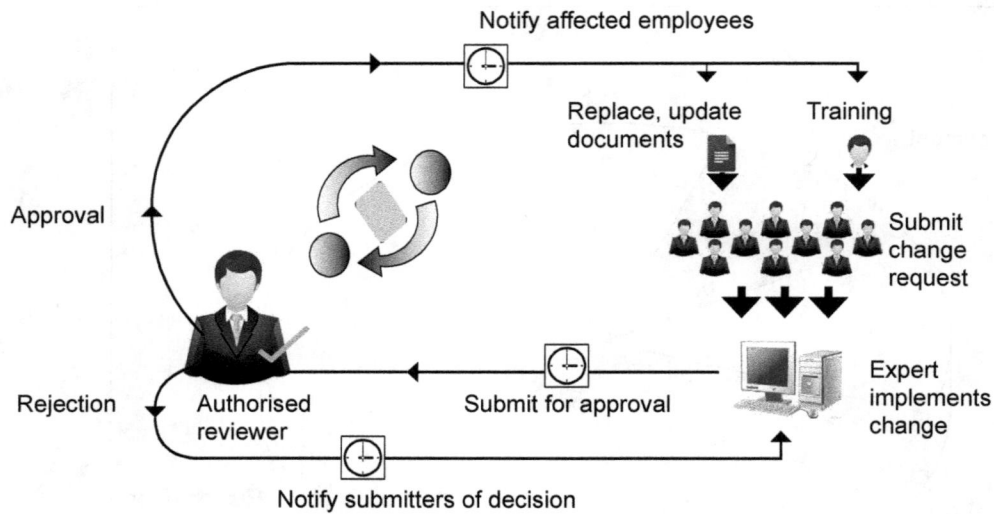

The change control pathway

The Standard Operating Procedure flowchart consists of two interlinking loops as per the above diagram. Through constant feedback of these interconnected loops, SOPs are constantly revised, upgraded and implemented through adherence to the Change Control SOP.

Recommended reading

1. College of Engineering, Texas A&M University, http://engineering.tamu.edu/

2. Environmental Health & Safety (administrative department), University of Washington, http://www.ehs.washington.edu/

3. ICH GCP 2.13: Systems with procedures that assure the quality of every aspect of the trial should be implemented.

4. Standard Operating Procedures for Good Clinical Practice at the Investigative, Centerwatch, http://www.ccrp.com/sop.shtml

5. Standard Operating Procedures for Good Clinical Practice, University of Washington, http://www.crc.washington.edu/Resources/GCPSOPInvSites.aspx

6. Standard Operating Procedures of the National Cancer Institute, https://cabig.nci.nih.gov/workspaces/CTMS/Meetings/SIGs/Best_Practices/SOPs/SOPs

7. United States Environmental Protection Agency, https://www.epa.gov

Self testing multiple choice questions

1. An operation standard describes the _____ way to carry out an operation.
   a. Quickest
   b. Best
   c. Easiest

2. To systematically draft standards the first thing you need to decide on is:
   a. Basic format and style.
   b. Categories for each standard.
   c. A work instruction.

3. An operation standard should contain:
   a. Materials required.
   b. Bulletins and sidebar notes.
   c. Names of individuals performing the task.

4. An operation standard will also contain:
   a. Alternative methods.
   b. An operation report.
   c. Methodology.

5. A work instruction should include:
   a. Names of the employees.
   b. Results of operations.
   c. Procedures for drafting instructions.

6. Components of an SOP that should go in the work instruction sheet include:
   a. Any abnormalities.
   b. Materials and equipment required.
   c. The final values of the test.

7. Product standards:
   a. Define the price that should be charged for the treatment.
   b. Define the quality requirements that a product must satisfy.
   c. Define the number of hours a technician is needed to create products.

8. Raw material standards:
   a. Make clear the quality characteristics of raw materials.
   b. Define the purchasing price for raw materials.
   c. Specify the brand names for incoming raw materials.

9. Inspection standards describe:
   a. Making on-site improvements.

b.   Evaluation standards.

c.   Who are the inspection staff.

10. Standards for maintaining equipment describe the methods for:
   a.   Purchasing equipment.
   b.   Using equipment for alternative purposes.
   c.   Preserving equipment.

11. Standards for maintaining measuring devices describe the _____ methods for measuring devices.
   a.   Identification
   b.   Maintenance of accuracy
   c.   Calibration

12. Packaging standards describe:
   a.   The specifications for packaging materials.
   b.   The sterilisation methods for packaging materials.
   c.   The purchasing requirements for packaging materials.

13. To check the extent to which standards are being put into practice you should focus on the questions:
   a.   Have the standards been re-classified?
   b.   Have newly established standards been placed on hold?
   c.   Have standards been established for all operations?

14. To keep standards up-to-date you have to:
   a.   Implement and maintain them.
   b.   Establish and implement them.
   c.   Maintain and revise them.

15. Revise standards whenever out-of-control events occur:
   a.   Because standards are not followed.
   b.   Even though standards are followed.
   c.   Unexpectedly and have no relationship to revision.

16. Revise standards when:
   a.   Raw materials are changed.
   b.   Equipment is cleaned.
   c.   The quality targets are changed.

17. Carry out periodic reviews:
   a.   Even when there are no out-of-control events.
   b.   Only when there are no out-of-control events.
   c.   Whenever there are out-of-control events.

18. Put a person in charge of revising each standard who:
    a. Is a department manager.
    b. Is a quality manager.
    c. Has practical experience.

19. Each standard should include a section that provides the history of _____ the standard.
    a. Establishing and revising
    b. Drafting and signing
    c. Drafting and operating

20. Which of the following items could be included in these history notes?
    a. A detailed explanation of items specified in the standards.
    b. Relationship with external standards.
    c. Issues that have to be decided in the future.

21. To use the standards to check that operations have been carried out properly you will need:
    a. Rules for handling things that do not go according to plan.
    b. Methods for checking and confirming that operations have been carried out correctly.
    c. Rules for revising quality characteristics.

22. It is not the purpose of a quality check for SOP performance that:
    a. Operations have been carried out according to the standards.
    b. The results of an operation satisfy the standards.
    c. Operations have been carried out as directed by the supervisor.

23. An SOP will not need to be revised if:
    a. The results of an operation satisfy the standards.
    b. The results of an operation satisfy the supervisor.
    c. Operations have been carried out according to the standards.

24. When abnormal quality characteristics arise you should first:
    a. Revise the standard.
    b. Investigate the cause.
    c. Check the quality.

25. In carrying out these checks:
    a. Decide who is least qualified to carry out the check.
    b. Choose the sampling and measurement equipment to be used.
    c. Understand in principle the proper performance of the SOP.

26. The items that have been checked in these checks should be summarised in operations reports and:
    a. Confirmed by management.
    b. Confirmed by technician who was evaluated.
    c. Sent to the patient.

27. If standardisation is to be successful employees should be educated and trained in:
    a. Those standards not applicable to their job areas.
    b. The history of quality consciousness.
    c. How to carry out the operations described in the standards.

28. If employees are to appreciate and use standards they need:
    a. To understand the important role that standardisation plays in maintaining quality.
    b. To realise that operations must never be carried out according to the standards.
    c. To understand the standards and work instructions that are at cross purposes.

29. The first step in skills training should be to:
    a. Assess the current skills that operators have.
    b. Set the target level to be reached.
    c. Aim for the highest level of skills.

30. Always retrain operators whenever:
    a. Operation methods are reviewed.
    b. Operation methods are changed.
    c. Out-of-control events occur.

31. Giving operators systematic and continuous guidance in operations will ensure that:
    a. Nothing will go wrong in the production process.
    b. Quality is maintained and improved.
    c. Equipment is updated.

32. One thing that cannot be achieved through quality consciousness is to:
    a. Produce only quality products.
    b. Recognise the important quality characteristics of their products.
    c. Recognise that rejected products destroy customer/patient confidence.

33. If operations have not been carried out according to the standards which of the following steps may have to be taken?
    a. Remove the standards.
    b. Retrain the employees.
    c. Review the training programme.

34. Organising and classifying standards systematically throughout a company will:
    a. Prevent standards being duplicated.
    b. Prevent standards contradicting each other.
    c. Both a and b.

35. Rules for product and manufacturing standards should specify:
    a. Quality level and inspection items for products and raw materials.
    b. How to promote the use of statistical methods.
    c. Criteria, sequences, procedures and methods in design, manufacturing and inspections.

36. Company-wide assessment should be carried out at least:
   a. Once every three months.
   b. Once every six months.
   c. Once a year.

37. Each department to be assessed should:
   a. Be informed in advance of important assessment items only.
   b. Not be informed in advance of any assessment items.
   c. Be informed in advance of all assessment items.

38. The ultimate purpose of standardisation is:
   a. Removing abnormalities.
   b. Creating a quality system.
   c. Achieving improvement.

39. Managers should create an environment where employees:
   a. Will feel able to report problems.
   b. Will recognise problems and cover them up.
   c. Will solve problems independently.

40. The effect of improvement activities in each workplace should be summarised and recorded in order to:
   a. Provide a basis for giving employees awards.
   b. Revise standard operations.
   c. Demotivate employees from seeking further improvement.

41. A Standard Operating Procedure has _____ main purposes that it serves.
   a. Three
   b. Five
   c. Seven

42. The section labelled PURPOSE in an SOP is to:
   a. Define the standard being written to.
   b. Explain "Why" the procedure is being carried out.
   c. Explain "What" the procedure applies to.

43. The reason why SOPs are best if they are written short and simple is:
   a. Most technicians have short attention spans.
   b. Less detail permits greater flexibility in performance.
   c. Too much in the document lowers productivity.

44. A Work Instruction is:
   a. A simple description of daily production plans and related information based on the SOP.
   b. A detailed, complex description of daily production plans based on the SOP.
   c. Identical to the SOP but is located at the site where the work is being performed.

45. In TQM, the 4M1E short form refers to:
    a. Money, machinery, materials, methods and equipment.
    b. Men, machinery, materials, methods and equipment.
    c. Men, machinery, materials, methods and environment.

46. When SOPs are revised or updated:
    a. Ensure that all old copies are removed from the workplaces.
    b. Ensure that every version remains at the workplace for ease of reference.
    c. Ensure that they are archived along with the previous versions.

47. There are _____ main reasons why employees do not follow SOPs.
    a. Three
    b. Five
    c. Ten

48. When performing a quality check on the performance of an SOP, QA will be satisfied with a minimum performance score of:
    a. 80%.
    b. 90%.
    c. 100%.

49. There are _____ main reasons why technicians are not consistent in performance.
    a. Three
    b. Five
    c. Ten

50. The rules for formatting SOPs include details regarding the use of:
    a. Font size, word count and symbols used.
    b. Numbering systems, symbols and terminology used.
    c. Subscripts, superscripts and italics used.

Chapter 12.
Precision measuring equipment

Metrology and calibration

Metrology and calibration is essential as it ensures that equipment being used is accurate in its measurement capacity, as most final products have to be within a specified range of acceptability prior to use. The acceptable range of precision is the basis upon which the quality characteristics of the product have been determined. Therefore failure to accurately measure the products as being within these required parameters results in failure to meet the required domestic and/or international standards.

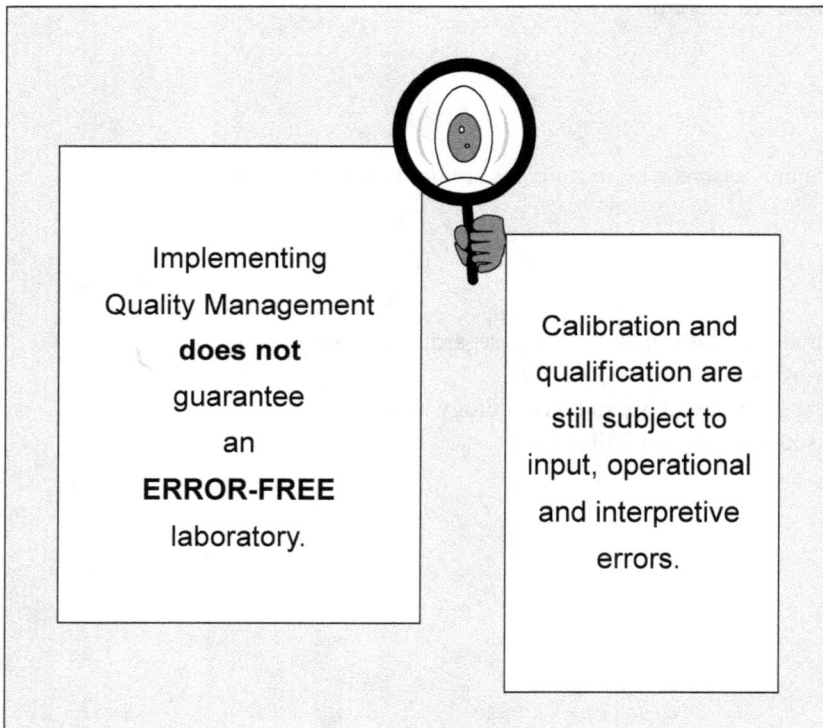

Implementing Quality Management **does not** guarantee an **ERROR-FREE** laboratory.

Calibration and qualification are still subject to input, operational and interpretive errors.

12.1 Performing proper calibration

The first step in ensuring that any measurements are correct is to use qualified measuring equipment, whether the metrology is performed in-house or by a contracted agency. The equipment used for calibration must be validated for accuracy within an acceptable range of precision, and verified as suitable to a range of conditions in which the company's products are manufactured and meet their quality characteristics. Therefore every measuring device must have a recent certificate of verification of its accuracy under normal use condition.

To ensure proper calibration:

a. Use contracted companies that have national certification.
b. Use only equipment that:
    i. Is appropriate for the manufacturing conditions and product quality characteristics.
    ii. Has the specifications that it is supposed to have.
    iii. Has the required precision.
    iv. Measures as easily as possible.
    v. Measures with a minimum of error.
    vi. Is cost-effective when used.
c. Get a clear idea of the procedures and tools for using the equipment correctly if performing the metrology in-house.
d. Companies that perform their own calibrations:
    i. Need to avoid damaging the equipment, especially needles and other fine apparatuses and only use equipment that is properly maintained.
    ii. Use equipment where the measuring error is minimal and within the specified range.
    iii. Have the analysers and calibrators adjusted and verified on a regular basis determined by the frequency of use of the production equipment.
    iv. Need to have checks and verification of calibration equipment to be signed off by QA.
    v. Store the calibration equipment properly.
    vi. Maintain a record log of servicing for all metrology equipment.
    vii. For in-house calibration, must implement a training programme for operators.

12.2 Calibration procedure requirements

Procedures are required for:

a. Calibrating, maintaining precision, storing and using the measuring equipment.
b. Calibration methods, and how to record the measurement results.
c. Storage of the calibration equipment in a way that will preserve its precision:
    i. Ensure that it is protected from dust, humidity, rust, and other harmful factors.
    ii. Ensure that it is wrapped up and stored carefully, even when used daily.
d. Proper labelling of calibration equipment and maintaining a master file as to equipment locations.
e. Maintaining the precision and functioning of the calibration equipment, and for inspecting, and repairing it.
f. Training for the proper use of equipment, and how to recognise any alerts for errors that may emerge either in the equipment or in the measurements.

Responsibility and authority for measurement control:

a. Clarify the responsibility and authority of the department in charge of measurement control which is most often shared between Engineering, QC and QA.
b. Document the company's system of measurement control.
c. Appoint someone to be responsible in each department, and clarify the scope of this person's authority.

A company will often need to change the specifications of measuring equipment when there are changes in manufacturing conditions. If the quality characteristics change, then results outside the range are indicative of a problem with the product. If the frequency of use of the equipment changes and results in measurements outside the range, then the equipment likely has to be recalibrated. To interpret the significance and impact of such variation of measurement precision after such changes, take into account the purpose, frequency, durability and reliability of the equipment before drafting and implementing any change control or non-conformance document.

Establish master files that document the purchase, calibration, repair and disposal of measuring equipment. These will provide records that make it easy to find the date of calibration and whether the equipment passed the calibration criteria.

Each unit of measuring equipment should have its individual file, marked with its unique identification number.

Relevant files to be included in an equipment file:

a. Operator's manual and specifications.
b. Acceptance inspection certificate and calibration certificate.
c. History of the item, including purchase, repairs, etc.

In-house operators of calibration equipment:

a. Must be qualified and validated to perform the metrology tasks.
b. Receive individual training programmes prior to calibrating equipment.
c. Use control check data in order to self assess performance.
d. Have training records that are kept up to date.

12.3 Calibration of measuring equipment

It is necessary to calibrate all equipment involved in production that provides a measurement. Determining the appropriate intervals is based on international standards. Calibration refers to testing the measuring equipment against specific reference points on a calibrator or analyser and certifying they are within acceptable limits. The calibration work may be performed in-house as long as a company has qualified personnel for performing such tasks. If not, then it is best to hire a qualified outside agency.

It is essential to:

a. Establish calibration procedures.
b. Mark the calibration expiry dates.
c. Issue calibration certificates.

If the equipment to be measured is used in the conduct of a Good Laboratory Practice (GLP)/GTP procedure, then the equipment must be calibrated by a GLP registered company that is qualified to do so.

When conducting in-house calibration:

a. Draft your calibration plans in advance.
b. Decide on a date to calibrate the equipment, after consulting the related departments.
c. Record the calibration results in calibration certificates and report them to the departments.
d. Consult the departments about what measures to take if deficiencies are found.
e. Recalibrate the equipment when deficiencies have been corrected.

Clearly mark the calibration expiry date on each piece of equipment:

a. Use labels that are easily identifiable by colour and shape.
b. Use a permanent ink that will not be easily erased.

It is important to show that the equipment has been calibrated by either an in-house, or external contracted service in case a problem is reported in the future.

CALIBRATION

ID:
DATE:
BY:
DUE:

Example of calibration label

The calibration certificates include:

a. The calibrating service or department.
b. The standard to which the unit was calibrated.
c. The calibration date.
d. The expiry date.
e. The person performing the calibration.
f. What exactly is being certified.

12.4 Establishing measurement control at associated facilities

It is important that any companies closely associated with the company in the treatment of patients adhere to the same level of metrology involving the calibration of instruments. In this case, all hospitals with which the company is affiliated must also achieve this standard.

This will require:

a. The company has working knowledge of the status of measurement control in the hospitals.
b. Standardising the procedures for using and calibrating the equipment in these associated facilities, and possibly training their personnel in these procedures if they are not performing to an equal level.
c. Providing the associated facilities with the acceptable ranges for your products.

Partnered companies must also carry out measurement control. In fact their overall quality assurance system should be of a similar level to that of the company's, since their level of quality often determines the quality in respect of efficacy of the products.

The following steps should be taken:

a. The department in charge of measurement in a company, in cooperation with its affiliates, carries out measurement control in the partnered companies if they are not using a contracted service.
b. The company's QA will periodically audit the measurement control system in the partner companies.
c. The departments involved in metrology at a company should provide guidance and assistance on measurement control when deficiencies emerge at its partner companies, or when requested to do so by these companies.

12.5 Review of measuring equipment

To review your measuring equipment and method:

a. Review and improve:
    i.   The measurement precision of drawings and specifications.
    ii. The precision of measuring equipment.
    iii. Measurement conditions.
b. When the measurement values are abnormal, check the methods and equipment.

When values are abnormal, then the equipment and/or methods may be unsatisfactory for technical or procedural reasons:

a. Technical:
    i. Inappropriate equipment has been selected.
    ii. The design or manufacture of the equipment is inappropriate.

iii. The precision of the equipment has deteriorated.

b. Procedural:

i. Operators lack measurement skills.

ii. The measurement operation standards are inappropriate.

Customer information:		**Equipment information:**	
Contact name _____		Name and description _____	
Department name _____		Model _____	
Phone number _____		Serial number _____	
		State and ID number _____	

Calibration log

Identification		Inspection completed	
Model		Technician	
Serial #		Serial #	
Range		Date	
Number of channels			
Test conditions			
Balance serial #		Balance model	
Sensitivity		Balance calibration date	
Correction factor			
Air temperature			
Barometric temperature			
Relative humidity			
Control data		**Test results**	
Calibration date		Selected volume: 200	Pass / Fail
Technician		Nominal volume: 1000	Pass / Fail
	Selected volume:		Nominal volume:
	Selected volume:		Nominal volume:
	Selected volume:		Nominal volume:
	Selected volume:		Nominal volume:
	Selected volume:		Nominal volume:
	Selected volume:		Nominal volume:
	Selected volume:		Nominal volume:
	Selected volume:		Nominal volume:
	Selected volume:		Nominal volume:
	Selected volume:		Nominal volume:
		0		0
Measured volume	0	Measured volume	0

Example of calibration form

12.6 Measuring equipment qualification

Qualification of a measuring device involves the verification that the measurements are precise and accurate over the scope or system range of the device, remaining within acceptable limits. For example a thermometer can be calibrated that it is precise and accurate against the calibrator or analyser for specific temperature points at 25°C or 37°C. But since temperatures can actually be anywhere between or outside those two parameters, then qualification is required that the thermometer still retains an acceptable level of precision and accuracy across the entire range or scope of the thermometer.

12.7 Measuring equipment validation

Equipment does not stand alone nor does it function all on its own. Measuring equipment is usually part of a system or process that works in conjunction with other equipment, environments and operators. Under perfectly normal conditions, equipment can usually maintain its precision and accuracy but if the environment, or the operator, or the timing mechanism are slightly different from the norm, how well that equipment can still remain precise and accurate is the basis for performing validation.

12.8 Analytical Instrument Validation (AIV)

Just as no two analysts are identical and therefore their performance may also not be identical, regulatory agencies such as the FDA have come to the realisation that no two pieces of equipment, especially those from different manufacturers may be identical. That being the case, then equivalence between pieces of equipment must also be proven since they may be measuring identical samples by completely different processes, using different algorithms, and arriving at differing results, creating the issue as to which unit is correct.

The major problem in the industry is that equipment manufacturers are filling the market with equipment for the same purpose but which operate differently. There are different algorithms, electronics, circuitry and software, etc. but each still claims to provide the best results. If the results differ, how can the end user know which one really is the best. Therefore, the goal of Analytical Instrument Validation is to set performance parameters that will result in quality data and confidence that the unit provides a universally acceptable result.

Therefore, calibration, qualification and validation are used to describe the overall process of determining whether or not equipment is functioning for its intended use. Since most equipment providers have their own sets of standards, then proving equivalence between units is essential, especially if multiple units from different manufacturers are used in the same laboratory for providing the same information.

In order to avoid issues with equipment from different manufacturers:

a. Companies should buy identical units for use in the same laboratory.
b. Companies should always purchase from the same manufacturer.
c. Companies should obtain a statement from the manufacturer that it has tested the validity of all its models and found them to be equivalent.

If different analytical units from different manufacturers are being used for the same product then:

a. A comparative validation of the units will need to be performed according to the FDA suggestions regarding Analytical Instrument Validation.
b. Qualified personnel will have to provide a scientific argument for any differences if the results are to be used for the process validation report.
c. Manufacturers will need to provide any information they already had of comparative studies they performed.

Recommended reading

1. Abernathy, R.B., et al. *Measurement Uncertainty Handbook*. Research Triangle Park, Instrument Society of America, 1980.

2. American National Standards Institute (ANSI) and the National Conference of Standards Laboratories (NCSL). ANSI/NCSL Z540-1: General requirements for calibration laboratories and measuring and test equipment, Final Draft, 29 October 1993.

3. Belanger, B. *Measurement Assurance Programs, Part I: General Introduction*, National Bureau of Standards (NBS) Special Publication 676-I, May 1984. Gaithersburg, MD: National Bureau of Standards. And, Croarkin, C. *Measurement Assurance Programs, Part II: Development and Implementation*, NBS Special Publication 676-II, April 1985.

4. Castrup, H.T. Calibration requirements analysis system, paper presented at the 1989 NCSL Workshop & Symposium, Denver, 9-13 July 1989.

5. Eisenhart, C. Realistic evaluation of the precision and accuracy of instrument calibration systems, *Journal of Research of the National Bureau of Standards-C. Engineering and Instrumentation*, 67C(2), April-June 1963.

6. Juran, J.M., ed. *Juran's Quality Control Handbook*, 3rd edition. New York: McGraw-Hill, 1974.

7. Trigg, G.L., ed. Calibration and maintenance of test and measuring equipment to collective phenomena in solids, *Encyclopedia of Applied Physics*, Vol. 3, Wiley-VCH, 1992.

Self testing multiple choice questions

1. The first step when you are going to purchase new equipment is to:
   a. Select equipment that has the required precision.
   b. Acquire the catalogues from all the equipment sales companies.
   c. Establish your User Requirement Specifications (URS).

2. The first thing to do after purchasing the equipment is to:
   a. Perform an acceptance inspection.
   b. Set up a log book.
   c. Record it in a company assets register.

3. Prior to writing the SOP for a particular item of equipment with measuring capabilities:
   a. Determine which calibration methods are appropriate.
   b. Determine which errors are typical for the unit.
   c. Determine who will be operating the equipment.

4. Store calibrated measuring equipment where:
   a. In high traffic areas with easy access by everyone.
   b. The environmental conditions are least likely to interfere with its precision.
   c. Under controlled access, so that technicians cannot make adjustments.

5. To record the implementation of control procedures use recording methods which make it _____ to rewrite records.
   a. Difficult
   b. Easy
   c. Impossible

6. Establish procedures for changing the specifications of measuring equipment when:
   a. There are changes in manufacturing conditions.
   b. There are changes in the quality characteristics.
   c. The frequency of use of the equipment changes.

7. The master file for each piece of equipment should include:
   a. The vendor's catalogue.
   b. The acceptance inspection certificate.
   c. The list of competitive pricing.

8. When conducting in-house calibration:
   a. Draft the calibration plans in advance.
   b. Decide on whom in the related departments will do the calibration.
   c. Know in advance how to deal with any calibration errors so they are not recorded.

9. Clearly mark the calibration expiry date on each piece of equipment:
   a. In Arabic numerals.
   b. In ink that can be easily erased and rewritten as necessary.
   c. On any label that can be easily removed.

10. Calibration certificates should show:
    a. The name of the calibrating department.
    b. The standards used.
    c. The labels used.

11. Unsatisfactory measurement results may be due to:
    a. Selecting the appropriate equipment.
    b. Deterioration in the precision of the equipment.
    c. Using an equipment of too high a level of precision.

12. The acceptable range of precision is the basis upon which the quality characteristics of the product:
    a. Have been determined.
    b. Are refined.
    c. Are adjusted.

13. Failure to accurately measure the quality characteristics of products results in:
    a. Having to repeat tests so that an acceptable mean range can be determined.
    b. Variability of the treatments when infused into customers/patients.
    c. Failure to meet the required domestic and international standards.

14. Implementing quality management:
    a. Guarantees an error-free laboratory.
    b. Does not guarantee an error-free laboratory.
    c. Guarantees an error-free laboratory only if it is Total Quality Management.

15. Calibration is subject to:
    a. Input, operational and interpretive errors.
    b. Mechanical errors only.
    c. Errors resulting from routine maintenance.

16. Precision measuring equipment must be _____ as suitable to the conditions.
    a. Verified for accuracy and validated
    b. Qualified for accuracy and verified
    c. Validated for accuracy and verified

17. Every acceptable measuring device must have:
    a. A recent certificate of verification of accuracy under abnormal conditions.
    b. A recent certificate of verification of accuracy under normal conditions.
    c. Proof of a certificate of verification even if it is expired.

18. Companies that perform their own in-house calibration need to have:
    a.   Analysers and calibrators adjusted and verified on a regular basis.
    b.   A service contractor monitor their performance for accuracy.
    c.   A wider range of precision and accuracy than a service contractor.

19. Verification of calibration equipment is signed off by:
    a.   QC.
    b.   QA.
    c.   The operator.

20. In-house operators of calibration equipment need to be:
    a.   Trained only to the in-house performance expectations.
    b.   Senior technicians with long records of experience.
    c.   Qualified and validated to the regulatory standards.

21. If a company uses measuring equipment for a GLP/GTP procedure then calibration must be performed by:
    a.   Someone from QA.
    b.   A GLP registered company.
    c.   The senior company engineer.

22. When deficiencies in the equipment have been corrected then:
    a.   Recalibration of the equipment is required.
    b.   The equipment can be used immediately.
    c.   Recalibration must be performed by a service contractor.

23. Identifying on the label that the equipment has been calibrated in-house or by an external contracted organisation is important.
    a.   True.
    b.   False.
    c.   Depends on the equipment type.

24. Affiliated companies cooperating with the manufacturer in treatment or production must:
    a.   Be included in the metrology certification of the therapeutic manufacturer.
    b.   Adhere to the same level of metrology independent of the manufacturer.
    c.   Not need calibration of their instruments because they are not the manufacturer.

25. The manufacturer of the product must provide the affiliated company or hospital with:
    a.   The calibration and analysing equipment for their metrology.
    b.   The acceptable ranges for any measurements of the product to be used.
    c.   The SOP for working with the product when outside the acceptable ranges.

26. Analytical Instrument Validation (AIV) was developed because:
    a.   The same equipment from the same manufacturer can have different results.
    b.   Equipment for different purposes can be calibrated and qualified to be identical.

    c.   Equipment for identical purpose from different manufacturers is not identical.

27. The goal of AIV is to set performance parameters:
    a.   That allow different equipment manufacturers claim to be the best.
    b.   That result in quality data and confidence of a universally accepted result.
    c.   That permits different results from different equipment to be accepted as true.

28. To avoid issues of AIV, companies should:
    a.   Always purchase from the same manufacturer.
    b.   Never purchase the identical machine as manufacturers always update the models.
    c.   Always purchase from different manufacturers creating a wider acceptance range.

29. The first step in ensuring measurements are correct is to:
    a.   Always have the same person do the measurements.
    b.   Use qualified measuring equipment.
    c.   Purchase equipment from a reputable vendor.

30. If precision and accuracy is vital then a company should use:
    a.   Equipment borrowed from another department or company.
    b.   Equipment known to operate with a minimum of error.
    c.   Equipment that has been in storage and untouched for a long time.

31. Storage and wrapping of metrological equipment is essential:
    a.   Even if the equipment is used frequently.
    b.   Only when the equipment is being used sporadically.
    c.   When the equipment is being moved into long term storage.

32. The departments most often responsible for measurement control are:
    a.   QA, Production, and QC.
    b.   QC, Engineering, and Production.
    c.   QC, QA, and Engineering.

33. When there is a change in manufacturing conditions, then:
    a.   Unacceptable changes occur in the specifications of the measuring equipment.
    b.   There should be no resultant impact on the specification of measuring equipment.
    c.   A change in specifications of measuring equipment may be necessary.

34. If the quality characteristics of a product result in a change in precision measurements, then there is:
    a.   Likely a problem with the product.
    b.   Likely a problem with the equipment.
    c.   A need to change the specifications of the measuring equipment.

35. When the frequency of using the measuring equipment change resulting in variation of the precision, then:

a. Purchasing of new equipment is required.

b. Recalibration is most likely required.

c. Change of the specification of the measuring equipment is required.

36. The master file for equipment should include information regarding:
   a. The operating procedure for the specific equipment.
   b. The disposal of specific equipment.
   c. The purchase negotiations of the equipment.

37. Every unit of measuring equipment must have:
   a. A specific SOP that corresponds to the individual piece of equipment.
   b. A unique colour coding.
   c. A unique identification number.

38. The repair history of the measuring equipment should be kept in the equipment file.
   a. True.
   b. False.
   c. Depends on the equipment.

39. When new equipment is purchased, the in-house calibration personnel must be:
   a. Able to perform the calibration immediately.
   b. Trained to that equipment after they calibrate it.
   c. Trained to that equipment prior to calibration.

40. It is a requirement to calibrate _____ of the metrological equipment used in production.
   a. Some
   b. All
   c. None

41. It is necessary for a company to periodically audit the measurement control system of affiliated (subcontracted) companies.
   a. True.
   b. False.
   c. Infrequently.

42. To review measuring equipment properly, Engineering should:
   a. Examine the drawing and specifications of the equipment.
   b. Disassemble the equipment on a routine basis for examination.
   c. Return it to the manufacturer for complete evaluation.

43. When the precision values of measuring equipment are abnormal, there are _____ possible technical reasons.
   a. Three
   b. Five
   c. Seven

44. When considering the purchase of equipment from a different manufacturer, one should:
    a. Obtain a guarantee vendors are responsible for any precision variation.
    b. Obtain proof vendors have tested against units like you have and they were equivalent.
    c. Obtain a waiver that permits you to have variation from the machines in your laboratory.

45. If using a different machine which is providing varying results, then:
    a. As long as the use of the machine is approved, then it does not matter.
    b. A scientific argument has to be provided why the results should be accepted.
    c. Immediately rerun the tests on the machine normally used.

46. Calibration refers to the verification of measuring equipment to be accurate and precise:
    a. Against specific reference points on a calibrator or analyser.
    b. Against an entire range of values on a calibrator or analyser.
    c. Even when the system characteristics are being changed.

47. Qualification of measuring equipment refers to the ability of the unit to be accurate and precise:
    a. Against specific reference points on a calibrator or analyser.
    b. Even when the system characteristics are being changed.
    c. Against an entire range of values on a calibrator or analyser.

48. Validation of measuring equipment refers to the ability of the unit to be accurate and precise:
    a. Against specific reference points on a calibrator or analyser.
    b. Even when the system characteristics are being changed.
    c. Against an entire range of values on a calibrator or analyser.

49. Calibration can only be performed in-house as long as there are:
    a. Specific people designated to perform calibration.
    b. Specific people certified to perform calibration.
    c. No restrictions placed on which personnel can perform calibration.

50. Biotechnology companies can still be GMP accredited without calibration performance.
    a. True.
    b. False.
    c. In certain countries.

Chapter 13. Disposal and storage

Waste and
surplus materials

It is an established fact that laboratories that are neat and well organised tend to be more efficient and happier environments for technicians because confusion is reduced to a minimum and interruptions due to distractions from the primary functions are absent. This chapter presents a number of actions that can be undertaken to achieve this as part of the TQM implementation.

13.1 Remove unnecessary items from the workplace

Workplaces often leave things lying around that are not needed and only interfere with the workflow and corridors are often used as storage areas. There are two categories of unnecessary items:

a. Those that will never be needed again.
b. Those that will be needed again but not until well into the future.

The basic rule to follow is to dispose of those that will never be needed, and store those that will be needed. This will save space, allow ease of movement for employees and equipment, make operations more efficient, help prevent accidents, and create a more comfortable working environment for everyone. A system is still required to remove unnecessary items from the workplace.

Take the following six steps:

a. Establish judgment criteria for the necessity of items.
b. Identify and label items that will be needed again, and those that will not.
c. Designate one person in charge of dealing with all these items.
d. Establish storage places for items that will be needed again with appropriate standard procedures to govern the storage areas.
e. Establish standard procedures for disposing of items that will never be needed.
f. Keep a record of items that have been disposed of with all required details.

Establish a universal criteria policy for judging how necessary things are otherwise individuals or departments will make their own judgment based upon their own situation or feelings and this will actually cause further confusion in the workplace.

Unnecessary items can be found in a lot of different categories:

a. Finished or semi-finished cell products, dead inventory, prototypes, product accessories, components. (Dead inventory are items that have not been used or used beyond their expiry date.)
b. Raw materials and indirect materials made obsolete by changes in design.

Items may become unnecessary when:

a. Equipment is replaced.
b. Production methods are changed.
c. Product specifications are changed.
d. The customer/patient requests treatment changes.
e. An item reaches its expiry date, or its post-use stage.

After you decide whether an item which is unnecessary at present, will be needed again, mark it so that people can see this immediately. Use of coloured stickers such as green for available to use, yellow for under repair and red for disposal are common to the industry.

There are different classifications that you can use for marking:

a. Label defects, items to rework, items to re-sort, items to dispose of, etc.
b. Label items that are damaged, expired, or no longer used.
c. Label items in batches of a certain size, e.g. batches of 10, 50 or 100.
d. Indicate the name of the section or person who is responsible for them.
e. Indicate items' shelf life or expiry date.

There are various labelling systems:

a. By colour, number, or pattern label.
b. By separation of specific storage areas for certain products.
c. By product name or identity label.

13.2 Remove rejected goods from the workplace

If rejected items are left in the workplace they most likely will cause problems. Not only will they take up valuable space, but they can easily get mixed up with non-rejected items thereby putting any production at risk. This also makes it difficult to maintain accurate stock levels. The risk that rejected therapeutic products, chemicals or reagents being infused into customers/patients is always present when rejected goods are left in the workplace.

Take the following three steps to deal with rejected goods:

a. Establish the acceptance criteria that specify the quality features of materials, goods, and/or product, so that rejected goods can be easily identified.

b. Set up a system for removing rejected materials from the workplace as soon as possible.

c. Establish procedures for dealing with any abnormalities that emerge.

Everyone should know which features of an item they should be examining for defects. These are known as the quality features. Establish clear acceptance criteria that specify the quality features to be examined in finished products, WIP, indirect materials, and tools. In fact, everything involved in the production process from raw materials to finished products.

Examples of quality features to be examined are:

a. The material and its composition.

b. Morphology, number of cells, and accuracy.

c. Appearance, giving consideration to any deformations, misshapen, peculiarities of the cells.

The standards should include within an SOP that specifies the testing evidence to be used to set the acceptance criteria and to indicate what defects are being specifically looked for.

When discovered, there must be an immediate notice of occurrence of abnormality (Non-conformance Report) generated by the laboratory.

As well as dealing with rejected therapeutic products, you also need to have a system for dealing with abnormalities. An abnormality is any item or event in the production process that is different from what it should be. Not all abnormalities are defects, but mostly they are, and they must always be seen as indicating that something might be seriously wrong. It is essential to establish Standard Operating Procedures to identify and deal with any abnormalities that may occur. Following are the specific instructions required for dealing with any non-conformances and items out of specification.

SOPs that will be required:

a. A procedure for identifying an abnormal product.

b. A procedure for checking the quality of an abnormal product: identifying the quality features that should be examined, such as cell shape, size, accuracy of expansion, and appearance.

c. A procedure for reporting abnormalities: the steps to be taken by the technician who sees and reports the abnormality, and the steps to be taken by the laboratory manager who receives the abnormality report.

d. Procedures for:
   i.   Checking the machine or production process where the abnormality occurred, and verifying product quality.
   ii. Checking, when necessary, the production processes immediately before and after the process in which the abnormality occurred.
   iii. Checking initial production after the manufacturing process has started again.

e. A procedure for restarting operations again, and returning to the normal production process:

i. The instruction that is to be given to return to normal production.
ii. The method of informing the production processes immediately.
iii. The method of verifying that operating status and quality are now OK.

f. A procedure for correctly labelling rejected products and materials so that they can be handled appropriately and stored or disposed of according to their specific requirements.

Label defective products and waste with correct and appropriate warnings.

To establish procedures for disposing of items that will not be needed again:

a. Classify such items according to type and characteristic.
b. Request approval for their disposal from the departments concerned.
c. Confirm the procedures and criteria for disposing of these items.
d. Disable semi-processed goods or finished components before disposing of them so that they will not be accidentally reused.
e. Keep a record of the items disposed of as a document trail is required.

A record of items that have been disposed of will ensure that assets are controlled correctly. Regulatory bodies are primarily concerned with the demonstration that a company has control. Although an item may no longer be needed, it must be dealt with properly as a company asset, following the normal accounting procedures.

a. Decide what to keep records of:
    i. Machinery and equipment: sale, transferral to an outside company or to other departments, or disposal.
    ii. Raw materials, inventory, work in progress (WIP), packing materials, and components: transferral or disposal.
b. Decide on the format of an Asset Disposal Request Form:
    i. For equipment.
    ii. For raw materials, inventory, WIP, a final product, packing materials, and components.
c. Carry out disposal in accordance with the request form as follows:
    i. The department making the request fills in the form.
    ii. It gets the approval of related departments.
    iii. It carries out the procedure in accordance with the contents of the form.

13.3 Set up a storage system

A lot of time can be wasted when employees have to search for the items they require. When a company has an efficient storage system, it has everything well arranged, with signage to indicate where the different items are stored. By using a first-in, first-out system, it ensures that older items do not remain in storage while newer items are taken out and used.

It is essential to develop SOPs to cover the following:

a. Establishment of criteria for deciding how different items will be stored.
b. Establishment of rules and procedures for arranging and managing items in storage.
c. Development of a first-in, first-out (FIFO) system.

These SOPs should contain the following:

a. Explanation of the FIFO system to ensure that older items do not remain in storage while newer items are taken out and used.
b. Identification of which items require storage.
c. Determination of how many are to be stored at any one time at any one place.
d. Establishment of the required storage conditions.
e. Map of the locations as to where they are to be stored.
f. Conditions of how long specific items are to be stored.
g. How quality is to be maintained during storage. This SOP must contain information regarding auditing procedures of the warehouses.
h. Identification of who is to be in charge of storage as in regards to position. Ensure that the job description matches up with the SOP.
i. Timetables as to when inventories are to be taken (i.e. checks of inventory contents). Identification of the inventory procedure and which department performs it.

Storage locations to identify:

a. Seldom used items, ie. less than once a year. Dispose after use as will likely exceed expiry date.
b. Occasional use, ie. every six months. Store in a secure location outside the laboratory.
c. Normally used every 1 or 2 months. Possibly store within the laboratory.
d. Often used, ie. weekly. Store at a designated location close to the workbench in the laboratory.
e. Very often used, ie. daily. Store at the workbench.

A designated person should be placed in charge of removing items that are not needed, and for making sure that everyone follows the procedures. His/her name should be posted clearly in the storage areas so there is no doubt as to who the responsible person is.

This person should decide:

a. Which items are not needed.
b. Which items to store.
c. Which items to dispose of.
d. How to classify them into groups.
e. Which items should go in which storage places.

Store items that will be needed in the future separately from finished goods, WIP, and research items.

These will come in two main storage categories:

a. Items held on a company-wide basis and therefore intended for different locations. These may be located at the head office rather than at the intended usage sites. The SOP must ensure that the logging in and logging out procedures for a central storage are strictly adhered to and that a record of time, dates, location for distribution, purpose, item identification, quantity, and who has entered or released the material is identified.
b. Items held by different departments, divisions, sections, sub-sections, teams, e.g. research prototypes, rejected goods, equipment, parts, and indirect materials. Each area must maintain log books for the entry and release of these items.

Remember to develop the system so that people who are completely unfamiliar with the storage system are able to find what they want right away. Prepare a warehousing manual containing the rules, that also contains the procedures and standards for arranging and managing the different items in storage.

This manual should specify:

a. How to arrange storage space.
b. The purpose for the different storage areas.

c. Designation of shelves for certain items.
d. The numbering or identification system used for all items.
e. A general statement on the use of the FIFO system.
f. References to the SOP on how to dispose of unnecessary and non-compliance items.

The warehouse or storage personnel or those doing the actual work should be responsible for writing up the standards and manuals.

Justification for the FIFO system:

It is especially important to organise storage so that items that were produced or purchased first are used first. This is the first-in, first-out (FIFO) system.

This will:

a. Prevent items being stored until they are too old and therefore wasted.
b. Ensure that items are stored in their proper lots and are easily traceable.

To set up a FIFO system:

a. Designate someone as being in charge.
b. Establish a procedure to ensure that items entered into storage first are also the first to leave.
c. Clearly label items with their serial number, part number, production date, quantity, and date of entry into storage.
d. Make clear the order in which items have arrived. For example, classify goods by putting them in boxes according to the order received.
e. Clearly record the item's storage location and position. Maps of the storage area are a good idea showing locations of various items.
f. Establish a separate entrance and exit for raw materials if space permits. This prevents any crossover contamination that might occur.
g. Place heavy goods in rows rather than stacking them.
h. When storing important components, number each component within a box or container sequentially and use them in that exact numeric order.

13.4 Set up a good inventory system

A good inventory system is essential if a workplace is to run smoothly and efficiently. The inventory is the quantity of final products, reagents, WIP, indirect materials, raw materials, packaging materials, etc. that are kept in stock. Maintaining the right level of inventory means having whatever is needed for the production process when needed but not having far more than is actually required.

Chapter 16 on Production Control provides more specific guidelines on maintaining a product inventory, components and raw materials inventories, and a WIP inventory.

To set up a good inventory system take the following steps to be included in SOPs:

a. Decide on appropriate inventory levels, and set up the necessary Standard Operating Procedures to maintain these levels.
b. Carry out regular reviews that these inventory levels continue to be appropriate.
c. Carry out periodic inventories (i.e. checks of what is actually in the inventory) and keep records.
d. Set up an efficient storage system for the inventory.
e. Ensure that inventory levels can be seen at a glance.
f. Establish a convenient system for moving items into and out of the inventory.

To decide on and maintain appropriate inventory levels:

a. Keep a schedule of when items will be required by departments, customers/patients, internal process points, etc.
b. Awareness of changes in use: daily, seasonal, or by product.
c. Time between ordering and delivery date.
d. Expected Time of Arrival (ETA log of all items).
e. Approval of secondary suppliers for receiving alternative items if the required items are not available from the primary supplier.
f. Establish standards specifying the right levels.
g. Decide on a method to order refills or replacements, either the regular ordering method or the fixed-point ordering method. (Regular ordering method: goods are ordered regularly at set times. Fixed-point ordering method: if the inventory falls to a pre-determined level, then a purchase order for a fixed amount is automatically generated.)

Production conditions and customer/patient requirements are always changing, and therefore the inventory levels may also need to be changed too. This requires that you review them regularly.

To do this, take the following steps:

a. Appoint at least two people responsible for carrying out inventory checks:
    i. First person to be in charge of inventory review: this person will observe changes in treatment demand.
    ii. Second person to be in charge of inventory control: this person will control what is actually in the physical inventory.
b. Decide the timing of the review. This may be:
    i. When each production of a cell therapy lot is completed.
    ii. When a shipping plan is decided upon with regular reception dates.
c. Review regularly the criteria used to calculate maximum and minimum inventory levels. These will fluctuate as new technologies replace old cell therapy techniques.
d. Give special attention to reviewing costly inventory levels. Some items are best to be ordered on an as needed basis because of cost of maintaining a set inventory of the items.

e. Respond immediately to changes in demand. Progress usually moves in one direction. As technologies change, so do the required materials and it is unlikely that replaced materials will ever feed back into the chain again.

f. Remind the laboratories of the importance of maintaining appropriate inventory levels, and keep them informed of factors that have an impact on those inventory levels.

There may sometimes be discrepancies between the quantity of goods actually held in the inventory, and what is shown in the accounting records. Carry out periodic inventories (checks of the inventory) to find any such discrepancies. This is all part of the Quality Control process.

Periodic inventories should be taken at the following intervals:

a. Laboratory equipment, and accessories for equipment: about once a year, using the asset control ledger.

b. Finished products, WIP, and product accessories: once or twice every fiscal year, using the production control ledger.

c. Raw materials or components such as reagents: monthly or according to a regular cycle, using the production control ledger if needed more often.

d. In addition to the periodic inventories, take an inventory of equipment any time something is replaced, or something is newly installed, and keep a record of this. This keeps the inventory up to date and ensures whatever has brought in new is entered on to the ledger properly.

Before carrying out these periodic inventories decide on the following:

a. Which items to include:
   i.   Equipment.
   ii. Finished products, WIP, and product related materials.
   iii. Components.
   iv. Indirect materials and tools.
b. The inventory documentation for each item:
   i. Ledger (or control list).
   ii. Inventory control tags.

The periodic inventory should include the following actions:

a. Conduct the inventory according to who does what, where, when, why and how.

b. A specialist with technical knowledge takes the inventory at the job site using the asset control ledger.

c. If a discrepancy is discovered, contact related departments immediately and direct them to take appropriate actions.

d. Set up a system for keeping records of the periodic inventories. These records should include inventory control tags which can be used to track (by type, name, number, or component type) any discrepancy between the quantity in the ledgers and the quantity actually found in the inventory.

13.5 Set up an efficient inventory storage system

To ensure that shortages and excesses do not occur, display information about the inventory in such a way that people can recognise the inventory levels at a glance.

To establish such a system:

a. Decide which items in the inventory are to be managed in this way, and the quantities.
b. Decide what inventory data should be displayed so that people can quickly see:
    i. The standard amount, and the minimum and maximum amounts.
    ii. The amount used and the balance remaining.
    iii. The refill orders that have been sent off, and the date of the expected arrival.
c. Decide on a method for displaying this data:
    i. Electronically.
    ii. Signs, standing plates, tags on articles.
    iii. Using different colours.
d. Develop and standardise detailed measures for setting up and running the system.

Establish procedures for making sure that the right amount of items go to the right place at the right time. This includes materials, components, WIP, accessories, packing materials, etc.

Decide on:

a. The method for bringing items into the inventory: it is best to use a continuous supply method based on the production plan.
b. The point at which items are to be brought into the inventory.
c. The method by which the next production process comes to collect items.
d. The method for submitting requests for items from the inventory.
e. The times when items will be supplied to the workplace and the amount.
f. Standards for the supply of packaging and indirect materials based on the standardised inventory.

13.6 Mark passageways

It is important to clearly mark passageways in the workplace, both for use by employees and for moving goods. The reasons for doing so are:

a. Ensure the safety of employees when they are moving about.
b. Allow a smooth flow of goods and equipment in the workplace.

Decide which types of passageways you need:

a. Passages for employees between equipment.
b. Passages between workstations.
c. Passages for transport of materials.

Use the following guidelines to measure passages:

a. The passages for employees between equipment should be at least 800 mm wide.
b. The main passages between workstations should be at least 1.5 metres wide.
c. Passages for transport of materials should be at least 2.5 times the width of the containers that use them.
d. The lines on floors that mark passages if used should be approximately 100 mm wide.
e. The corners of the hallway intersections should be rounded off by at least 500 mm coving. The points where floors meet walls and walls meet ceiling should be rounded off by at least 500 mm coving.
f. For ease of flow, mark the passages clearly with coloured lines, and paint the surfaces a different colour from the workplace floor, using a non-slip epoxy coating.

13.7 Pack and move goods carefully

The company needs to write a series of SOPs covering the proper packaging and transporting of unfinished and finished products. These SOPs safeguard the quality of the product while at the same time ensuring there is no risk to those handling the products.

Pack and move finished products and WIP carefully in order:

a. Not to damage the therapeutic products (ie. cell rupture).
b. Not to include items that are not meant to be included (contamination).
c. Not to risk any danger to the employees handling them.

There are three key decisions to make:

a. How to pack goods.
b. How to move goods.
c. How to handle goods.

On the subject of how to pack goods:

a. Pack individual items in a transport bag or do not pack them at all.
b. Use a container: cardboard box, Styrofoam cooler, basket, etc.
c. Use pallets for large items such as equipment that may require a forklift to move.

On the subject of how to move goods:

a. Automatic conveyor used between machines.
b. By hand or using mobile equipment, e.g. trolley, pushcart, etc.
c. Conveyor (belt, roller, etc.)
d. Motorised vehicle, e.g. motorised truck, forklift, etc.
e. Manually by one or two individuals.
f. At the most suitable times when the product is least likely to be damaged, contaminated, or compromised by the move.

On the subject of how to handle goods:

a. Place goods on sufficiently large, flat surfaces to prevent items from falling down.
b. Limit the height of stacking levels to ensure a clear view.
c. Keep to weight limits for items that individuals move alone, so as to avoid back strain and fatigue.

Stacking criteria specify the stacking height for product boxes and pallets. Their purpose is to protect both products and workers who stack boxes and pallets, and people who work near them, from falling objects. Adhere to these guidelines within the SOPs.

The criteria are based on two key considerations:

a. The height should ensure that workers have an unobstructed view even when carrying boxes and pallets or transporting them with a forklift.
b. If pallets are stacked, they should not be stacked at such a height that workers have to work in a high location to check their contents from above.

13.8 Encourage everyone to follow the rules

If all the systems for keeping the workplace neat and well organised are to be effective, everyone, both supervisors and employees, must follow the rules. This requires rules that employees can follow, and manuals that describe them clearly.

The rules are:

a. Organising and arranging items throughout the work process.
b. Storing materials and finished products.
c. Removing environmental factors that may adversely affect product quality.

It is particularly important to educate employees about storage of materials when:

a. New employees arrive in the workplace.
b. New equipment is installed.
c. The workplace becomes disorganised.

Once trained, ensure that all employees that have undergone the training are signed off for the appropriate SOPs in their training records. Only those employees who are signed off as being qualified to perform the SOP are permitted to work in the storage and warehousing areas.

Recommended reading

1. Containment of bulk hazardous liquids at COMAH establishments containment policy: Supporting guidance for secondary and tertiary containment and implementation principles for regulators, U.K. http://docplayer.net/12029220-Containment-of-bulk-hazardous-liquids-at-comah-establishments-containment-policy.html, accessed 19 September 2016.

2. Drury, J., & Falconer, P. *Building and Planning for Industrial Storage and Distribution*. London: Architectural Press, 2003.

3. Environment, Health and Safety Office, California Institute of Technology. Hazardous waste management reference guide.

4. Environmental Association for Universities and Colleges. EAUC Waste Management Guide.

5. European Commission Guidelines of 5 November 2013 on Good Distribution Practice of medicinal products for human use, *Official Journal of the European Union*, C 343/1, 23 November 2013.

6. European Parliament and Council Directive 94/62/EC of 20 December 1994 on packaging and packaging waste. *Official Journal of the European Communities*, No. L 365, 31 December 1994, pp. 10-23.

7. GMP warehouse mapping: Step-by-step guidelines for validating life science storage facilities, brochure of Vaisala. http://www.vaisala.com/Vaisala%20Documents/White%20Papers/CEN-LSC-AMER-GMP-Warehouse-Mapping-White-Paper-B211170EN-A.pdf, accessed 31 August 2016.

8. Management Sciences for Health & Euro Health Group. *Managing Drug Supply: The Selection, Procurement, Distribution, and Use of Pharmaceuticals*. Sterling, VA: Kumarian Press, 1997, pp. 11-26.

9. Rushton, A., Croucher, P., & Baker, P. *The Handbook of Logistics and Distribution Management*. London: Kogan Page, 2006.

10. The Chartered Institute of Purchasing and Supply. How to develop a waste management and disposal strategy, 2007.

11. U.S. Food and Drug Administration. 21 CFR 820.150: Storage.
(Note: CFR stands for Code of Federal Regulations.)

12. WHO Technical Report Series, No. 961, 2011, Annex 9: Model guidance for the storage and transport of time- and temperature-sensitive pharmaceutical products.

Self testing multiple choice questions

1. Having a storage space for items that are unnecessary now but may be needed in the future will:
   a. Allow ease of movement for employees and equipment.
   b. Create a healthier but not necessarily happier working environment.
   c. Make operations less efficient.

2. Items may become unnecessary when:
   a. The storage system is changed.
   b. Product specifications are changed.
   c. Machinery or equipment remain in use for a long time.

3. Items do not need to be labelled if they include:
   a. Items to be replaced.
   b. Items to be reworked.
   c. Items should always be labelled according to their category.

4. Items may be considered marked:
   a. By tying them together with ropes.
   b. By attaching labels or stickers to them.
   c. By storing them in a certain place together.

5. One person should never be given the responsibility for deciding:
   a. Which items are not needed.
   b. Which items can be reworked.
   c. Whether unneeded items should be put in storage or disposed of.

6. There are two main storage categories:
   a. Items maintained by each department and items stored outside the building.
   b. Items maintained on a company-wide basis and on a departmental basis.
   c. Items located outside the building and inside the building.

7. Procedures for disposing of items that have been judged unnecessary include:
   a. Classifying items according to type and characteristic as some might get reused.
   b. Classifying items according to their cost so that the cheapest are disposed first.
   c. Rendering semi-processed goods unusable before disposing of them.

8. To record the disposal of unnecessary items:
   a. Provide minimal information and avoid forms if possible.
   b. Use the same request form for everything.
   c. Carry out disposal in accordance with the SOP and regulations.

9. If rejected products are put together with non-rejected products it can be difficult to maintain:
    a. Correct cost levels.
    b. Correct stock levels.
    c. Supply chains.

10. An example of a quality feature is the _____ of the item.
    a. Appearance
    b. Purpose
    c. Value

11. Signs used to indicate defects include:
    a. Standing signboard.
    b. Rust.
    c. Transparent tape.

12. A system for removing rejected items from the workplace should include:
    a. Separating or disposing of rejected goods as soon as they are identified.
    b. Discolouring rejected goods so that they look different from normal goods.
    c. Immediate incineration if it is cell based product.

13. Abnormalities are _____ defects.
    a. Always
    b. Often
    c. Seldom

14. The procedure for reporting abnormalities in storage should include the steps to be undertaken by:
    a. Any employee.
    b. The superior who receives the abnormality report.
    c. The person in charge of storage.

15. A procedure for dealing with abnormalities should include checking:
    a. Production immediately before the occurrence of the abnormality.
    b. Production immediately after the occurrence of the abnormality.
    c. Production before the equipment has a malfunction.

16. It is not necessary that the SOP for storing items should cover:
    a. The type of forklift required in the storage area.
    b. The time when items are brought into storage.
    c. A method for maintaining quality during storage.

17. The storage area for components, materials, tools, WIP, etc. most often will be:
    a. Off site.
    b. Within the workplace.
    c. In the inventory.

18. Designate a storage area for finished products according to:
    a. Its ability to sustain a quality environment for the finished product.
    b. Its proximity to transportation to the final destination.
    c. Its ability to hold a large quantity of items well in excess of current production.

19. The method of marking storage areas, locations and shelves may include:
    a. Marking with letters.
    b. Post-it notes.
    c. Using arrows that show a general location.

20. The storage system should:
    a. Provide space for discarded items.
    b. Provide space for reassembling lots made up from different containers.
    c. Allow verification through the first-in, first-out method.

21. To set up a "first-in, first-out" system the most important information on items is the:
    a. Production date.
    b. Serial number.
    c. Shipment arrival date.

22. Maintaining the right level of inventory means that:
    a. Whatever is needed for production is always in stock.
    b. There are always plenty of extra parts in stock.
    c. Parts can only be ordered when the production process requires them.

23. To decide on standard inventory levels:
    a. It is not necessary to remain informed of changes in the use of items.
    b. Keep a note of when incoming items are expected.
    c. Always assume that the order will be cancelled.

24. It is recommended that a company carries out a periodic review of inventory levels:
    a. Each time an incoming order arrives.
    b. Routinely on either a six-month or one-year basis for type of material.
    c. Depending on the type of material anytime from monthly to yearly.

25. Periodic inventories for finished products, WIP, and product accessories should be taken:
    a. Every six months.
    b. Once or twice every fiscal year.
    c. Once a year.

26. The procedure for carrying out an inventory should include the following:
    a. Relying solely on the departments to feed the correct information.
    b. A specialist with technical knowledge to take the inventory at the job site using the asset ledger.
    c. Contacting related departments on a daily basis to inform them about discrepancies.

27. To display inventory levels so that people can see them at a glance, show:
    a.   The amount of inventory used and the balance remaining.
    b.   The maximum and minimum amounts that should be in the inventory.
    c.   Only the dates of the past deliveries.

28. The most important reason passageways need to be marked out in the workplace is to:
    a.   Ensure the safety of employees moving about.
    b.   Ensure the speedy shipment of goods.
    c.   Allow a smooth flow of goods and equipment in the workplace.

29. One of the main types of passageways is:
    a.   Passages for driving motorised vehicles into the storage area.
    b.   Passages for inspectors/auditors to watch without being seen.
    c.   Passages for employees between equipment.

30. In marking passages:
    a.   Mark with coloured lines.
    b.   Round-off corners by at least 1500 mm.
    c.   It is mandatory to install mirrors to ensure safety.

31. The appropriate measurements are:
    a.   The main passages should be at least 1800 mm wide.
    b.   The lines that mark passages should be 100 mm wide.
    c.   The passages for employees should be at least 800 mm wide.

32. There are _____ steps to develop a proper storage and dispersal system.
    a.   Four
    b.   Six
    c.   Eight

33. It is necessary to set up a system for removing and disposing:
    a.   All rejected materials.
    b.   All materials with abnormalities.
    c.   All materials that are no longer currently needed.

34. When items are used very often, for example daily, then they should be stored:
    a.   At the central storage unit.
    b.   Securely at a site outside the laboratory.
    c.   At or close to the workbench.

35. The Warehouse Manual does not need to include:
    a.   How to arrange the items in storage.
    b.   A historical justification of the FIFO system.
    c.   An explanation of the FIFO system operation.

36. People cannot be expected to follow the rules:
    a.   Because the criteria and rules for storage are impossible to follow.
    b.   Because they are naturally resistant to taking instruction.
    c.   If the manuals are difficult to read.

37. The concept of the FIFO system is:
    a.   What comes in first, goes out first.
    b.   What comes in last, goes out first.
    c.   What comes in first, goes out last.

38. The normally practised methods for ordering replacements are:
    a.   At regular and preset intervals and when stock has run out.
    b.   At regular and preset intervals and a fixed point when the levels reach an established minimum.
    c.   At irregular intervals and whenever stock is needed.

39. Every inventory system should have:
    a.   Everyone participating in reviewing the inventory and controlling it.
    b.   A person to be in charge of review of inventory and market trends and another person to be in charge of inventory control.
    c.   One person to review and control inventory as well as check trends in the market.

40. Raw materials and reagents should undergo periodic inventory review at least:
    a.   Yearly intervals.
    b.   Semi-annual intervals.
    c.   Monthly intervals.

41. Before carrying out a periodic inventory it is important to decide:
    a.   Which items to include.
    b.   Which weekend is best to conduct the inventory.
    c.   Which products to include but not any related equipment.

42. If a discrepancy is discovered on a periodic inventory audit, then:
    a.   Ignore it unless it occurs again.
    b.   Contact the related departments immediately and let them deal with it.
    c.   Contact Senior Management immediately and let them deal with it.

43. Passageways for transporting materials should be at least:
    a.   1500 mm.
    b.   2500 mm.
    c.   2.5 times of the width of the material using the passageway.

44. Workplace flooring, including storage should be painted with:
    a.   A non-slip surface coating.
    b.   A smooth and polished surface coating.
    c.   A roughened and pitted surface coating.

45. When it comes to packaging, there are three decisions to make, which are:
    a.   How to pack, move and handle goods.
    b.   How to pack, stack and weigh goods.
    c.   How to move, transport and ship goods.

46. The ETA of goods and materials stands for:
    a.   Established Transportation Agenda.
    b.   Estimated Time of Arrival.
    c.   Enhanced Therapeutic Activity.

47. When establishing approved vendors it is important to have:
    a.   A guaranteed secondary supplier.
    b.   Contacts with at least three vendors that sell the same product.
    c.   Capability to manufacture yourself in case of an emergency.

48. Employees should undergo Storage and Disposal of Materials training:
    a.   Whenever the workplace becomes organised.
    b.   When equipment is removed from the laboratory.
    c.   When they are new to a company.

49. If space permits, a company should establish:
    a.   A separate doorway/passageway for incoming and outgoing goods.
    b.   A lounge for working staff far away from the storage area.
    c.   A common area for all materials to be collected and then dispersed.

50. The Asset Disposal Request Form is necessary for:
    a.   WIP and product development plans.
    b.   Equipment and raw materials.
    c.   Finished products and clothing.

Chapter 14.
Healthy and clean laboratories

Personnel environments

Everyone should be guaranteed to work in a comfortable and healthy environment as part of corporate policy. This proves to be the most productive environment for technicians. There are five sets of actions that a company can take to keep workplaces healthy and comfortable.

14.1 Keep your workplaces clean

To keep your workplaces clean:

a. Set up a system for collecting, storing and disposing waste.
b. The laboratories and workplaces are to be cleaned everyday.
c. Collect debris on an ongoing basis.
d. Establish Standard Operating Procedures for all cleaning activities.
e. Validate any cleaning operations to meet the regulatory standards.

To set up a system for collecting, storing and disposing of waste:

a. Write the SOP on waste disposal.
b. Keep garbage bins in a fixed location and mark them so that everyone can recognise them.
c. Separate usable items from those that can be thrown away through the use of separate containers.
d. Collect and store waste by type. Prepare separate bins for infectious waste that needs to be incinerated, biohazardous waste, sharps and glass waste that can be collected. Train everyone so that they automatically separate the waste at the end of each task.
e. Introduce a regular day and time for arranging collection of waste by type.

To have the workplace and laboratory cleaned everyday:

a. Write a Daily Cleaning SOP for the laboratories and workplaces.
b. Specify what is to be cleaned. This may include: equipment, walls, floors, ceilings, measuring devices; shelves, cabinets; storage areas, warehouses, restrooms.
c. Specify the type of cleaning agent that is required, ie. virucidal, bactericidal, bacteriostatic. Ensure that all cleaning agents are government approved. Write SOPs for those cleaning materials that have to be prepared in advance of use.
d. Specify who is responsible for cleaning which items and areas.
e. Draw up a cleaning schedule and associated forms that are checked off daily to

demonstrate what has been cleaned. Employees have to initial upon completion of each task.

f. Inspect the condition of equipment within the laboratory at time of cleaning.

g. Establish the SOP so that it is clear how well things have to be cleaned and identify what will need to be confirmed on an audit.

Give special attention to collecting debris:

Establish standards for procedures to keep the workplace clear of debris.

a. Debris includes dust, particles, minute fragments.

b. Vacuuming routinely is one way to reduce this small particle debris even if HEPAs exist in the facility.

Establish standards to ensure that all these cleaning activities are carried out in an organised way:

a. Specify what is to be cleaned.

b. Specify responsibility – who cleans where.

c. Include a schedule of when the cleaning is to be done.

d. Include criteria for evaluating the cleaning.

14.2 Maintain appropriate levels of lighting, temperature and humidity

In order to maintain workplaces comfortable and healthy, it is necessary to define the factors that are required in order to create such an environment. Once identified, then it is necessary to establish standards in order to sustain and maintain these factors. These include not only cleanliness, but also the right levels of lighting, temperature, and humidity.

The standards should be based on the type of work being performed, and recognise that certain areas such as laboratories have such standards defined by the regulations which must be met. Include these regulatory standards in the SOP.

Desirable levels of lighting, temperature and humidity in office and administration areas will be different from regulatory governed areas and more open to variability. The methods used to control these levels will also vary. In some work locations that are exposed to the external environments, or high heat and humidity areas such as equipment wash areas, it will be virtually impossible to provide a comfortable environment, so instead you should regulate the amount of time that any employees have to spend in such locations.

Some environmental factors, such as temperature and relative humidity, can be measured numerically, while others, such as odour and vibration, either cannot be measured or are often ambiguous. Because of this, it is important not to get too caught up in numbers. Rather, set standards that relate to the way one actually feels in a particular environment. In laboratory areas it will be necessary to enforce strict specific standards for minimum and maximum brightness, temperature and relative humidity and therefore these areas are closely monitored and daily data must be recorded.

Lighting. While lighting requirements for places such as laboratories and inspection locations may be as much as 600 lux or in some cases even more, it is more common to simply specify "appropriate lighting" and set suitable lighting levels for those areas outside regulatory requirements. (Lux is a unit of illumination equal to 1 lumen per square metre.) Generally most other areas operate around 300 lux.

Temperature. Temperature control is a very important factor in keeping the workplaces comfortable. Depending on the materials being worked with in the laboratory, there is sometimes a very narrow permissible range for temperature. Even if an environment must be kept cool, then appropriate clothing should be provided to the technicians so that they can remain in a comfortable working environment. Limit working time in any extreme environments. Apart from special cases however, it is not critical to specify the temperature range which is usually set between 18°C and 23°C.

Noxious factors. A healthy and comfortable workplace must have low levels of noise, odour, vibration, and dust. Establish standards for procedures to keep these down. Keep in mind too that products and tools should be stored in suitable environmental conditions. Remember that workplaces that produce noise, odour, or vibration must counter their impact in the local community just as much as in the workplace. Regulations have specific limits for noxious factors and these must be adhered to. For example, 60 decibels and above is considered an environment that can induce hearing loss if exposed for a long period of time. Limits to gases such as ammonia are also covered by the regulations and therefore these noxious factors must be monitored at regular intervals established by a company specific to the work being done.

Humidity. While humidity is recognised as an important element in health, appropriate humidity levels are usually not specified in numeric terms, except for special types of work. Normally it is left to nature to determine humidity levels in most facilities. Variations due to seasonal differences, such as the wet and dry seasons, and to the location, are not unusual. However it is important to prevent excessive humidity or dryness, since these can damage health and affect product. In cases where certain products do have humidity critical limits, then it will be necessary to ensure the humidity stays relatively constant through the use of humidifiers and dehumidifiers.

14.3 Prevent environmental pollution: Treat laboratory waste

Laboratory waste can be a cause of environmental pollution. When contaminated water and contaminated air are released from laboratories and workplaces in sufficient quantity, then they can represent a hazard.

This has led to a new appraisal of a company's obligation to society. A company is now seen as having as much responsibility for avoiding environmental pollution outside the laboratory as for maintaining cleanliness inside. Companies must establish standards for disposing of their waste in a way that will not cause environmental pollution.

Laboratory waste includes trash, chemicals, biological material, adventitious agents, waste water, and exhaust gas. There are a variety of contaminants that can impact on product quality.

A partial list includes but is not limited to:

a. Chemical contamination
    i.   Product residues
    ii. Decomposition residues
    iii. Cleaning or disinfecting agent residues
b. Microbiological contamination
    Bacteria, moulds, spores, viruses and pyrogenic antigens
c. Unintended materials
    i.   Airborne (particulate) matter
    ii. Lubricants, residues from cleaning agents

Standard Operating Procedures should:

a. Specify the procedures to be followed before disposing of any waste.
b. Specify the conditions for disposing of it. Follow international and regulatory guidelines which apply to certain waste products.
c. Separate the waste in the SOP by type, and standardise the disposal methods utilised for each type.
d. Identify any designated equipment to process waste that requires treatment prior to release; ie. incineration, chemical decontamination, filtration, etc.
e. Identify biohazardous or chemical waste that must be sent for treatment to the proper agency.
f. Write a manual for processing this waste before it is passed to the agency. Follow this manual carefully.
g. Establish standards for disposal: for example, an impurity concentration rate percentage, or pH for treated water, and ppm of various elements and metals, or the quantity of particles contained in a gas. These standard values should not contradict those in government laws and regulations.
h. Ensure that equipment for the treatment of biohazardous waste, water purification equipment, and equipment for measuring the amount of waste particles in the air, etc. adheres to official standards for the prevention of contamination within the laboratory and the outside environment.
i. Measure the waste upon disposal at designated intervals, and keep records. Assign a Quality Control person to be in charge of treating laboratory waste and possibly a second person to control the overall waste disposal process. The controllers should take appropriate action when the measured value does not meet the standards.
j. Educate employees in the importance of these standards and encourage them to cooperate in implementing them.

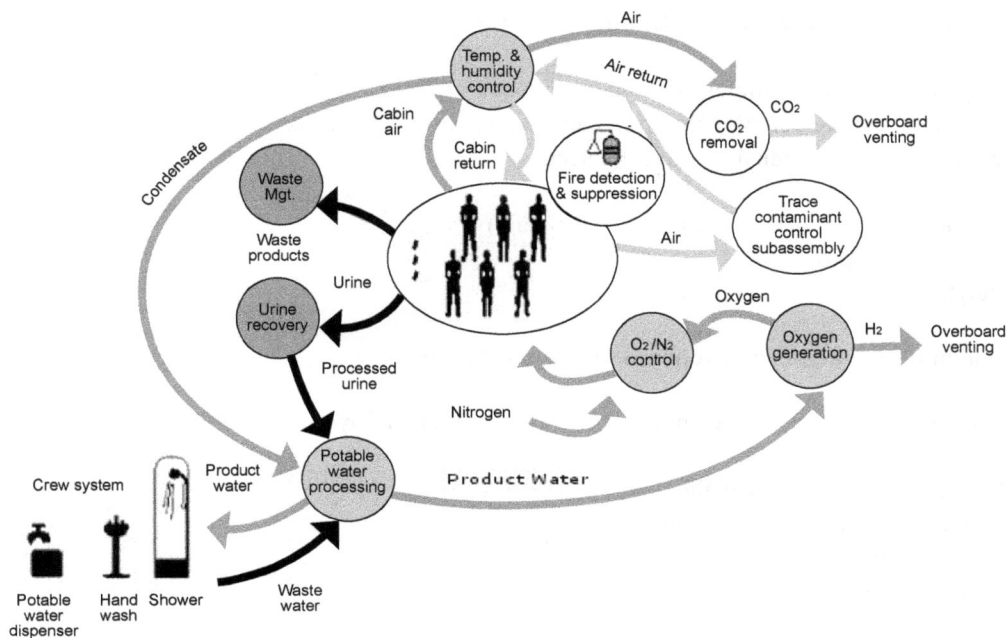

Environmental control is essential in the immuncell therapy facility.

14.4 Validation of the cleaning programme

It is important to know that the cleaning programme adopted by a company is effective in achieving its objectives. Under the cGMP/GTP requirements, a company must demonstrate that the cleaning methods are designed to ensure that any finished product is free from contamination and that the environment eliminates the risk of transference of contamination from one product to another, from the use of cleaning agents to the product, from product to the technicians, as well as the reverse situation from the technicians to the product. Therefore it is essential that a company establishes criteria by which it will test to see that the cleaning programme is achieving its objectives.

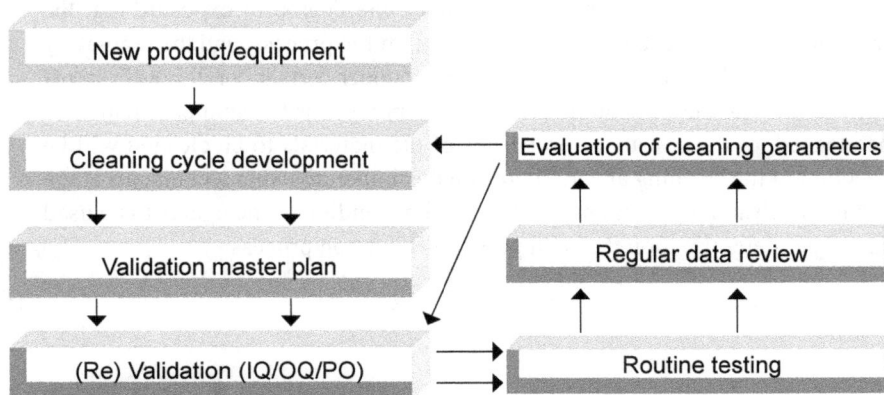

Cleaning validation life cycle management

Protocol elements to consider in cleaning validation:

a. Design/construction complexity of the equipment
Generally the more complex the equipment the harder it is to clean.
b. Characteristics of residuals/product to clean
c. Cleaning agents
d. Type of cleaning process (automated vs. manual)
e. Manufacturing process
f. Analytical methods and their sensitivity

Equipment design and construction is important because:

a. Adequate design/structural complexity and configuration allows cleaning.
b. Material and surface
    i. Non-reactive and cleanable
    ii. Compatibility with detergents
c. Stage of manufacturing process
    i. Upstream
    ii. Down stream
    iii. Drug product
Increased risk the further down stream in the manufacturing process.
d. Structural/design complexity
Size and process piping configuration for Cleaning in Process (CIP)
    i. Potential dead leg/space
    ii. Adequate turbulence
    iii. Adequate slope
    iv. Nozzle design and locations
    v. Branch piping orientation

14.5 Cleaning agent selection

In order to select the appropriate cleaning agent, it is first necessary to identify the degree of risk that contamination represents to the product, the degree of exposure that the product will have to the cleaning agent as any particular step of the process, and the regulatory requirements for determining the appropriate cleaning agent under certain conditions. The pH of the cleaning agent is a major determinant in the selection process as the acid/base nature of the material to be cleaned, or the resistance properties of the materials to be cleaned will be affected by the selection of the cleaning agent. Therefore it is imperative that technicians at the company, involved in the cleaning process have a basic understanding of the agents to be used in order to make a proper determination that will pass the validation process.

	Acidic	Alkaline
Advantages	• Inorganic soilings are more soluble in acidic detergent • Works well as blends of several acids	• Make organic soiling water soluble via the change of chemical/physical nature of organic soilings • Easy to choose
Disadvantages	• Equipment corrosion • Difficult to choose • Less compatible with other cleaning components	• Precipitation of water hardness • No cleaning effect on mineral residues • Limited to organics • Very concentration dependent

Acidic v.s. alkaline cleaning agents

Some basic rules for selection of cleaning agents:
a. For the cleaning of organic materials, alkaline cleaners are the best.
b. Blending of other cleaning components with alkaline detergent will enhance the cleaning effects.
c. Sequestering agents may be necessary to prevent scaling within equipment.
d. Corrosion inhibitors such as silicone may be required for certain equipment.
e. Complexing or building agents are added to enhance the cleaning effect but by themselves are poor cleaners.

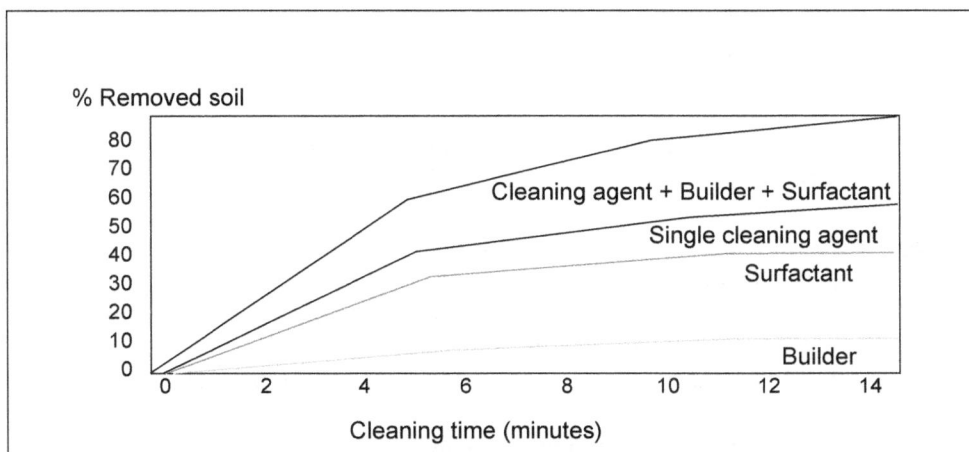

Cleaning agent selection

Some basic rules for the use of a cleaning agent:

a. There is no reuse of cleaning agents in the biological/pharmaceutical industry. Once a specific area, workbench, or piece of equipment is cleaned then the detergent used is to be disposed of.

b. Validation of the cleaning agent's effectiveness is done across the sequence of its use and not at the individual step. If found to be ineffective then the procedural steps are examined to identify the source of the non-conformity.

c. Normally a pre-rinse is performed prior to the use of the detergent using clean water at a neutral pH.

d. Different types of soilings will require different exposure times to the detergent. Therefore knowledge of the specific exposure times is required in advance.

e. Since the detergent itself can prove to be a source of contamination, then a post cleaning rinse is required in order to remove any detergent residues. Post rinse is performed using clean water of neutral pH.

f. Determining the effectiveness of the cleaning agent is a combination of certain factors, including pH, temperature, concentration and exposure time.(See chart below.)

pH

		5.5	6.0	6.5		
Temperature (°C)	30				10	Concentration (%)
	40				20	
	50				30	
		10	20	30		

Exposure time (minutes)

The specific parameters to be examined to determine the appropriate selection of the cleaning agent are:

a. Specificity. Is the cleaning agent able to perform its required action when the contaminating agent is being masked by other materials and solutions?

b. Sensitivity. What are the Limits of Quantitation and Detection before the cleaning agent is no longer effective in eliminating the contaminating agent?

c. Precision. Can the cleaning agent consistently provide reduction of organisms to be within the acceptable limits?

d. Accuracy. Can the cleaning agent completely eliminate a specified undesirable organism?

Cleaning procedures should be drafted or established before validation and once completed then the validation protocol should include:

a. Description of the manufacturing process and the correlations to the precise procedures and cleaning agents selected.

b. The cleaning procedures should refer to all relevant SOPs.

c. Details of the sampling requirements in order to establish the effectiveness of the cleaning procedures.

d. The various test methods that will be implemented in order to ensure ongoing quality checks by the Quality Control units. These test methods can include those listed in the following chart but are not limited to only these tests. (See table below.)

Specific methods	Non-specific methods
• Bioassay	• TOC
• ELISA	• pH
• HPLC	• Conductivity
• Q-PCR	• UV
• SDS-PAGE	• Bioburden
• LAL	• Osmolarity

Recommended reading

1. ICH Expert Working Group. ICH Q7A: ICH Harmonised Tripartite Guideline: Good Manufacturing Practice Guide for Active Pharmaceutical Ingredients, 2000.

2. U.S. Food and Drug Administration. 21 CFR 211.65: Equipment construction; 21 CFR 211.67: Equipment cleaning and maintenance; 21 CFR 211.182: Equipment cleaning and use log.
   (Note: CFR stands for Code of Federal Regulations.)

3. U.S. Food and Drug Administration. Guide to inspections validation of cleaning process, July 1993.

4. World Health Organization. WHO Technical Report Series, No. 908, 2003, Annex 4 Good Manufacturing Practices for pharmaceutical products: Main principles.
   - 4.11: "It is of critical importance that particular attention is paid to the validation of analytical test methods, automated systems and cleaning procedures."
   - 16.11: the paragraph on cross-contamination.
   - 16.15: "Before any processing operation is started, steps should be taken to ensure that the work area and equipment are clean and free from any starting materials, products, product residues, labels or documents not required for the current operation."
   - 16.18: "Time limits for storage of equipment after cleaning and before use should be stated and based on data."

Self testing multiple choice questions

1. Different disposal containers should be used for separating waste by:
   a. Quantity.
   b. Age.
   c. Type.

2. Ways of minimising waste should be considered:
   a. From the start of production in the laboratory.
   b. When the product is being designed.
   c. When the first finished immunotherapy products appear.

3. During cleaning it is important to inspect the condition of:
   a. Cleaning materials.
   b. Measuring devices.
   c. Water leaks.

4. Checkpoints for cleaning machines include:
   a. No measuring tools exist.
   b. No gummy residues from tape.
   c. No residues of any kind present.

5. Which of the following should you include in creating standards for cleaning?
   a. Specify what is to be cleaned.
   b. Specify responsibility – who cleans where.
   c. Specify the cleaning materials.

6. Cleanliness in a laboratory means specifically to:
   a. Clean things according to the cleaning SOP.
   b. Keep things clean after they have been cleaned.
   c. Keep work uniforms clean.

7. The Laboratory Manager should make regular rounds to:
   a. Help raise employee awareness of cleaning requirements.
   b. Check that employees are cleaning correctly.
   c. Check that employees are comfortable enough.

8. A plan for monthly and daily activities should include:
   a. A programme for educating employees.
   b. A programme for disseminating employees.
   c. A list of items to be carried out within a certain period.

9. Establishing standards for maintaining a healthy workplace environment will involve:
   a. Having medical personnel in the workplace.

b. Defining what constitutes a healthy workplace.

c. Establishing specific standards for the features that make a healthy workplace.

10. Standards should be based on:
   a. The type of work being done.
   b. The size of the job site.
   c. The number of employees on the job site.

11. Desirable levels of lighting, temperature and humidity will vary depending on the _____ of the workplace.
   a. Size
   b. Purpose
   c. Location

12. Which environmental standards cannot specify numerical measurements? Those for:
   a. Temperature.
   b. Odour.
   c. Relative humidity.

13. Where exact measurements are not possible it is better to use standards that relate to:
   a. How people feel in a particular environment.
   b. How many hours people are working.
   c. How hard people are working.

14. For most areas, standards:
   a. Should specify lighting of 600 lux.
   b. Give no specification.
   c. Specify "appropriate lighting".

15. It is critical that standards specify the temperature level:
   a. Always.
   b. Never.
   c. In special cases.

16. Humidity levels should normally be:
   a. Left to nature.
   b. Specified exactly.
   c. Specified as "appropriate".

17. Noisy workplaces should be designed and insulated so that:
   a. No noise escapes.
   b. The level of noise that escapes is not noticeable in neighbouring areas.
   c. All noise in the workplace is eliminated.

18. Which of the following should be done to deal with odour in the workplace?
   a. Fix odour-eliminating filters to all equipment that produces odour.

    b.   Fix odour-eliminating filters at all doors and windows.

    c.   Filter all gas before it is released outside the facility.

19. Which of the following helps reduce the levels of dust generated in the workplace?
    a.   Install a dust catcher near the areas where dust is created.
    b.   Install measuring equipment and regularly take measurements.
    c.   Use fans to keep the dust from circulating.

20. When storing finished products, equipment, etc., ensure:
    a.   Their functionality is not reduced.
    b.   It is not more difficult for employees to handle them.
    c.   They are securely locked up.

21. Which of the following guidelines should be followed in order to store products, etc. in suitable environmental conditions?
    a.   Prevent leaks where rain might get in.
    b.   Do not store any items outdoors.
    c.   Choose the proper storage location for each item type.

22. Which further guidelines should be followed?
    a.   Mark gaps through which the wind might pass.
    b.   Mark part name, lot numbers, etc. clearly on locations and on items.
    c.   Clean items on a regular basis, and at the beginning and end of work.

23. A company is now seen as having _____ responsibility for avoiding pollution outside the factory as/than inside.
    a.   as much ... as
    b.   less ... than
    c.   more ... than

24. Standards for treating laboratory waste should specify:
    a.   Procedures to be followed before disposing of waste.
    b.   Conditions for actually disposing of it.
    c.   Procedures for informing employees that waste has been disposed of.

25. Guidelines for disposing of waste should include:
    a.   Write up a manual for the contracted disposal agency to use.
    b.   Separate waste by type and standardise disposal methods for each type.
    c.   Give waste that requires treatment to a proper agency.

26. Standards should be established for waste disposal that:
    a.   Are always in adherance of government laws and regulations.
    b.   Are almost as high as those in government laws and regulations.
    c.   Do not go against government laws and regulations.

27. Employees should be:
    a. Required to implement the environment standards.
    b. Encouraged to cooperate in implementing the environment standards.
    c. Educated in the loopholes of the environment standards.

28. Examples of procedures that can be followed to prevent pollution include:
    a. Process strong acids and alkaline in a neutralising tank, and neutralise the chemicals prior to disposal.
    b. Put carp in the waste water before discharging it.
    c. Put carp in the waste water after it leaves the factory.

29. Standards for preventing environmental pollution vary _____ from one industry to another.
    a. A great deal.
    b. A little.
    c. Not at all.

30. Under the cGMP/GTP regulations, a company must demonstrate its cleaning methods:
    a. Have a refreshing fragrance.
    b. Ensure the final product will be free of contaminants.
    c. Have been used by some other companies before.

31. When designing the cleaning validation protocol, it is important to consider:
    a. The manufacturing process.
    b. What colour the protective clothing should be.
    c. How quickly the procedure can be done.

32. When assessing equipment to purchase, a consideration for cleaning should be the _____ of the equipment.
    a. Material surface
    b. Brand
    c. Cost

33. When Cleaning in Process (CIP) it is important to consider:
    a. That there is a high degree of turbulence, slope and dead space to maintain.
    b. The cycle speed, nozzle design and the need for dead space.
    c. The turbulence, slope and orientation of the piping.

34. When selecting a suitable cleaning agent, you do not have to take into consideration:
    a. Residues.
    b. Exposure time.
    c. Country of origin.

35. Inorganic contaminants are more soluble in:
    a. Acidic detergents.
    b. Neutral pH detergents.
    c. Alkaline detergents.

36. Equipment is more likely to be corroded when using a detergent that is:
    a.   Alkaline.
    b.   Neutral pH.
    c.   Acidic.

37. Alkaline detergents are limited to contaminants that are:
    a.   Organic.
    b.   Inorganic.
    c.   No limitations.

38. When removing mineral deposits, the choice of detergents is usually:
    a.   Neutral pH.
    b.   Acidic.
    c.   Alkaline.

39. When alkaline detergents are mixed with other cleaning components the cleaning effects are:
    a.   Reduced.
    b.   Enhanced.
    c.   Depends on the components as they can reduce or enhance.

40. Silicone is a good:
    a.   Corrosion inhibitor.
    b.   Organic solvent.
    c.   Demineraliser.

41. By themselves, complexing agents or building agents are:
    a.   Excellent cleaners.
    b.   Poor cleaners.
    c.   Better than once mixed.

42. In the immunotherapy industry, it is permissible to use the same prepared cleaning agent on multiple surfaces in the laboratory.
    a.   True.
    b.   False.
    c.   Depends on the circumstances.

43. Effectiveness of the cleaning agent is a combination of:
    a.   pH, volume, time and viscosity.
    b.   pH, temperature, time and concentration.
    c.   Temperature, viscosity, concentration and volume.

44. Specificity is the measurement of the cleaning agent's ability to:
    a.   Perform consistently within acceptable limits.
    b.   Eliminate the unwanted organism or contaminant completely.
    c.   Perform even when the contaminating agent is masked by other materials or solutions.

45. Sensitivity is the measurement of the cleaning agent's:
    a.   Limits of quantitation and detection of the contaminating agent.
    b.   Performance within acceptable limits.
    c.   Ability to eliminate the contaminant completely.

46. The ELISA is an example of a _____ method to analyse.
    a.   Specific
    b.   Non-specific
    c.   Special

47. To set up a system for collecting, storing and disposing of waste, the recommended first step would be:
    a.   Mark the garbage bins.
    b.   Write the SOP on waste disposal.
    c.   Arrange for a regular pick up time.

48. In high heat and high humidity areas, a company should have:
    a.   Employees wear minimal clothing.
    b.   A limit on the time employees can work in the area.
    c.   Do not worry about it, as it is part of their job.

49. For general purpose areas, the lighting level would be around _____ lux.
    a.   50
    b.   300
    c.   600

50. Long term exposure to noise levels above _____ decibels will be harmful and likely result in hearing loss.
    a.   10
    b.   25
    c.   60

Chapter 15.
Safety and managing risk in the laboratory

Emergency situations

Each year hundreds of employees are killed or seriously injured working at laboratories. The vast majority of these injuries could have been prevented, and the severity of the injuries could be greatly reduced by installing general safety rules when working in the biotherapeutic industry. As more and more companies become involved using modified cell lines and viral vectors, the potential for serious injury or illness increases proportionally. But even the basics of general public health and safety have to be adhered to before more complex hazards can be controlled.

There are nine basic actions that can be taken to improve safety at a company.

15.1 Mark emergency exits and put up warning signs

The two safety priorities in the workplace are that:

a. People can leave the building quickly when there is an emergency: mark emergency exits clearly as laboratories can become very confusing places in the dark.
b. People do not enter dangerous areas unless they are intending to work there for a specific purpose: put up warning signs showing areas that are dangerous to enter. Ensure that only qualified personnel have access to any potentially dangerous locations.

Indicate emergency exits and evacuation routes clearly. The evacuation routes should lead to clearly identified evacuation sites where people can gather safely. When people are gathered in one place it is easier to see if anyone may still be in the building.

Take the following actions:

a. Put emergency exit signs on the doorways of all rooms, and illuminate these at night.
b. Put up signs to indicate how to follow evacuation routes, and illuminate these at night.
c. Ensure that all emergency exits and evacuation routes are kept clear of obstructions.
d. Install ladders and evacuation chutes, if the design of the building requires them.
e. Mark tools to be used for evacuation and keep them always ready for use.
f. Ensure that absolutely every employee knows where the evacuation sites are.
g. Use emergency power sources to illuminate the emergency signs in case of a power failure.

Clearly indicate areas that are dangerous for employees to enter or to approach.

These might include:

a. A specified area within a certain distance of operating machinery and equipment.
b. Sites where there are movable facilities and equipment.
c. Sites where harmful gases are stored or used.
d. Sites containing radioactive substances.
e. Areas where there are electrical hazards e.g. high-voltage power cables.
f. Sites where there are infectious agents cultured, harvested or used.
g. Sites where bio-organisms and GMOs are stored.

Fix danger signs where they can be easily seen – on walls, doors or suspended from ceilings. Use the same sizes, colours and words that the public authorities use.

The decision to designate certain areas as dangerous will be based on a code of practice for the biopharmaceutical industries.

15.2 Provide protective clothing and tools

Wherever necessary, protective clothing and tools to lessen the risk, and to protect employees from injury and death must be utilised.

A company must:

a. Write an SOP to cover the gowning and protective clothing requirements as well as the procedure in which to don this clothing. This SOP should include details on when and where this clothing and gear must be worn.
b. Provide protective clothing and tools wherever there is any risk to employees.
c. Check that these are worn and used properly.
d. Install devices that automatically halt operations or sound alarms when something goes wrong.

Employees who work in dangerous environments must use protective clothing and tools. These must conform to standards set by the government.

These will include:

a. Eyeglasses or face shields for operations that involve aerosolising infectious or potentially biohazardous materials.
b. Masks for all operations conducted within the laboratory to avoid inhaling or exhaling particles.
c. Gloves for handling any items for aseptic technique.
d. Protective uniforms or gowns when working inside the laboratory. Working with infectious materials may require double gowning procedures.
e. Protective shoes for safety and prevention of introduction of contaminants.

Check periodically that protective clothing and gear are being used properly, and keep a record of the audit results.

Follow these procedures:

a. Designate the operations and the employees that require protective gear at each hospital and laboratory site.
b. Keep a record of where this protective gear is distributed for joint use, and where for individual use.
c. Check that the protective gear conforms to the standards set by the government, regulators or industry organisations.
d. Keep records of periodic checks on the number of items of protective gear that are in use.
e. Record the periodic cleaning and maintenance of protective gear.

Patrol teams appointed by Quality Assurance should conduct periodic safety patrols and report when protective gear is not being worn. QA will act immediately on any out of compliance reports and undertake retraining and corrective actions.

15.3 Raise safety awareness

Employees must be taught to be safety conscious. They must be aware, directly and personally, of the importance of accident prevention in all the company's activities.

To achieve this, a company should take the following actions:

a. Assess the level of safety awareness throughout the company.
b. Promote company-wide safety awareness.
c. Provide formal safety education.
d. Train employees to recognise hidden dangers.

```
┌──────────────────────────────────┐
│  Name of material                │
│                                  │
│  ┌───┐                           │
│  │   │   HEALTH                  │
│  └───┘                           │
│  ┌───┐                           │
│  │   │   FLAMMABILITY            │
│  └───┘                           │
│  ┌───┐                           │
│  │   │   REACTIVITY              │
│  └───┘                           │
│  ┌───┐                           │
│  │   │   PROTECTIVE              │
│  └───┘   EQUIPMENT               │
└──────────────────────────────────┘
```

Safety label

Consider these requirements:

a. The company must give the safety issue high priority.
b. Managers must try to establish a relationship in which subordinates will tell them about potentially dangerous situations.
c. Employees must follow the safety rules and encourage others to do so too.
d. When something goes wrong with the safety systems it must be given priority and corrected immediately.
e. Signs are posted in obvious places to remind employees to be safety conscious.

To promote company-wide safety awareness:

a. Ensure every department has a copy of the safety SOP.
b. Examine safety issues periodically, make improvements and record the results.
c. Train employees to recognise potential dangers. Provide safety instructions and reminders to employees regularly.
d. Encourage employees to go to lectures and seminars where they can get safety qualifications such as CPR.
e. Organise activities at which safety posters and slogans can be displayed.
f. Encourage employees to participate in national safety events, and run in-house safety campaigns.

General safety courses should cover the following:

a. List of potential dangers at the work site.
b. Methods for operating equipment safely.
c. Methods and procedures for safety activities.
d. Methods for handling potentially dangerous materials.
e. How to report safety abnormalities.

Keep records of the safety education given to new employees, to employees changing jobs, and to employees receiving special purpose education. These records should include the time and dates of the courses, the length, the names of the employees and the instructors, and the subjects taken. These records become part of their personal file in HR.

15.4 Establish safety standards and regulations

Every company should have safety standards and regulations to prevent accidents at work, to clarify job responsibilities in relation to safety, and to encourage employee initiative in undertaking safety activities.

A company should:

a. Establish safety standards and regulations based on the law of the country, industry standards and/or international requirements.
b. Train employees in the knowledge and application of the safety standards.

c. Put instructions for operating equipment in writing to ensure that employees follow the standard procedures.

15.5 Ensure facilities and equipment are safe

Facilities and equipment are recognised sources of danger. It is a matter of identifying those areas with potential to cause accidents so that employees can be cautious when exposed to these sources.

Take the following actions:

a. Carry out a thorough safety inspection when you purchase or build new facilities or renovate existing ones.
b. Before using new equipment, discuss with the manufacturer potential concerns for safety. Those that make the equipment will be the ones most aware of its hidden dangers.
c. Establish standards for the management and maintenance of facilities and equipment and identify the potential risks in the SOPs under the Safety Section.
d. Watch out for abnormalities or anything unusual.

When something unusual happens, regardless of whether an injury occurred or not, always investigate why it has happened, and take whatever countermeasures are necessary to avoid its reoccurrence. There are many abnormalities concerning facilities and equipment that can indicate danger.

These are:

a. Unusual movement of machines and equipment.
b. Strange noises coming from facility mechanicals or equipment.
c. Heat or vibration not normally seen with the equipment.
d. Changes in pressure for those units having pressure sensors.
e. An unexpected stop in the production, cycle.

The most critical abnormalities are:

a. Sound or light alarms do not go off when they should, or when a production cycle stops.
b. Sound or light alarms go off when the machines and equipment are working normally.
c. Machines and equipment stop automatically or alarms activate even though values are within the designated levels.
d. Abnormalities in production cycles: Values cannot be set back to the original settings, alarms do not activate, etc.

15.6 Keep accident records

Records are a valuable way of learning from past accidents and preventing them recurring. Use an accident report form to give the causes of the accident, and the measures to prevent similar accidents recurring.

There are three main actions to take:

a. Fill in a full accident report form with all the details of the accident.
b. Classify and maintain these accident records.
c. Use the accident records to assess the accident rate.

First, write a quick report of the accident as soon as it happens. Then hold a meeting to identify the causes and decide on countermeasures to prevent similar accidents in the future. After this meeting write up a full accident report, using an in-house form.

The accident report form should have sections for the following information:

a. The circumstances at the time of the accident: include columns for what happened, and where, when, who, why and how it happened.
b. The causes of the accident: include columns for material causes (unsafe conditions), human causes (unsafe actions) and managerial causes (unsafe systems or procedures).
c. Any other factors that contributed to the accident.
d. Countermeasures aimed at improvement: include columns for what measures should be taken, who should take them, how they should do so, and when they should have them finished.
e. Confirmation by laboratory managers that countermeasures have been carried out, with their comments.
f. Confirmation and approval of the countermeasures by senior managers.

Keep permanent records of all accidents. These are valuable reference materials that lessons can be learned from. These reports can also be used to compile company statistics.

Classify the accident records as follows:

a. Incidents where there was not an actual accident but employees were concerned.
b. Extremely light accidents: injuries can be treated by nurses and do not impede work.
c. Light accidents that allow work to continue: injuries must be treated by doctors, or doctors diagnose the injuries as presenting no problem if work restrictions are observed.
d. Accidents that require work suspension: injuries are judged by a doctor to require additional rest. The doctor must issue a certificate for paid leave.
e. Accidents that require work suspension and cause after-effects: injuries are judged by the doctor in charge as causing permanent disability. This is confirmed by another doctor from the appropriate public authority.

15.7 Set safety targets

Accidents will always happen. The critical question for each company is how often they happen, and especially if they happen more often at the company than in other companies. Set safety targets to lower the accident rate.

There are three actions to take:

a. Assess the accident rate in the company.
b. Set safety targets to improve (lower) the accident rate.
c. Check that these targets have been achieved.

To assess the accident rate in a company, work out the accident ratio numerically.

Ratios may be calculated as follows:

a. Per 1,000: ratio of injuries and deaths per 1,000 employees per year
b. Frequency: ratio of injuries and deaths per 1,000 working hours

It is important to assess the progress that is being made towards achieving the safety targets. This will also allow every employee to see what is being done to make the workplace safer, and where problems may still exist.

15.8 Be prepared to deal with disasters

Disasters do not happen very often, however when they do happen, whether they be major accidents at work, fires in a company or the local community, or natural disasters like earthquakes, floods, etc., the results are devastating. A company should set up a Risk Management Plan to prepare systems and train employees to deal with potential disasters.

The Risk Management Plan will identify company officers who will be responsible for taking concrete action in any disasters that may occur, and for providing emergency training as required.

Appoint people to the following roles, at a company:

a. Emergency Chief is responsible for the training, instruction and supervision of fire-fighting leaders and their deputies. Each laboratory and worksite will need technicians trained in how to combat small fires.
b. Fire-fighting leaders need to know when the fire cannot be contained and evacuation has reached, and to report injuries and other information to headquarters.
c. Communication and liaison officers report disasters to local authorities such as police and fire stations as well as any provincial or national disaster-handling centres.
d. Evacuation assistance officers help all employees reach evacuation sites, confirm the numbers of registered employees evacuated, and report these numbers to regional authorities.

e. Transport officers remove important documents and objects (data disks and other items).

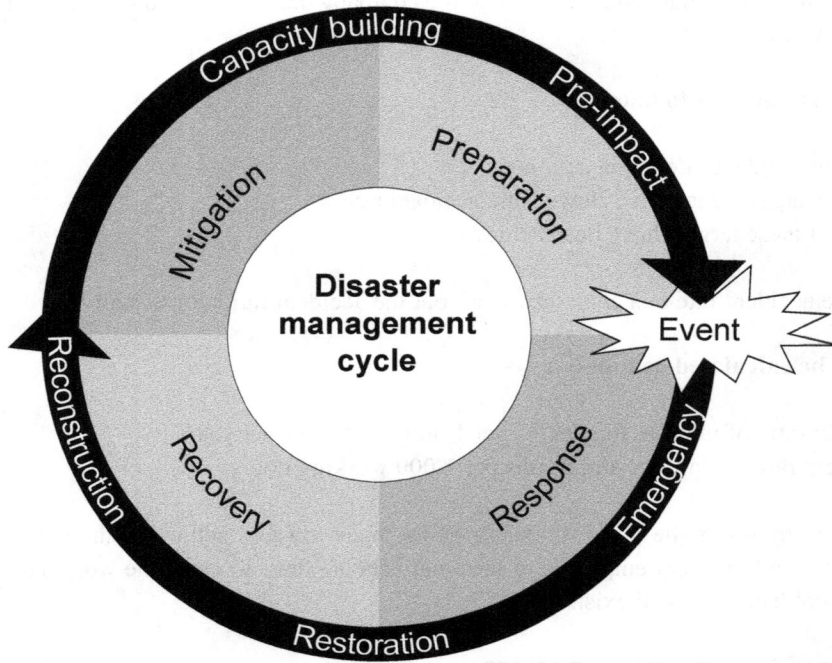

Disaster management cycle

Make an assessment of the company's readiness to deal with disaster in terms of the following five stages:

a. First stage: Establish company-wide goals, but virtually no self-directed activities.
b. Second stage: Recognise the issues that need to be addressed, but no self-directed activities have been undertaken.
c. Third stage: Allocated roles, and employees have taken on self-directed activities in accordance with divisional, departmental, and sectional plans.
d. Fourth stage: Allocated roles, and employees have taken on self-directed activities in cooperation with related departments and sections.
e. Fifth stage: Allocated roles, and company-wide activities have been undertaken vigorously.

**"Plan as you operate,
operate as you plan"**

Risk management plan before it happens

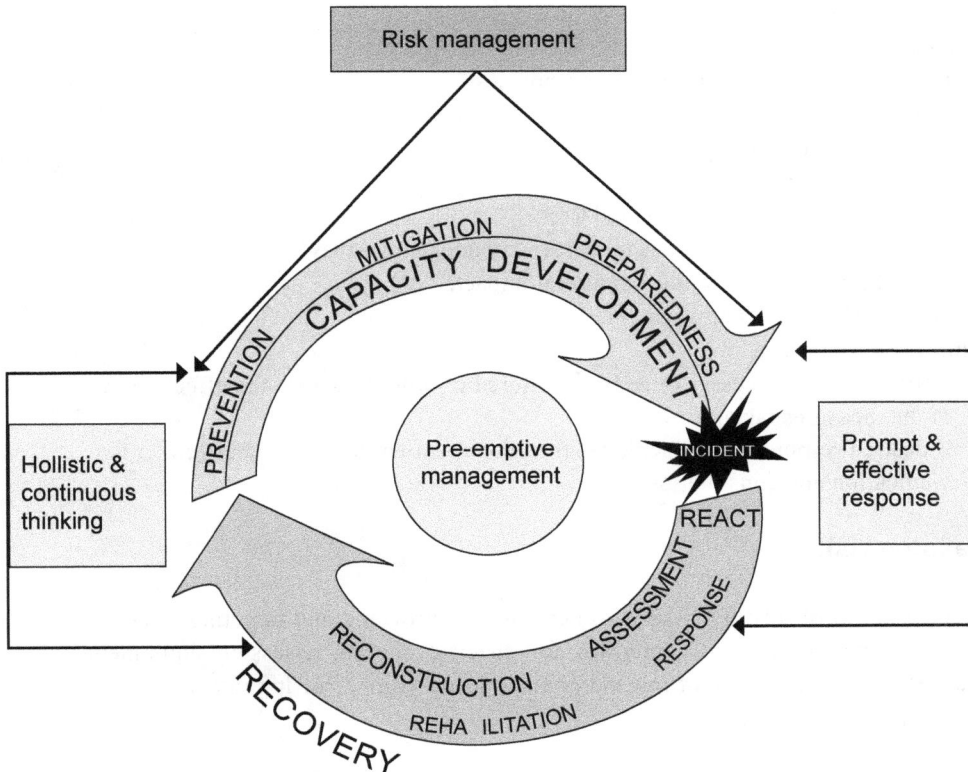

Pre-emptive management

15.9 Risk management

Risk management is the name given to a logical and systematic method of identifying, analysing, treating and monitoring the risks involved in any activity or process. It is a methodology that helps Quality Managers make best use of their available resources. Risk management is an integral part of biological/therapeutic planning and an essential component of GTP.

There are seven steps in the risk management process:

a. Establish the context.
b. Identify the risks. Define the type of risk.
c. Analyse the risks.
    i. Likelihood of the risk to occur (probability and frequency).
    ii. Impact, cost and consequences of an occurrence.
d. Evaluate the risks.
    i. Ranked according to management priorities.
    ii. Cost impact assessment.
    iii. Determine the inherent level of risk.
e. Treat the risks.
    i. Prioritise which has to be treated first.
    ii. Allocate sufficient resources to treat (personnel and money).
    iii. Accept those risks with no or minimal impact.
    iv. Document the risk management plan.
f. Monitor and review.
    i. Activities and processes must be monitored in order to assess the effectiveness of the measures taken.
    ii. If desired response is not achieved then the treatment has to be changed.
g. Communication and consultation.

15.10 Assessing risk

A common method of treating risks is to develop risk profiling and targeting systems. This requires selecting performance activities for specific checks, assessing equipment failure, contamination agents, origin of raw materials, transportation, staffing rate, routing of materials, value, etc.

Types of risk assessment:

a. Quantitative
    i. Quantification of hazard.
    ii. Exposure assessment.
    iii. Population at risk.
b. Qualitative
    i. Opportunity of risk.
    ii. Informed judgment.
    iii. Exposure magnitude.

Risk is measured in terms of a combination of the consequences of an event (Impact) and their likelihood (Frequency).

Impact →

	Extreme	Very high	Moderate	Low	Negligible
Almost certain	Severe	Severe	High	Major	Moderate
Likely	Severe	High	Major	Significant	Moderate
Moderate	High	Major	Significant	Moderate	Low
Unlikely	Major	Significant	Moderate	Low	Very low
Rare	Significant	Moderate	Low	Very low	Very low

Frequency

Evaluation of risk table

15.11 Control points

In order to determine potential risks, then it is necessary to establish a procedure by which controls can be applied to essentially prevent or eliminate any safety hazards or reduce them to an acceptable level by keying on specific areas where problems are likely to occur. Focusing on those specific areas which have the highest impact identifies what are referred to as the Critical Control Points. These CCPs can be found anywhere along the therapeutic chain, all the way from the initial design to the infusion of the product into the patient.

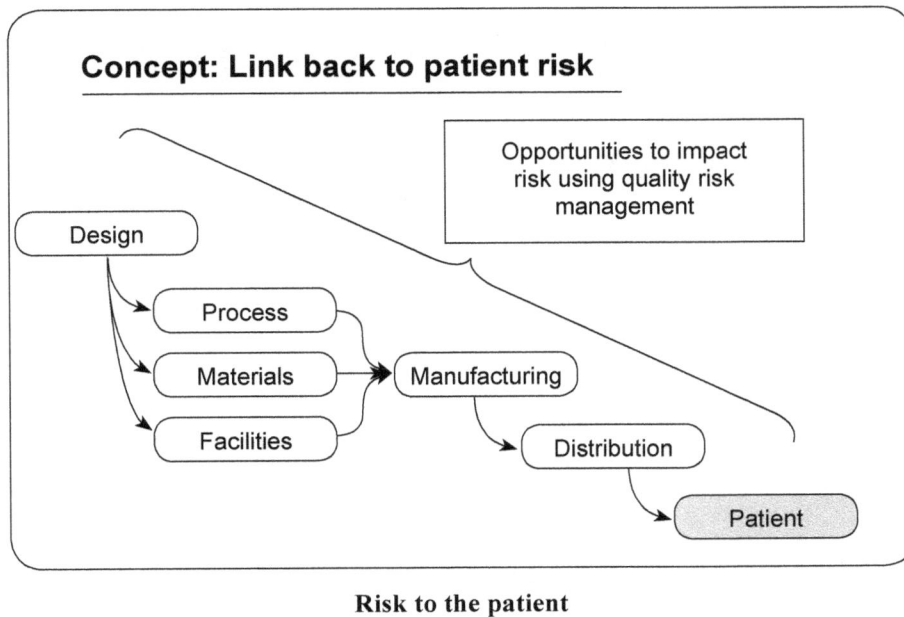

Concept: Link back to patient risk

Opportunities to impact risk using quality risk management

Design → Process, Materials, Facilities → Manufacturing → Distribution → Patient

Risk to the patient

Source: A presentation by H. Gregg Claycamp, Office of New Animal Drug Evaluation, U.S. Food and Drug Administration, at ICH Q9: Quality Risk Management, CDER Advisory Committee for Pharmaceutical Science (ACPS), 5-6 October 2006, Rockville, MD.

Recommended reading

1. AS/NZS 4360:2004, Risk Management Standard (Australian/New Zealand standard).

2. BS 31100 Code of practice for risk management.

3. Enterprise Risk Management, Committee of Sponsoring Organizations of the Treadway Commission (COSO), U.S.

4. ISO 31000:2009 Risk Management – Principles and guidelines.

5. ISO Guide 73:2009 Risk Management – Vocabulary.

6. Q9 Quality risk management, The International Council for Harmonisation of Technical Requirements for Pharmaceuticals for Human Use (ICH).

7. Risk Management Standard. U.K.: Institute of Risk Management (IRM), Public Risk Management Association (ALARM), and Association of Insurance and Risk Manager (AIRMIC), 2002.

Self testing multiple choice questions

1. For a proper safety plan it is important to set out and clearly mark:
    a. Bomb shelters.
    b. Evacuation routes.
    c. Emergency vehicles.

2. To indicate the emergency exits:
    a. Put up guide signs to indicate evacuation routes.
    b. Ensure that all emergency exits are securely locked and bolted.
    c. Install rope ladders and evacuation chutes on all floors.

3. Areas that are dangerous for employees to enter or go near are:
    a. The area outside a certain range of operating machinery and equipment.
    b. Restricted access sites.
    c. Sites with movable facilities and equipment.

4. Fix danger signboards on:
    a. Walls.
    b. Flush with ceilings.
    c. Chairs.

5. Match protective clothing and tools with the operations, or operation contents:

    i.   Eyeglasses   ① Working at altitudes, or with forklifts
    ii. Masks   ② Organic solvents
    iii. Gloves   ③ High temperatures
    iv. Helmets   ④ Noise
    v.   Earplugs   ⑤ Hot, sharp or pointed items
    vi. Protective uniforms   ⑥ Specific chemical substances
    vii. Heat-resistant uniforms ⑦ Cutting, grinding and chipping

    a. i⑤, ii⑥, iii②, iv④, v⑦, vi①, vii③
    b. i⑦, ii②, iii⑤, iv①, v④, vi⑥, vii③
    c. i⑦, ii⑤, iii②, iv①, v④, vi⑥, vii③

6. To periodically check that protective clothing and tools are being used properly:
    a. Record where protective gear is distributed for joint use, and where for individual use and observe that it is being used.
    b. Check that protective gear conforms to the standards set by the business facility.
    c. Do periodic checks on the numbers of items of protective gear that are released.

7. Examples of protective devices that halt operations or activate alarms include:
    a. Safety fences around apparatuses.
    b. The rotor blades of centrifuges.
    c. Systems that monitor environmental parameters.

8. Checkpoints for promoting safety awareness include:
    a. Supervisors encouraging employees to make their own safety rules by not performance.
    b. Permitting employees to have accidents so they can be used as examples of what not to do.
    c. Signs prompting employees to be safety conscious.

9. Actions to promote safety awareness include:
    a. Arrange study visits to companies with better safety records.
    b. Provide safety instructions and reminders to employees at regular meetings.
    c. Provide training on ways of increasing potentially dangerous situations.

10. Formal safety education must be provided when:
    a. Employees begin to work in a company.
    b. Employees are about to move to another company.
    c. Employees experience an accident.

11. General safety courses should cover the following points:
    a. Methods to avoid reporting accidents.
    b. Methods for handling materials.
    c. Methods for reporting on other employees.

12. Standard laboratory protective equipment (PPE) includes the following items in a level 2 (low risk) environment:
    a. Latex gloves, mask, gown, cap, safety glasses and shoes (shoe coverings).
    b. Mesh gloves, face shield, cap, and shoes (shoe coverings).
    c. Gown, cap, respirator, gloves and shoes (shoe coverings).

13. There are _____ stages in a company's readiness to deal with disaster.
    a. Five
    b. Seven
    c. Ten

14. Safety standards and regulations should:
    a. Be based on the law of the country and/or international regulations.
    b. Take into account the local culture.
    c. Take into account the history of the workplace.

15. Safety standards and regulations should be involved in:
    a. Operational procedures and methods.
    b. Travelling to and from work.
    c. Behaviours outside the workplace.

16. Written procedures for operating equipment should cover:
    a. The methods to tweak performance beyond specifications.

    b.   The names of the people to report an accident to.

    c.   The specific safety items to be noted during the operations.

17. Employees should receive general safety courses which include:
    a.   Potential dangers, handling dangerous materials, smoking risk.
    b.   Potential dangers, how to report safety abnormalities, handling dangerous materials.
    c.   Safety activities, handling dangerous materials, smoking risk.

18. To make training in the safety standards as effective as possible:
    a.   Restrict education to standards related to the operations concerned.
    b.   Do not use standards and regulations for individual units of equipment as teaching materials.
    c.   Keep records of education and training.

19. A meeting held after an accident occurs should include the following:
    a.   Ask the question "Why did it happen?" and establish root cause.
    b.   Focus primarily on the legal consequences to a company.
    c.   Gather only information relevant to the after-effects of the accident.

20. Safety committees only need to be established at:
    a.   Senior management levels.
    b.   Government organisations only.
    c.   All places of work.

21. A safety patrol should be carried out:
    a.   Daily.
    b.   Weekly.
    c.   Periodically.

22. When employees feel that the facilities, equipment and behaviour at the workplace are a source of danger to them, they may:
    a.   Request the safety committee to audit and make a diagnosis.
    b.   Diagnose the situation themselves and report to the safety committee.
    c.   Refuse to do any more work until the company deals with the situation.

23. Safety patrols should:
    a.   Be the responsibility of a single employee.
    b.   Be staffed by two or three employees.
    c.   Be any number that can make up a team.

24. Facility safety checks should be carried out at the following stages:
    a.   Only at the design stage.
    b.   At all stages from design through operations.
    c.   Only at the operational stage.

25. Establish standards that will protect equipment from:
    a. Avoidable damage.
    b. Functional deterioration.
    c. New employees.

26. It is important to keep maintenance records for each unit of equipment because:
    a. It looks impressive on a regulatory audit to have all that paperwork.
    b. It makes effective use of the data for preventive maintenance.
    c. It measures performance loss which is the most important factor.

27. Abnormalities that may have safety implications include:
    a. Unusual movements of operators.
    b. Usual noises from operation.
    c. Changes in pressure in all cases.

28. To correct abnormalities:
    a. Try to ask certified individuals to perform recovery and restoration tasks that require certificates.
    b. Place a card that says "power supply must be turned off' around switches when suspending the operation of machines and equipment.
    c. Report abnormalities to superiors at work whenever necessary.

29. An in-house accident report form should include the following two stages:
    a. The circumstances causing and the countermeasures implemented.
    b. The costs and delays caused by the accident.
    c. Confirmation that the accident has been recognised and someone will examine it.

30. Causes of an accident can include:
    a. Unsafe SOPs.
    b. Unsafe human factors.
    c. Unsafe records.

31. When setting safety targets:
    a. Set the target levels as high as possible.
    b. Aim for levels that are impossible.
    c. Set target levels at past achievement levels.

32. Match jobs with duties:
    i.   Fire-fighting leaders
    ii. Fire-fighting officers
    iii. Transport officers

    ① Remove important documents and data.
    ② Direct and control regional teams, confirm the stage of evacuation, report injuries to HQ.
    ③ Rush to the sites and perform early fire-fighting activities.

    a.   i②, ii③, iii①
    b.   i①, ii③, iii②
    c.   i②, ii①, iii③

33. Match stages of readiness to deal with disaster with content:

i.	First stage	①	Roles are distributed and self-directed activities are undertaken in accordance with divisional, departmental, and sectional plans.
ii.	Second stage	②	Roles are distributed and company-wide activities are undertaken vigorously.
iii.	Third stage	③	Issues to be addressed are put in order, but there are no self-directed activities.
iv.	Fourth stage	④	There are company-wide goals, but virtually no self-directed activities.
v.	Fifth stage	⑤	Roles are distributed and company-wide activities are undertaken in cooperation with related departments and sections.

    a.   i④, ii③, iii①, iv⑤, v②
    b.   i④, ii①, iii③, iv②, v⑤
    c.   i④, ii③, iii②, iv⑤, v①

34. The following is a potential safety concern:
    a.   Vibrations normally associated with the equipment.
    b.   Electrical hum.
    c.   An unexpected stop in the production cycle.

35. An example of an alarm abnormality is:
    a.   Sound and/or light alarms fail to go off when values are within accepted limits.
    b.   Sound and/or light alarms go off when the equipment is working normally.
    c.   Sound and/or light alarms go off when the production cycle suddenly stops.

36. The four major sections of the disaster management cycle are:
    a.   Preparation, mitigation, documentation and recovery.
    b.   Preparation, response, recovery and mitigation.
    c.   Recovery, restoration, reconstruction and restitution.

37. The person in the role of Emergency Chief is responsible for:
    a.   Training, instruction and supervision of fire-fighting leaders and their deputies.
    b.   Taking charge of the laboratory when an emergency (fire) breaks out.
    c.   Arranging to move all materials off site so they are untouched by the disaster.

38. A commonly used safety ratio is:
    a.   Injuries versus deaths of employees per year.
    b.   Injuries or deaths per 1,000 employees per year.
    c.   Injuries or deaths per number of people in the industry per year.

39. Risk management is:
    a.   An impossible dream since risk is uncontrollable and spontaneous.
    b.   A means by which the odds of process failure can be calculated.
    c.   A logical and systematic method of identifying, analysing, treating and monitoring the risks involved in any process.

40. The second step in risk management is to:
    a.   Identify the risk.
    b.   Evaluate the risk.
    c.   Monitor and review.

41. Analysing risk requires examining both:
    a.   Likelihood and frequency.
    b.   Probability and frequency.
    c.   Impact and likelihood.

42. Two types of risk assessment are:
    a.   Quantitative and qualitative.
    b.   Variable and sustainable.
    c.   Critical and casual.

43. Informed judgment is part of:
    a.   Quantitative risk assessment.
    b.   Qualitative risk assessment.
    c.   Critical Control Point establishment.

44. A very high likelihood of occurring with moderate impact on product risk is considered to be:
    a.   High.
    b.   Major.
    c.   Severe.

45. By identifying Critical Control Points, a company can:
    a.   Avoid a few problems from occurring.
    b.   Essentially prevent or eliminate any safety hazards.
    c.   Understand risk better but not eliminate it.

46. Critical Control Points can:
    a.   Exist anywhere along the therapeutic chain.
    b.   Only be identified at the design phase.
    c.   Only be identified following infusion.

47. The material causes of an accident are also referred to as:
    a.   Unsafe conditions.
    b.   Unsafe systems or procedures.
    c.   Unsafe actions.

48. In biopharmaceutical companies, the decision to designate certain areas as dangerous is based on:
    a.   An individual decision process within a company.
    b.   A code of practice for the industry.
    c.   A logical and systemic method of identifying risk.

49. Proper gowning procedures must be governed by:
    a.   An SOP.
    b.   Individual responsibilities.
    c.   Regulatory requirements.

50. In risk management, an event with low impact but almost certain to happen is classified as:
    a.   Significant.
    b.   Major.
    c.   Moderate.

SECTION FOUR
Production, materials and development

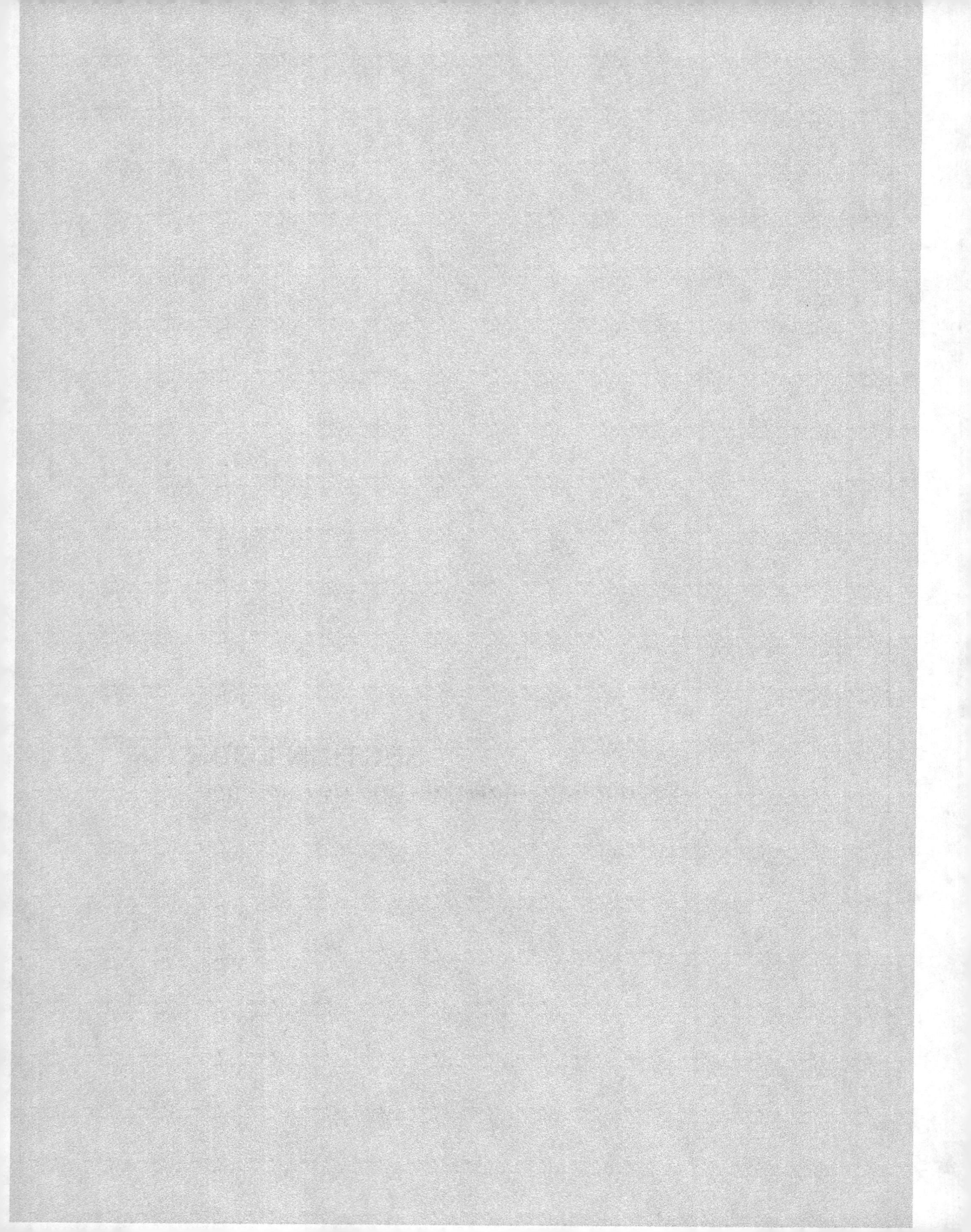

Chapter 16. Production control

Sustaining
quality in biotechnology

Production control is the management of the production processes to ensure that a company produces a therapeutic product of the quality that the market wants, in the right quantity, and ready for delivery at the right time, and that it continues to improve the efficiency with which it does so. This chapter will provide the key actions to be undertaken in order to achieve these goals that a business remains sustainable for the long term. There are six critical areas in which control has to be exercised in order for a company to achieve the goal of long term sustainability. It is the responsibility of management to see that each of these critical areas is being controlled properly. The failure to do so in any one of these areas will result in a chain of events that adversely affects all the remaining areas and results in a company losing control.

Design and production web

16.1 Prepare production plans (Design and Production Controls)

There are two key plans that the production laboratory must prepare in order to have effective production control:

a. An annual production plan, based on the marketing plan.
b. A daily or monthly production plan.

The daily or monthly plan is essential to ensure that production follows the annual plan. Because of the intermittent nature of immunotherapy companies, by which production is more to an on-demand basis corresponding to patient availability, then a monthly plan (expectation of production) is more suitable than a daily plan. The delivery plan ensures that cell products are delivered from the laboratory site to the infusion site on time and as required. Whereas, in most other industries this does not present too great a logistical problem, when dealing with immunotherapies, attempting to forecast the number of treatments, both initial and repeat, as well as which site, when dealing with multiple hospitals, becomes extremely difficult. The only means by which to do so is to look at trends from previous years for baseline determinations and then add on an incremental percentage increase based on the assumption that as a company's reputation grows, so will its customer/patient base.

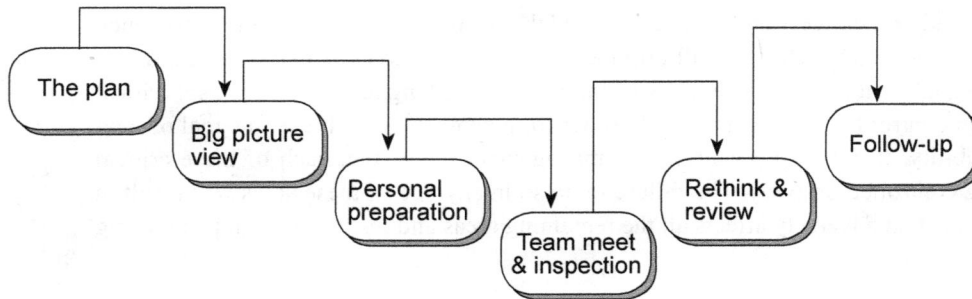

Planning steps

16.2 Annual production plan (Equipment and Facility Controls)

The annual production plan will cover the company's fiscal period, which will usually be the 12 months from tax time rather than the calendar year.

This plan will determine:

a. The type and quantity of products that will be produced during this period.
b. The production conditions: location, materials and personnel, and the internal production departments and external suppliers which will participate. It may also include other detailed plans that support this basic plan.

The annual production plan will be largely determined by your schedule of expected treatments plan. Before preparing it, look carefully at the scheduled treatments plan and then look at the status of your production processes:

a. Examine the amount of materials in the inventory, and the rate at which raw materials are being moved into and out of the inventory.
b. Check if the current production capacity can produce the quantity required by the projected sales plan.
c. If production capacity is insufficient, investigate what countermeasures can be taken in order to achieve the proposed level.

d. Investigate the time and expense needed to implement these countermeasures.

e. Finally, prepare your production plan based on the results of all these investigations.

The annual production plan may have to be radically changed if there are changes in either the market or in production conditions. When this happens, review the plan and make whatever changes are needed, whether to the marketing plan, or to the planned production quantity and methods, or to both.

Dealing with changes in the market will involve three actions:

a. Investigate the changed market needs.

b. Decide on the company's treatment strategy based on these needs.

c. Decide on the estimated therapeutic treatment quantity.

A good annual production plan will help to ensure that a company:

a. Produces the therapies that the market wants.

b. Delivers the right quantity of each type of treatment on time.

c. Continues to improve the efficiency of its production processes.

To ensure that production follows the annual production plan, it is important to also prepare daily or monthly plans, based on the annual plan.

These should specify:

a. How much is to be produced daily or monthly.

b. Efficient methods for producing this quantity.

c. How much this will cost.

d. The target level of quality.

You should have a production administrator who will assess the situation in your section, and prepare a plan that will ensure the required production quantity. This requires regular meetings with related departments or sections to discuss any problems they may have. Take these into account in preparing the plans.

In general, in those industries where they produce a repeatable, identical product such as the automotive industry, which includes assembly operations, the monthly and daily plans are based on a daily schedule. This schedule specifies the number of days required for each step, from receipt of the order through to the final product. Because biological products are neither identical nor consistent in their production and manufacture, any schedule or plan drawn on a short term basis is likely to undergo major changes.

Changes sometimes have to be made to the daily plan, for example when:

a. A product has to be produced out of the scheduled order.

b. Unscheduled work has to be done.

c. Design changes have to be made.

d. There are cancellations by either the hospital or patient.

Planning should allow for the occurrence of such events – it must take the realities of the job site into account.

16.3 Keeping to plan (Process Controls)

When the monthly production plan has been agreed, there are several important actions to take to ensure that production keeps to this plan.

a. Organise your production stream (flow) in the best possible way.
b. Set and keep to standard times.
c. Know your production capacity.
d. Manage each process precisely.

Organise the production in laboratories in the best possible way to carry out the monthly plan economically and efficiently.

Take the following steps:

a. Decide on the daily production quantity on the basis of the monthly plan.
b. Decide on a suitable speed to provide this production quantity.
c. Decide on the best way to organise operations for maximum potential.
d. Allocate the workforce according to this organisation of operations.
e. Arrange for the supply of components and materials to be on hand.
f. Give production instructions to the sections that produce related components, based on the cell expansion speed.

Hold regular meetings with sections related to the monthly production plan to review the plan, to investigate the problems in each section, and to decide on a strategy for handling these problems.

If the current production system is not capable of meeting the monthly production plan, it is better to review the equipment and the standards rather than to simply increase the workforce because the equipment and how it is operated is most often the limiting factor.

Standard time is the time required to perform a standardised operation. You need to manage production according to the standard times (based on history) in order to keep production stable and to stay within the targeted production costs and timeline.

To set and keep to standard times:

a. Check how long it takes to carry out the operation as specified in the operation standard, and then set this as the standard time.
b. Remember that the time required to complete a given operation may vary not only from one individual to another, but even when the same individual performs the same operation repeatedly.
c. Take into account the physical requirements.
d. Take into account the environment.

e. Set hours separately for skilled employees and for general employees.

f. Educate and train employees so that they can perform standard operations within the standard times.

g. When operation conditions change because of changes in design or equipment, respond immediately through Change Control, revise the operation standard and the standard time, and check the results for acceptance.

It is essential to know the precise production capacity of each process. This is the quantity that each process is capable of producing within a certain period.

There are a number of points to consider here:

a. A process may not actually produce to its full capacity because:
    i.   Of defects in the production process.
    ii. Of failure of machinery or equipment.
    iii. An associated process is producing at a lower capacity.

b. Where a lower capacity process is slowing down production, synchronise capacity among the related processes: increase the capacity of the processes with lower capacity by improving equipment or by increasing the number of operators, etc.

c. When starting production of a new immunotherapy, establish, in hard numbers, the production capacity of each process, and adjust the capacities of the different processes to create a balance in overall production.

d. Production capacity can always be improved by regularly checking how many products are actually produced. Too many cell products being produced in the laboratory simultaneously not only present a risk of cross-contamination but also administration of a particular product to the wrong customer/patient.

To manage each process properly you need to:

a. Prepare a draft process schedule (by process and by date).
b. Work with related sections to create an implementation plan based on this draft.
c. Check progress several times a day with reference to the implementation plan.
d. If there is a delay in a production process, immediately gather all the relevant information, check the situation, and decide on countermeasures. It is important to minimise any loss of production.

When your monitoring system indicates a delay, find out the causes and deal with them:

a. The reasons for failing to keep to the schedule are normally found in the "4M" – material, machinery, man, and methods of work. Examine these four areas to find the causes, and take temporary measures to deal with them.

b. Decide on countermeasures to deal with these causes, and implement these with the cooperation of other related sections.

c. Evaluate the results of these measures, and, if necessary, review the standards.

d. The standards that you establish for the countermeasures should be appropriate for the specific situation: measures may differ depending on the degree of delay.

e. Take measures to prevent delays recurring and check the effect of these measures.

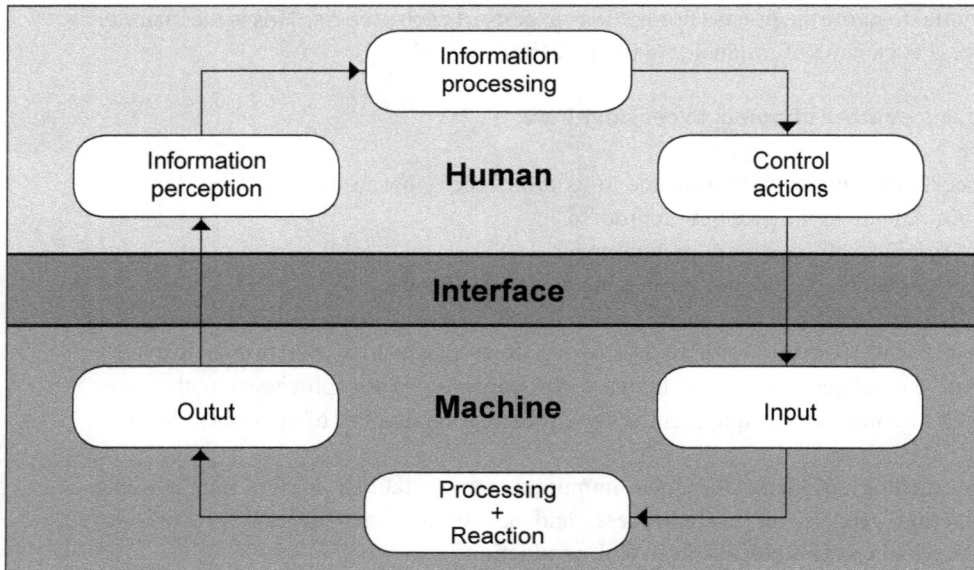

Human-machine interface

Remember that if production delays are not dealt with they will recur repeatedly and indices will show that your processes are out of control.

A critical factor in keeping to the production plan is the level of productivity of both employees and equipment. One measure of this is the number of hours operations are actually performed during the hours that they could be performed. This is the operating ratio. It can be put in percentage terms:

$$\frac{\text{Operating hours}}{\text{Actual working hours}} \times 100\%$$

To improve productivity, decide on effective operating ratios and establish standards for these using numbers as indices:

a. Decide which non-operational tasks can be counted as actual work.
   Examples are:
   i.   Going to pick up materials, shipments, etc.
   ii. Replacing parts on equipment and reagents.
   iii. Calibrating or adjusting equipment.

b. Identify lost time for which employees are not responsible.
   Examples are:

      i.   The power failed.

     ii.   Equipment malfunction or breakdown.

    iii.   Undergoing an audit or inspection process.

    iv.   Waiting for essential materials to be delivered.

c. Reduce indirect work – work not directly related to the production process and therefore only of secondary importance which can be done by someone else other than the operator.

d. Establish a realistic standard time for each operation based on historical data.

The scientific study of manufacturing operations is called industrial engineering (I.E.). Many companies overlook this function which is part of the Quality Operations portfolio for enhancement and improvement of processes.

Delay-free production depends on having a smooth supply of the various components and materials needed for daily production. Prepare a plan that will ensure this. Base your plan on the daily or monthly production plan and distribute it in advance to related sections.

Be aware of the following potential problems and seek preventive actions:

a. Rejected products may be produced if the components and materials in the production process have changed, even if that change is not anything more than changing the supplier.

b. Breakdowns may occur in machinery or equipment during the process, especially if the equipment is not being regularly inspected, cleaned, or adjusted after each process.

c. Transport problems may hold up the supply of components and materials and therefore ensuring sufficient stock is kept on hand for key components is essential.

To be prepared for these problems, standardise and implement a control system of the production process for each component and material. If a problem does occur, collect related information, and decide on recurrence prevention measures.

In many cases, companies hold a meeting when they are preparing the monthly plan in order to identify problems that might arise in supplying each component and material. They then make adjustments, determine a daily production plan based on the information they have gathered, and decide on the quantity of components and materials that should be on hand. If a defect is found in the components and materials that have been supplied, it is not unusual for companies to stop production until the cause has been found but this is not realistic when the product is an immunotherapy and required almost immediately. Therefore, part of the preventive actions should be having a list of secondary approved suppliers that can be contacted immediately.

16.4 Deal with fluctuations in production (Records-change Control)

Changes in patient demand may result in fluctuations in the required level of production. For example, a customer/patient may originally opt for four treatments and then change their mind and request additional treatments should the therapy be working well. Or the opposite may be true where multiple treatments are initially requested and this number is reduced at the request of the customer/patient. Furthermore, expansion of cells is not an exact science and there is no guarantee that you will reach treatment levels of $>10^8$ each time or that the cells expanded are precisely the immune specific ones required. This being the case, then repeated productions might be necessary for a single patient until the number or type of cells are correct.

To anticipate or respond to such changes you may have to:

a. Re-organise the production system.
b. Allow for fluctuations in production when negotiating contracts with suppliers.
c. Adjust the internal stock of components reagents and materials.
d. Develop multi-skilled employees who can be re-allocated to different jobs.

There are a number of ways that you can re-organise the production system:

a. Change the number of technicians.
b. Allocate employees to different jobs.
c. Change from a single-shift system to a multi-shift system, or vice versa.
d. Reduce the production quantity by stopping production of a certain process or reducing the operational hours.
e. Adjust the speed of machinery and equipment.

It is important to:

a. Be flexible when making changes to processes: relate changes to the size of production fluctuations.
b. Be careful when making changes: Large fluctuations in production often are the source and major cause of defects.

When drawing up a contract with an external supplier it is essential to agree about what will be done if there are fluctuations in production as a result of changes in patient and market demand. Such changes will obviously result in changes to supply requirements.

Note the following:

a. When arranging a contract, make a clear allowance in the contract for both increases and decreases in the quantity to be supplied without penalty.
b. When a major increase in the required supply quantity causes difficulties for the primary supplier, have a secondary supplier already under contract.
c. It is important to clearly reserve the right in the contract to make adjustments to the supply amount within a certain limited range of fluctuation, and to request a separate contract to handle changes beyond this limit.
d. Consider the capacity of your suppliers to respond to fluctuations when preparing the monthly production plans.

It is important for a company to be able to adjust the inventory of components and raw materials in response to any fluctuations in production.

In order to do so:

a. A company must constantly monitor and manage the maximum and minimum stock levels for the components, reagents and raw materials needed for a particular therapeutic product.
b. A company must institute a mechanism to ensure that there will always be enough components, reagents and materials available if production has to be increased.
c. The ability to handle such changes in production indicates flexibility in production capacity.
d. It is important to minimise the gap between maximum production capacity and actual production capacity if a company is operating proficiently.

It is a necessary practice to be able to re-allocate technicians to different jobs or projects when there are fluctuations in production. Training employees in multiple skills is one of the best methods for responding flexibly to fluctuations in production.

This can be achieved in several ways:

a. Educate and train employees systematically so that they will be able to perform a broader range of functions, and therefore support re-allocation.
b. Introduce multi-skill training through an in-house qualification programme.
c. Supervisors recommend specific employees to obtain various licences.

16.5 Maintenance of inventories (Material Controls)

Elimination of high storage costs of raw materials and reagents is essential if a company is to practise TQM.

Plan and maintain several inventories:

a. Finished processed cell product inventory.
b. Components and raw materials inventory.
c. WIP (work in progress) inventory.

Storage systems are appropriate if they:

a. Allow staff to see the inventory levels at a glance.
b. Provide a first-in first-out (FIFO) system for delivering and collecting items. See previous chapters for more general guidelines on storage and maintaining inventories.

Excessive product can become a serious problem, resulting in:

a. Additional storage costs.
b. Products deteriorating, aging, or expiring.
c. Technologies changing making products obsolete or redundant.

For these reasons it is important to plan the product inventory carefully. Decide on the right level, and maintain this level by adjusting production.

Take the following steps:

a. Decide how many treatments and therapies have to be over a specific period of time, and incorporate into the monthly plan.
b. Take into account any possible changes in the treatment plan, and establish a standard inventory quantity that includes some extra inventory.
c. Extra inventory quantities take into consideration the following:
    i.   Production capacity.
    ii.   The production period per patient versus time.
    iii. Deterioration of the product and raw materials over time.
    iv. Market trends in regards to treatment type.
    v.   The product's competitiveness against other therapies.
    vi. Inventory costs.
d. Maintain the calculated volumes in the inventory at all times.

A WIP (work in progress) inventory is essential to provide the right quantity of partially processed goods for all the final production processes.

To establish such a system:

a. Decide which items in the inventory need to be managed and the quantities required at all times.
b. Decide what inventory data should be displayed for people to quickly see:
    i.   The standard amount, and the minimum and maximum amounts.
    ii.   The amount already used and the balance remaining.
    iii. Any refill orders that have been submitted, and the date of the expected arrival.
c. Methods for displaying this data are:
    i.   Electronic format.
    ii.   Signs, standing plates, tags on articles.
    iii. Log books.
d. Develop an SOP that governs the operation of the system.

This computer/electronic system can be used to manage all inventories: finished products, partially finished products, components, reagents and raw materials.

In many cases, companies check raw materials weekly or monthly, or when they receive these materials. It is still recommended to regularly conduct a visual on-site check of the actual goods, materials rather than depend solely on the accuracy of a computer system. Computer systems are only as good as the technician loading the updated figures into the system.

Items received first should be used first to prevent their deterioration. "First-in" refers to the order of receipt into the storage area. See notes in Chapter 13 on the FIFO system.

16.6 Inspect finished products, deal with abnormalities and pursue continuous improvement (CAPA Controls)

There are additional actions to be taken in order to improve production control.

These include:

a. Inspect finished products with reference to the product standard.
b. Define what constitutes an abnormality in a product, and standardise the countermeasures to be taken when one is identified.

Inspect finished products with reference to the product standard or the accepted limits for product specifications.

Take the following steps:

a. Establish an inspection standard that corresponds to the theoretical or desired standard of the therapy product.
b. Attach full certification when the product is being shipped to the centre where the infusion will take place.
c. Put an identifying mark on the infusion bag or another therapy container to indicate that it has passed inspection and is acceptable for use.
d. When a product does not pass the inspection, identify the problem and any obvious root cause. Notify QA to conduct an investigation by submitting a Non-conformance form or Out of Specification form. If QA investigates and finds no issues then continue to process the product.
e. Keep the inspection results for a certain period to satisfy the quality assurance requirements, and regulatory requirements, usually a minimum of seven years.
f. Following the non-conformance investigation of the therapeutic product, measures must be instituted to prevent recurrence.

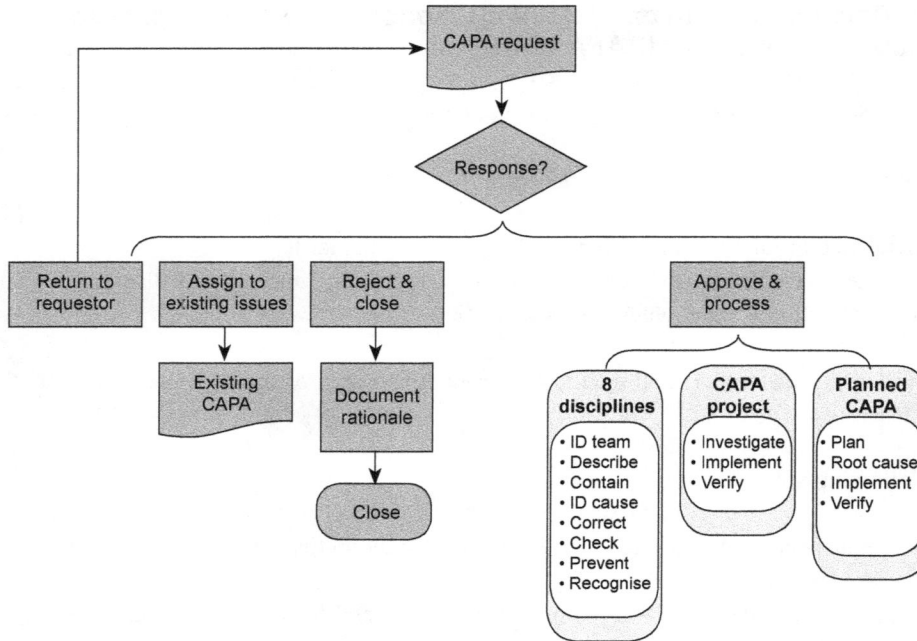

The CAPA process

To prevent reoccurrence, root cause must be established. When trying to identify the root cause, it is necessary to focus on the following possibilities.

Root causes which result in abnormalities may stem from:

a. Quality problems.
b. The deterioration of products and semi-finished products as a result of temperature, light, or time.
c. Damage to container or packaging of the product.

Take the following steps:

a. Establish Critical Control Points (CCPs) for examining progress in the production processes, and collect information on any abnormalities which appear in both quantity and quality occurring at those points.
b. CCPs are determined by understanding of the process and recognising there is a high probability at this particular juncture that things could go wrong.
c. Standardise any procedures for dealing with abnormalities.
d. Decide on which countermeasures will bring the process back into normal limits.
e. Standardise the countermeasures.

Example: Infusion bags that were found to be contaminated.

The first step was to collect information on possible contamination sources of the packages.

These include the following CCPs:

a. Problem of contamination at the manufacturer. (Supply)
b. Contamination at time of receipt. (Unpacking)
c. Contamination when moved into the laboratory. (Aseptic technique)
d. Contamination at time of filling. (Mechanical)
e. Contamination when storing/transferring the final product. (Logistics)

Employees found that there was a 20% occurrence rate of this contamination in the final process. When bags were tested for contamination while being removed from their packages there was no evidence of any contamination. Working back from the final process to find the point that was generating the contamination it was discovered that in this particular instance it was the filling process that was causing the problem. The seal was leaking on the plasmaphoresis unit that filled the bags, allowing room air to enter into the product. Since filling is performed in a Class C environment, the air is not filtered any higher than 10,000 particles per cubic foot.

Corrective measures:

a. Replace the gasket where the silicone tubing attaches to the unit.
b. Review the standards for maintaining and checking the plasmaphoresis unit.
c. Add antibiotics to the product suitable for intravenous injection and without interference with cell expansion.
d. Standardise this procedure after changes.

Recommended reading

1. Åström, K.J., & Murray, R.M. *Feedback Systems: An Introduction for Scientists and Engineers*. Princeton and Oxford: Princeton University Press, 2008.

2. Camacho, E.F., & Bordons, C. *Model Predictive Control in the Process Industry: Advances in Industrial Control*. London: Springer, 1995.

3. Hill, T. *Production Operations Management: Text and Cases*. Prentice Hall, 1991.

4. Morari, M., & Zafiriou, E. *Robust Process Control*. Englewood Clis, NJ: Prentice Hall, 1989.

5. Ogunnaike, B.A., & Ray, W.H. *Process Dynamics, Modeling, and Control*. Oxford University Press, 1994.

6. Payant R.P., & Lewis, B.T., *Facility Manager's Maintenance Handbook*. McGraw-Hill, 2007.

7. Rivera, D.E., "An introduction to mechanistic models and control theory," tutorial presentation at the SAMSI Summer 2007 Program on Challenges in Dynamic Treatment Regimes and Multistage Decision-making, 18-29 June 2007. http://csel.asu.edu/controleducation (item 9), accessed 31 August 2016.

8. Saunders, M. *Strategic Purchasing and Supply Chain Management*. Pitman, 1994.

9. Seborg, D.E., Edgar, T.E., & Mellichamp, D.A. *Process Dynamics and Control*. John Wiley & Sons, Inc., 1989.

10. Slack, N., Chambers, S., Harland, C., Harrison, A., & Johnston, R. *The Operations Management*. Pitman, 1995.

Self testing multiple choice questions

1. The annual production plan determines:
   a. The type and quantities of products that will be produced during this period.
   b. Which internal production departments and external suppliers will participate.
   c. The range within which the sale price of the product should be set.

2. The annual production plan is determined largely by:
   a. The annual plan.
   b. The scheduled treatments plan.
   c. The market plan.

3. Which of these is not a control requirement to meet expected patient requirements:
   a. Check if the current production capacity can produce the quantity required.
   b. Examine the amount of materials in the inventory, and the rate at which products are being moved into and out of the inventory.
   c. Prepare an R&D plan based on new technologies.

4. The production will not have to be radically changed if there are changes in:
   a. The supply of raw materials.
   b. Research methods.
   c. Production conditions.

5. An annual production plan is a valuable means of ensuring that a company:
   a. Produces the products that the customers/patients want in advance of sales.
   b. Guarantees to reach a certain income for the year.
   c. Continues to improve the efficiency of its production processes.

6. The precise production capacity for each process is the quantity that each process:
   a. Is capable of producing.
   b. Is capable of producing within a certain period.
   c. Produces within a certain period.

7. When drawing up a contract with an external supplier it is essential to agree about supplier responsibilities if there are fluctuations in production due to:
   a. Employee absenteeism.
   b. A breakdown in equipment.
   c. Changes in market demand.

8. If a contract has already been made with an external supplier which does not include an agreement on how to handle fluctuations in production, then:
   a. Conditions must be decided each time such a change takes place.
   b. The contract must be immediately changed to include such an agreement.
   c. A new supplier should be found.

9. To determine the proper WIP inventory keep the following points in mind:
   a. The more inventory the better.
   b. When you create a new inventory plan because of changes in production, establish the changed inventory as the standard inventory level.
   c. The standards are unchangeable even if there is a change in production.

10. Inventory management at a glance is the best way to control inventory precisely, because:
    a. Nobody is needed to manage the inventory.
    b. People can immediately see the level of the inventory.
    c. It allows a first-in, first-out system.

11. "First-in" refers to:
    a. The order of arrival in the inventory.
    b. The order of production or purchase.
    c. The order of anticipated usage.

12. The first-in, first-out method may be difficult to use because:
    a. It is very complicated.
    b. It is often time-consuming.
    c. It requires a lot of storage space.

13. There are _____ critical areas in which production control must be exercised.
    a. Four
    b. Six
    c. Ten

14. The critical areas of production control do not include:
    a. Process and production.
    b. Materials.
    c. Personnel.

15. There are _____ key plans that a production lab must prepare in order to have control.
    a. Two
    b. Three
    c. Five

16. Because of the individual basis of applying immunotherapy, the appropriate plan is:
    a. Daily.
    b. Weekly.
    c. Monthly.

17. If production capacity is insufficient, then:
    a. Produce what you can and postpone the remainder.
    b. Pressure the workforce to increase their output to meet the remainder.
    c. Investigate what countermeasures allow achievement of the proposed level.

18. Countermeasures before implementation have to be weighted against:
    a.   Employee needs for time off and holidays.
    b.   The time and expense required in order to implement.
    c.   Profit and loss estimates.

19. Changes within the market will require _____ responsive actions.
    a.   Three
    b.   Five
    c.   Seven

20. The first action when responding to a change in the market is to:
    a.   Decide what will be a company's treatment strategy be based on the change.
    b.   Investigate what the change means in terms of marketing.
    c.   Decide on the estimated quantity of new treatments.

21. The most common limiting factor when production plans are not met is:
    a.   Size of the workforce.
    b.   How the equipment is being operated.
    c.   How the standards are being followed.

22. Standard time is:
    a.   Based on the 24-hour clock.
    b.   Based on an 8-hour work day.
    c.   The time required to perform a standardised operation.

23. Standard times are set by:
    a.   Industry standards.
    b.   Examining historical records from a company.
    c.   Backtrending from the time when the product is required.

24. In order to perform within the standard time, it is necessary to:
    a.   Use only experienced staff and technicians.
    b.   Educate and train employees so that they can perform within standard time.
    c.   Terminate any staff who cannot meet the standard time criteria.

25. Multiple cell therapy products prepared in the laboratory simultaneously are:
    a.   A time and labour saving practice that is encouraged.
    b.   An indicator of successful control practices.
    c.   At risk of cross contamination and infusion into the wrong patients.

26. Successful practice of TQM involves:
    a.   Elimination of high storage costs of new materials.
    b.   Maximising product in storage several times the actual need, so shortages do not occur.
    c.   Elimination of the need for storage by using all product and materials at each run.

27. A cell therapy company should have several inventories. These include:
    a. Raw materials, obsolete products and waste materials.
    b. Final products, obsolete products and work in progress.
    c. Final cell products, components and raw materials, and work in progress.

28. The Operating Ratio is a percentage calculated from:
    a. The number of hours available to work divided by the hours actually worked.
    b. The number of hours actually worked divided by the hours available to work.
    c. The number of hours the technician worked divided by equipment operation hours.

29. When calculating the Operating Ratio it is important to:
    a. Count total down time whatever the cause as actual work.
    b. Count those hours spent on unrelated non-operational tasks as actual work.
    c. Count those hours spent on related non-operational tasks as actual work.

30. The scientific study of manufacturing operations is called:
    a. Industrial Engineering.
    b. Scientific Engineering.
    c. Constructive Engineering.

31. In order to avoid delays caused by a breakdown in raw material supply it is important:
    a. To have penalty clauses to be paid by the supplier in such an occurrence.
    b. To have penalty clauses paid by a company to the supplier in such cases.
    c. To have secondary approved suppliers readily available and under contract.

32. It is important to minimise the gap between maximum production capacity and _____ production capacity.
    a. Theoretical
    b. Actual
    c. Historical

33. One of the best methods for responding to fluctuation in production is:
    a. To train employees in multiple skills.
    b. To contract out production needs to other companies.
    c. To increase or decrease the staff size accordingly.

34. As part of the CAPA control in production, it is important to define in advance:
    a. Why errors happen and how they best can be hidden.
    b. Which department is likely to blame and how they will be disciplined.
    c. What constitutes an abnormality and what countermeasures will be required.

35. When a product does not pass an inspection, it is important to immediately send a report to:
    a. QC.
    b. QA.
    c. Senior management.

36. In a non-conformance report, the term OOS refers to:
    a.   Out of Standardisation.
    b.   Out of Selection.
    c.   Out of Specification.

37. The inspection and product reports are usually kept a minimum of _____ years.
    a.   7
    b.   10
    c.   20

38. The three actions to be undertaken when a non-conformance is identified are:
    a.   Containment, investigate Root Cause, and suggest Corrective Actions.
    b.   Investigate Root Cause, identify who is responsible, begin disciplinary action.
    c.   Immediately undertake corrective actions, validate those actions, write a report.

39. The term CCP refers to:
    a.   Common Correction Points.
    b.   Closed Circuit Programming.
    c.   Critical Control Points.

40. CCPs are defined as those points where there is:
    a.   A construction fault that results in recurrent problems.
    b.   A high probability at that juncture that something could go wrong.
    c.   An intrinsic fault in a system that cannot be corrected.

41. Recommendation for specific personnel to be licensed for certain technological skills should be made by:
    a.   Their direct supervisors.
    b.   Human Resources.
    c.   Their colleagues.

42. Large fluctuations in production are often the major sources of:
    a.   Unrecoverable costs.
    b.   Overtime.
    c.   Defects.

43. One of the ways to re-organise the production system is to:
    a.   Fix employees to specific jobs.
    b.   Adjust the speed of the machinery and equipment.
    c.   Keep the employee numbers static.

44. A change in supplier can result in:
    a.   Rejected product.
    b.   Increased cell production yields.
    c.   Less product variability.

45. In order to prevent breakdowns in machinery and equipment a company should perform inspection, cleaning and adjustment:
    a. After each use.
    b. Monthly.
    c. Annually.

46. To improve productivity:
    a. Assign unrelated tasks on a regular basis.
    b. Provide unrelated tasks to those employees in production to avoid boredom.
    c. Remove unrelated tasks from those employees working in production.

47. If production delays are not dealt with decisively then they will:
    a. Occur repeatedly.
    b. Tend to disappear on their own accord.
    c. Not be a problem as delays are minor.

48. The 4M reasons for failure do not include:
    a. Manpower.
    b. Malfunctions.
    c. Materials.

49. Production capacity can _____ be improved.
    a. Never
    b. Always
    c. Sometimes

50. Training can be judged as successful when employees can:
    a. Perform all procedures but not necessarily within standard times.
    b. Perform some procedures within the standard times.
    c. Perform standard procedures within the standard times.

Chapter 17. Process control

Maintaining
quality

Process control is about making certain that the manufacturing processes produce cell products of the required quality and quantity are in a continuous and stable manner. In reference to immunotherapy, that means that the products are tested as being safe and efficacious for each situation and respective client. Moreover, dealing with biological products means that a degree of variability must be permitted but not to the extent that consistency is compromised. There are several mechanisms for evaluating and maintaining process control.

17.1 Process Control Plan, and Process Capability Study

The manufacturing of a product involves numerous and different processes. If any of these is not functioning properly, then the quality of the product will be directly affected. Since a process usually consists of many different factors such as employees, equipment, materials, facilities and methods as discussed previously, then all of these can impact on the quality of the product. Because there are so many different factors then it can also be quite difficult to check for them unless a company has a prepared plan for doing so. A Process Control Plan is the device by which companies can check all of these factors, and identify where any problems may exist. In addition, a Process Capability Study can demonstrate if the process is actually capable of producing products of the required quality consistently.

This text describes the use of a QC process chart as an effective Process Control Plan.

A QC process chart has two functions:

a. Confirmation of results: To confirm that acceptable products have been produced, using control charts, graphs, check sheets, etc.
b. Process or factor analysis: To provide feedback for the process when abnormal values occur. In doing this it uses characteristic diagrams, scatter diagrams, stratification and other control techniques which show the cause and effect relationships. Factors that can be analysed include personnel, materials, equipment, operation methods, and environment. (See Chapter 8 on Statistical Process Control for guidelines on using these techniques.)

To prepare a QC process chart you need to take four actions:

a. Briefly describe the manufacturing process. Specify what constitutes quality in the process – what you regard as the right level of quality either in how the process functions or in the product that the process produces.

b. Establish control criteria (the range of permissible limits to quality) to measure the level of quality achieved – decide on control points in the process which can be examined to see if production is going as it is supposed to, and in a continuous and stable manner.

c. Choose a method of inspection to check that the right level of quality is being achieved, and the specific items that will be inspected or measured.

d. Prepare a written procedure to carry out this inspection.

One of the most critical aspects of process control is to ensure that a process actually has the capability to produce products of the required standard in a stable manner. Use a histogram to evaluate this. Collect statistical distribution data (mean value and dispersion) on the quality characteristics of the products when the process is operating in stable conditions. This data will allow you to estimate the probability that the process is meeting standard values.

To carry out a Process Capability Study, first exclude dispersion due to abnormal causes. Use an $\bar{x}$-R control diagram, process capability diagram (transition graph including standard values) and histograms to show the process capability graphically. Use Process Capability Index C_p or C_{pk} to indicate quantitatively the capability of the process to meet given standard values.

17.2 Application of process capability

Many companies insist on assessing the process capability in order to assess how close the process performs within the specification limits. Using standard formulas works well with most companies but with biological products, the probability of each process or production lot being consistently close to the production mean is unlikely due to the natural variability of the product. Therefore such measurements as C_p (Process Capability) and C_{pk} (Process Capability Index) have to be modified if they are to be applied to biological products. It is this adjustment which provides us with P_p (Process Performance) thus providing a useful measurement for non-centred distributions. In order to demonstrate statistical control, C_{pk} must be greater than 1.5. This is difficult with most biological products.

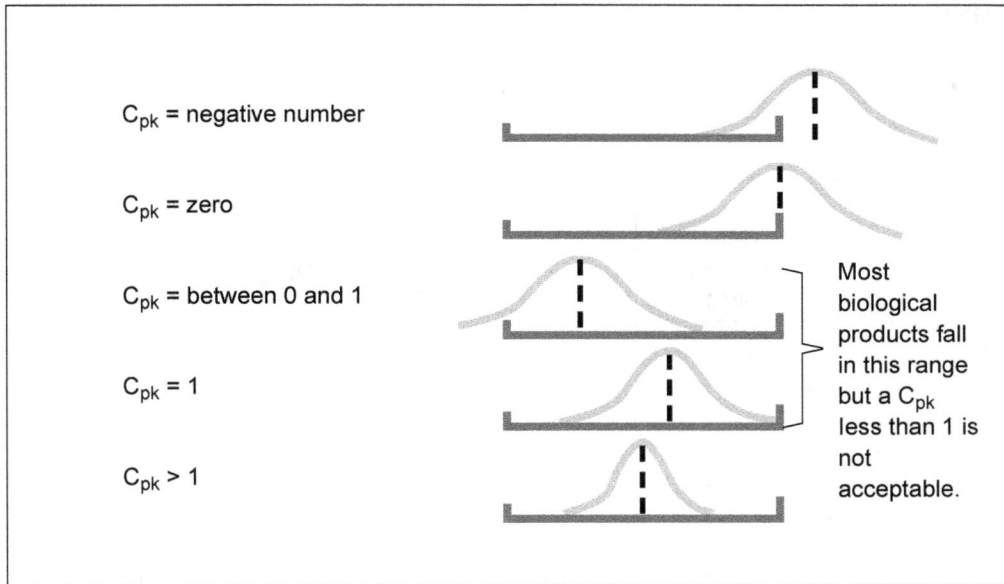

Interpreting C_{pk}

According to the C_{pk} calculations the biological production process from our example in the preceding diagram failed. It is viewed as being non-sustainable and prone to variability, therefore considered as being out of control. According to the Performance Capability Index for most biological companies, even though statistically it has been proven that 92% of their production lots would be within acceptance limits with most lying in a narrow range around the mean, their processes would still be considered to be out of control or not capable.

Therefore, when dealing with biologics, a formula that takes into consideration the fact that regulatory authorities' requirements often exceed current capabilities is required if those companies are to have a measure of process control and process capability. Since most product titres are at the low end of the limits, setting alarm and action levels will be different at either end of the spectrum unlike normal product companies that manufacture very standardised and uniform products. Process capability in a biological company is a function of stability, efficacy and safety which C_{pk} and similar calculations were never intended to assess and this has to be taken into account.

A new theorem is required:

a. Taking into account the non-centered distribution of biological products and increased variability, then a specific formula is necessary for products of this nature.
b. Since a formula for biological processes, representing yields, titres, potencies, etc. does not exist, then one had to be developed for this purpose.
c. For lack of a better name I refer to this as the Goldenthal Theorem.

STEP ONE:

$$Z_1 = \left(\dfrac{\overline{X} - \text{Lower specification limit}}{\sigma} \right) \qquad Z_2 = \left(\dfrac{\text{Upper specification} - \overline{X}}{\sigma} \right)$$

In Step One, I refer to the calculation of a **Z** factor. **Z** is the individual factor that takes into consideration that actual values for biological products may have a mean that is often lower than the acceptance range mean. The acceptance range mean is usually determined by National Regulatory Authorities that have a fixed range that is theoretically maximised for potency and often unattainable in real-life production. As such, this results in most biological products having a much smaller range to the left of the theoretical (acceptance mean) than on the right. By dividing these separate ranges by the standard deviation I can calculate the number of deviations possible within the shifted curve on each side of the mean. This factor will be smaller on the left side of our curve for most biological products than the right which acknowledges the much tighter limitations on biological product values at the lower end of the acceptance criteria.

STEP TWO:

Potency Capability Index
(Goldenthal Theorem)

$$T_{pk1} = \left[(C_p * Z_1) \right] \qquad T_{pk2} = \left[(C_p * Z_2) \right]$$

In Step Two the **Z** values are multiplied by the Process Capability Factor and resulting values are then read no differently from the process capability score. This Potency Capability Index tells us that when we achieve values below the mean for the product the process is capable of controlling its variability but is even better when we produce a lot above the mean value, which is a more realistic representation of biological products. This process control determination can be applied to products where potency, yields and titres are the determining factors of whether or not a company can produce a sustainable product.

17.3 Operation standards

Two factors are essential if a manufacturing process is to achieve the target quality efficiently. There must be operation standards that describe the best way to carry out the operations that make up the process, and there must be knowledgeable, skilled and motivated operators who will follow the standards.

This brief text highlights:

a. The need to review standards when there are changes in the production process.
b. The importance of training the operator.

When you change any part of the production process, review the operation standards and, if necessary, revise them:

a. Stipulate and document the procedures to be followed, because there are so many factors that can influence quality.
b. Ensure that the change will not cause any unexpected problems.
c. After the change has been made, gather and analyse the relevant data in order to check if the goals of the change have been reached, and how quality and productivity have been affected.
d. Remove any operation standards that are no longer being used. These should be archived properly according to the SOP on archiving documents.
e. Conduct periodic reviews to see how the new standards are being implemented, and how effective they are.

One of the factors in the production process which greatly affects quality is the operator. Because the operator is human, there will always be some variance in operations. An operator may carry out the same operation a little differently each time, and when the same operation is carried out by different operators, there is yet more risk of variance. What is a priority is to keep this variance to a minimum. Managers can try to minimise this variance in three ways – by educating, training and motivating operators to remain consistent.

17.4 Dealing with out-of-control events and non-conforming products

A primary purpose of process control is to identify anything that may be going wrong in the manufacturing process, and correct it.

Problems may be indicated by:

a. Out-of-control events: Processes that are not functioning as they are intended to. The problems may lie in operators, equipment or materials.
b. Non-conforming products: Products, reagents or raw materials that are not of the required quality.

The source of any out-of-control events may be related to operators, equipment or materials. Basically they are any events in the process, or outputs of the process, that do not meet the specifications (or criteria) that govern the process. If the process goes according to plan there should be no occurrence of "out-of-controls". The function of process control is to detect them, investigate why they occur, and make whatever improvements are necessary to prevent them occurring again.

To detect and deal with out-of-controls take the following actions:

a. Take emergency action to deal with the immediate situation.
b. Investigate the root cause of the out-of-control event.
c. Inform related departments of what is happening.
d. Take action to prevent further out-of-control events occurring.
e. Confirm that this action has been effective in resolving the problem.

Control chart: An essential tool in investigating the cause of out-of-controls is a control chart. Use control charts to examine the process and determine whether it is in a stable condition in terms of both quality and quantity.

Out-of-control report. Write up reports on out-of-controls and pass them onto Quality Assurance and senior management:

a. Describe the process situation and give details of:
    i. The out-of-controls.
    ii. The investigation that was carried out.
    iii. The causes that have been identified.
    iv. Whatever countermeasures have been taken and their effectiveness.
    v. Any actions that are still to be taken.
b. Record details of the actions and opinions of the department in charge of handling of out-of-controls.
c. Specify the date and person in charge of implementation at each step, from detection of out-of-controls through preventive action and confirmation of its effectiveness.
d. Establish handling criteria for the report.
e. Ensure that the non-conformance report has its own identification number.
f. After countermeasures have been taken the process averages or dispersion may change. If this happens, revise the control characteristics and control lines.

Non-conforming products or non-conformities, are products, parts, and materials, both finished and unfinished, which are found, usually on inspection, not to meet the required quality criteria.

To deal with non-conformities take the following actions:

a. Clarify who has the responsibility and authority for handling them.
b. Segregate them as soon as possible from conforming products and identify/mark them appropriately. Then remove them from the manufacturing process.
c. Record their occurrence.
d. Check whether there were any problems with previous manufacturing lots.
e. Decide how to dispose of them. Depending on the nature of the non-conformity, they may be used as they are, reworked, regraded, or scrapped. This decision should be taken in accordance with a predetermined procedure by those with responsibility and authority. Depending on the circumstances, a review committee may discuss the issues involved.
f. Once this decision has been made, act on it as quickly as possible.
g. When non-conforming products have been reworked they must be re-inspected.
h. Check both the inspection items where the original problem arose, and any inspection items that could have been affected by the reworking.
i. Implement recurrence prevention measures.
j. Record the results of these measures in a report. These records can then be used as basic data to help analyse the cause of any future non-conforming products and decide on appropriate action.

To ensure that effective countermeasures (and, when necessary, emergency actions) are taken to eliminate out-of-controls and non-conforming products, it is important to have all the relevant data on hand and to manage it properly.

When collecting, recording and using this data you need to:

a. Specify the objectives that the data is to be used for. Different objectives require different types of data and different methods of collection.
b. Plan a system for collecting and documenting the data.
c. Decide where each type of data is to be gathered and how often.
d. Assign people to collect, record and process the data.
e. Summarise the data in a chart.
f. Use a QC process chart to clarify the cause and effect relationships in the data.
g. Keep data which indicates the time sequence so that you can identify when the non-conforming product or out-of-control occurred and what caused it. This includes such details as material lot, equipment and personnel.

Data from inspections and quality checks can be shown to auditors from outside a company as evidence that improvements have been implemented, when customer audits or product liability issues arise.

17.5 Early control system and foolproof operations

Out-of-control events and non-conforming products are most likely to occur when:

a. New equipment and techniques are introduced, often at the start of production of new products. An early control system will detect many such problems.
b. Operators make careless mistakes. Foolproofing operations can reduce operator mistakes significantly.

An early control system will detect a variety of problems at an early stage in the production of new products, help to solve them quickly, and stabilise the new manufacturing process. Fortunately, there are a variety of computer software programmes available that perform this early control system function, raising alarms and flags when a process trend is detected that suggest a problem is developing. These automated control process programmes produce a variety of charts and it is up to the individual company to decide which software suits their needs best.

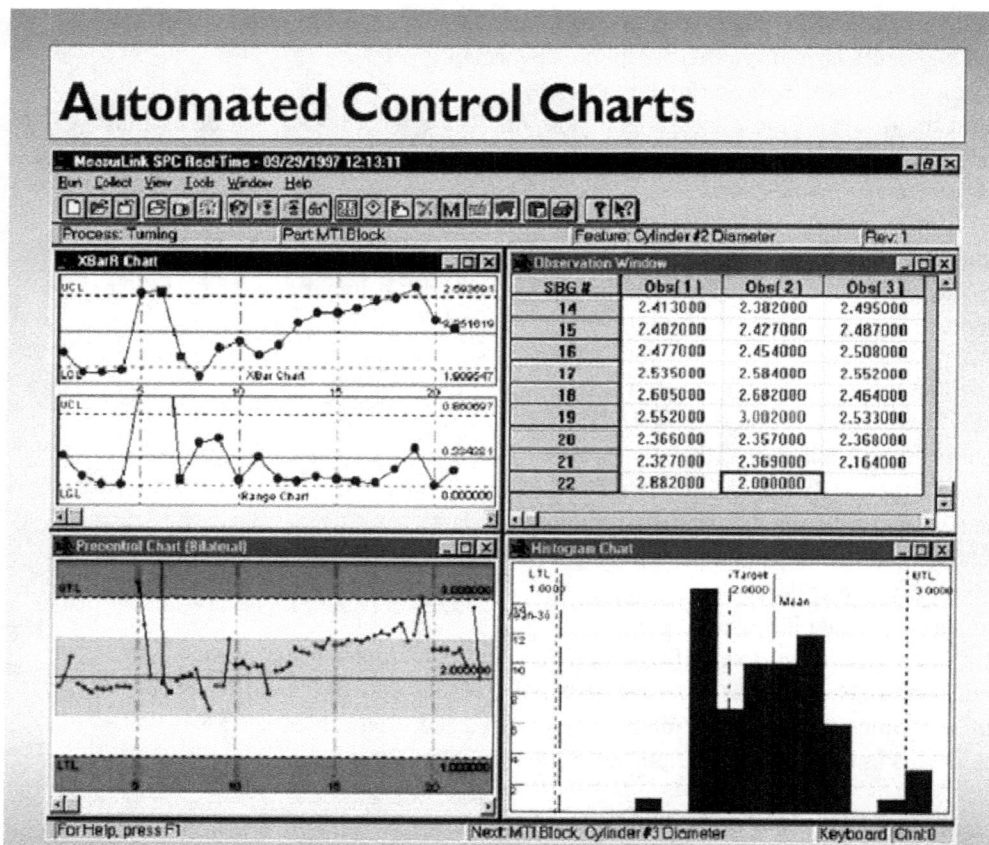

To establish an early control system take the following actions:

a. Decide in advance a period during which countermeasures may be introduced to correct any problems in the new process. The length of this period will depend on the product, the equipment being introduced, and how new the manufacturing technique is.

b. Analyse in detail any factors in the current situation that could affect the new process and product quality, so that latent defects can be recognised and removed as soon as possible.

c. Form a project team to quickly solve any problems. Decide how many people are needed and what skills they should have.

d. Use marks and other symbols to differentiate problem operations and equipment from other processes, and focus control measures on them.

The best way to foolproof operations is by establishing a process in which mistakes are less likely to occur: make various adjustments to operation methods, and to the way that parts, materials, equipment, and tools are handled.

There are two ways to foolproof operations:

a. **Preventive:** Design operations so that there is no possibility of mistakes being made no matter who is the operator. This method may include the elimination of certain tasks, replacing them with risk-free procedures, or making them easier to perform.

b. **Reductive:** Try to spot mistakes as soon as they occur, so that their effects can be contained or eliminated. This method involves two stages: defect detection where mistakes are discovered, and effect mitigation where the effects of the mistake are minimised or eliminated.

Recommended reading

1. Coughanowr, D.R. *Process Systems Analysis and Control*, 2nd ed. New York: McGraw-Hill, 1991.

2. Goldenthal, A.E. Setting alert and action limits for vaccine titres in the biologics industry, 2012. https://www.academia.edu/5308442/, accessed 29 August 2016.

3. Luyben, W.L., & Luyben, M.L. *Essentials of Process Control*. McGraw-Hill, 1997.

4. Seborg, D.E., Edgar, T.E., & Mellichamp, D.A. *Process Dynamics and Control*. John Wiley & Sons, Inc., 1989.

5. Smith, C.A., & Corripio, A.B. *Principles and Practice of Automatic Process Control*, 2nd ed. New York: Wiley, 1997.

Self testing multiple choice questions

1. The key phrase in process control is to:
    a. Produce only quality.
    b. Produce in a continuous and stable manner.
    c. Produce quantity only.

2. Which of the following is a function of QC process chart?
    a. Confirmation of plans.
    b. Confirmation of results.
    c. Confirmation of financial loss.

3. Control points will show whether production is:
    a. Continuous and flexible.
    b. Stable and controlled.
    c. Continuous and stable.

4. To prepare a QC process chart you will need to:
    a. Establish control criteria to measure the level of quality achieved.
    b. Choose a method to confirm that quality has been achieved.
    c. Choose a system for changing the process.

5. When dealing with biological products:
    a. Variability cannot be tolerated.
    b. A degree of tolerance of less than 1% is permitted.
    c. A degree of variability must be permitted.

6. A Process Control Plan allows a company to:
    a. Check all equipment and materials used only.
    b. Check all factors and identify where problems may exist.
    c. Check all control levers employed by operators only.

7. Which of the following can be used to confirm that acceptable products have been produced?
    a. Cost analysis.
    b. Control charts.
    c. Marketing graphs.

8. Which of the following are used to provide feedback for the process when abnormalities occur?
    a. Characteristic diagrams.
    b. Multiplication.
    c. Stratification.

9. Key factors that can usually be analysed to provide feedback when abnormal values occur include:
    a. Personnel and equipment.
    b. Equipment and cost cutting.
    c. Personnel, especially management.

10. A Process Capability Study is used to evaluate whether a process has the capability to produce products of:
    a. The required standard in the fastest time possible.
    b. The highest standard whatever time it takes.
    c. The required standard in a stable manner.

11. A Process Capability Study collects and uses statistical distribution data on the quality characteristics of the products manufactured when the process:
    a. Is operating in stable conditions.
    b. Is stopped.
    c. Is operating under maximum pressure.

12. The purpose of this study is to estimate the probability that the process meets:
    a. Company expectations.
    b. Standard values.
    c. Operator values.

13. Control charts are valuable because:
    a. They prevent operational changes.
    b. They can confirm that acceptable products have been produced.
    c. They provide marketing information that satisfies customers.

14. There will always be some variance in operations because:
    a. Operators are difficult to train.
    b. Operators are human.
    c. Operators are not motivated.

15. Factors that can be analysed using control charts include:
    a. Personnel, materials, equipment, methods and environment.
    b. Environment, suppliers, costs, equipment and breakdown.
    c. Methods, personnel, environment, regulations and methodology.

16. A way of assessing whether a process has the capability of producing stable product is:
    a. Use of a pie chart.
    b. Use of a histogram.
    c. Use of Venn diagrams.

17. A control chart is used to examine a process and determine whether it is in a stable condition terms of:
   a.   Quality and efficiency.
   b.   Quality and quantity.
   c.   Quality and process.

18. The use of C_p and C_{pk} is to indicate that the process meets the standard values:
   a.   In a qualitative manner.
   b.   In a quantitative manner.
   c.   Both a and b.

19. Actions for the control of non-conforming products, do not include the following:
   a.   Separate non-conforming products as soon as possible from conforming products and identify them appropriately.
   b.   Confirm that there are no problems in the inspected items and therefore all other products must be acceptable.
   c.   Record the results of any control measures taken in a report.

20. C_p and C_{pk} do not work well with biological products because of the:
   a.   Inherent variability of biological products.
   b.   Inherent instability of biological products.
   c.   Impact of temperature on biological products.

21. Which of the following actions for collecting, recording and using data is incorrect:
   a.   Specify in advance the objectives for which the data will be used as different uses requires different collection methods.
   b.   Stipulate the critical points where data is to be collected, once and by whom.
   c.   Keep data indicating time of collection so that operators can be identified at the time any non-conforming product was produced.

22. In terms of standard process control, variability of the product is viewed as:
   a.   Being out of control.
   b.   Being completely acceptable as uniformity is not a requirement.
   c.   Being out of control unless it is a biological product.

23. In comparison to regulatory standards, most biological products are at the _____ of the acceptable limits.
   a.   Mean
   b.   Low end
   c.   High end

24. There are _____ ways by which to foolproof operators.
   a.   Two
   b.   Three
   c.   Five

25. Operations need to be foolproofed in order to prevent:
    a. Interference by other departments.
    b. Careless mistakes by operators.
    c. Malfunctioning of machinery.

26. One of the ways of foolproofing operations is "preventive" which includes:
    a. Eliminating certain non-essential operations.
    b. Reducing the effects of mistakes.
    c. Making operations more difficult to carry out.

27. Any improvements that are made should be implemented:
    a. Only in the workplace where the problem arose.
    b. Only in the department where the problem arose.
    c. Throughout a company.

28. To achieve the targeted quality efficiently, there are _____ factors essential to the manufacturing process.
    a. Two
    b. Three
    c. Five

29. Operational standards that are no longer in use should be:
    a. Destroyed immediately.
    b. Archived.
    c. Left in the laboratory for reference.

30. One of the major factors affecting quality is:
    a. Instruction from senior management.
    b. Interference from other departments.
    c. The ability of the operator/technician.

31. Managers can attempt to reduce the variance of operators in _____ ways.
    a. Two
    b. Three
    c. Five

32. The factors to reduce variance of the operators do not include:
    a. Motivation.
    b. Training.
    c. Penalisation.

33. Options for any products that are identified as non-conforming include:
    a. Reworking.
    b. Being automatically discarded.
    c. Ignoring the non-conformance as it is likely a one-off.

34. Out-of-control events are most likely to occur when introducing:
    a.   New equipment but the same technicians.
    b.   New technicians but the same old equipment.
    c.   New equipment and new technicians.

35. QC process chart has _____ functions.
    a.   Two
    b.   Three
    c.   Five

36. To prepare a QC process chart it is necessary to take _____ actions.
    a.   Two
    b.   Four
    c.   Six

37. The Goldenthal Theorem takes into consideration:
    a.   The right of centre distribution of biological products.
    b.   The non-centred distribution of biological products.
    c.   The left of centre distribution of biological products.

38. The Goldenthal Theorem can be used for biological processes measured in:
    a.   Yields, titres and potencies.
    b.   Titres, percentage weights, potencies.
    c.   Yields, cells per field, titres.

39. If a process goes according to plan there should be:
    a.   Only the occasional occurrence of out-of-controls.
    b.   Only technician sourced out-of-controls.
    c.   No occurrence of out-of-controls.

40. Every non-conformance report must have:
    a.   Its own unique identification number.
    b.   No more than one event recorded per report.
    c.   At least one week grace period before being submitted.

41. After countermeasures are undertaken, control charts may need to be revised because:
    a.   It makes it much easier to see when the countermeasures were implemented.
    b.   Each change in process should be accompanied by a new chart.
    c.   The process averages or dispersions may have changed.

42. When non-conforming products are reworked, they must be:
    a.   Held in quarantine until senior management instructs their release.
    b.   Re-inspected as if they were new products.
    c.   Released but do not require retesting.

43. Control criteria are based on:
    a. The range of production lot quantities
    b. The range of permissible limits to quality.
    c. The range of allowances for error.

44. The conduct of an inspection for non-conformance requires:
    a. There be an office of non-conformance established.
    b. Surprise inspections that are randomly set.
    c. There be a written procedure in place first.

45. In order to demonstrate statistical control for biological C_{pk} must be:
    a. Equal to 1.5
    b. >1.5
    c. <1.5

46. When the calculated mean of the product is far to the right of the acceptable range:
    a. Then C_{pk} will be a negative number.
    b. Then C_{pk} will be between 0 and 1.
    c. Then C_{pk} will be immeasurable.

47. The Z factor of the Goldenthal Theorem is an acknowledgement of:
    a. The inability to apply process control techniques to biological products.
    b. Much looser limitations of the biological product values over the entire range of the acceptance criteria.
    c. Much tighter limitations of the biological product values at the lower end of the acceptance criteria.

48. For a company to have successful process control, operators must be:
    a. Knowledgeable, skilled and motivated.
    b. Fast, consistent and compliant.
    c. Achievers, available and accurate.

49. Non-conforming products are:
    a. Both finished and unfinished products found not to meet the quality criteria.
    b. Both finished and unfinished products, as well as parts, found not to meet the quality criteria.
    c. Both finished and unfinished products, as well as parts and materials, found not to meet the quality criteria.

50. Data from a process control inspection should be shown to regulatory inspectors:
    a. Any time they request it.
    b. Only to provide evidence that improvements and corrective actions have been implemented.
    c. Never, as this is internal Quality Assurance information and they are not entitled to see it.

Chapter 18. Auditing external suppliers

Vendor selection

The quality of the product being manufactured is directly influenced by the quality of a company's external suppliers. Therefore it is imperative that the parts, reagents, and raw materials provided by these external suppliers is to a standard or level of acceptance that supports the quality performance standards attained by the company.

18.1 External suppliers

Even if there are only a few raw materials, reagents and components received from external suppliers, these will have a major impact on the quality and performance of a company's products.

Choosing an external supplier:

a. Selection of an external supplier must follow established procedures defined in an SOP without exception.
b. Preference should be given to suppliers with GMP accreditation.
c. Preference should be given to suppliers that have a respected track record in the industry.
d. An in-depth audit of suppliers must be initially performed, again at the time of acceptance and then every two years subsequently.
e. All terms of the supply arrangement are to be drawn up in a contract.
f. Identify and approve a secondary supplier for backup and emergency situations as defined in the SOP.

18.2 Auditing the external supplier

Before the final selection of an external supplier, carry out an in-depth audit to ensure that they have the capability to deliver goods to the agreed quality standard, on time, and have the ability to respond to any changes that may be required.

Use quantitative assessment methods to assess their quality capability. Check that they can respond immediately to any changes that you may make in your product specifications or design. Ensure that the supplier can respond to quality audit requests, that their employees have the necessary job training and that they continually increase their technological base of manufacturing.

Basic procedures for selecting an external supplier:

a. Collect information about potential external suppliers, both domestic and foreign.
b. Confirm that the supplier can meet the target purchase price and this price is sustainable for the long term.
c. Ensure that the supplier meets any domestic or international regulations that are required.
d. Prepare a questionnaire/checklist and have candidate companies complete the list and submit any written documents that describe their operations.
e. Carry out an in-depth audit.
f. Ensure that a company is capable of delivering what it claims.

The audit process:

a. Production resources: Their equipment, personnel, warehouse space and layout, processing methods, etc.
b. Quality control: Their inspection and control systems must include:
    i.   Their system for inspecting their own incoming supplies.
    ii. Their lot control and inventory control.
    iii. Their testing equipment and their management of it.
    iv. Their control of the manufacturing process.
    v.   Their finished goods inspection.
c. Investigate any prior deviations or audit findings and assess their capability to handle abnormalities through CAPA and internal change control:
    i.   Assess their ability to detect abnormalities.
    ii. Assess their ability to respond immediately to abnormalities.
    iii. Assess their reporting of abnormalities.
    iv. Assess their ability to identify and record causes of abnormalities.
    v.   Assess their ability to decide on countermeasures, how they check the results of those countermeasures.
d. Check their organisational chart showing the relationship between various individuals, management controls and whether they have enough people in the right places.
e. Generate a QC process chart showing how their production process is controlled.
f. Be aware of their Process Capability (C_p) Index for important quality characteristics.
g. Clarify that their Quality Assurance department is in charge of handling quality issues and changes in the production process and is independent of management.
h. Analyse the production activity and traceability of raw materials.

18.3 Inspect incoming supplies

It is essential that incoming supplies are always examined at the time of delivery to ensure they have not been damaged and still meet the company's requirements and specifications:

Required SOPs and forms:

a. Establish an SOP for performing the incoming inspection.
b. Prepare an inspection evaluation form for delivered items.

To establish the SOP:

a. The Quality Control department produces a draft plan for standards for incoming inspection.
b. The Quality Assurance and Purchasing departments meet to discuss the plan and make any necessary changes.
c. The final SOP is submitted and signed off for approval.
d. The results of the incoming inspections are monitored and, if necessary, the standards are revised as necessary.

The form should include:

a. The inspector's name and qualifications.
b. The name of the person in charge.
c. The date of inspection.
d. The inspection data: Compliance with the required specifications, quantity, selection method for inspection, and evaluation results.
e. The inspection level (ie. cursory, limited, full inspection).

This information should be entered in the form at the time of delivery and not afterwards. Any quality checks must be performed in real time in order to be valid. If any non-conformities are found, then the involved departments and the supplier are to be contacted. Keep the records on file so that any changes in quality while the items are in storage can be traced.

18.4 Contacting the supplier

Times to contact the supplier:

a. Communicate directly with the person in charge of quality at the supplier.
b. Keep the supplier notified about any changes in the allocation of work, changes in the allocation of responsibility, and technological development.
c. Inform the supplier regarding any analysis of defects performed on their products at the company's laboratories.
d. Inform the supplier of any changes in the specifications of raw materials.

18.5 Monitoring external suppliers

Use the supplier's SOPs and QC process charts to confirm that they conduct each process correctly and meet the company's requirements. Establish quality verification points (predetermined points where quality can be verified) when auditing the supplier.

The procedure is as follows:

a. Confirm the supplier is using the most recent operation standards and the QC process chart.
b. Confirm that each process and each piece of work is being carried out as specified in the most recent SOP version.
c. Match the actual work to the standards and the chart, and report as abnormalities any deviations and differences.
d. Have the supplier take corrective actions wherever and whenever appropriate.
e. Confirm that the production process is in accordance with the flow in the QC process chart.

Evaluation of goods delivered:

a. Regularly evaluate the level of quality (acceptance rate by lot, defect rate by lot), the rate of late delivery (number of days late by lot), and the quantity delivered (by lot and total).
b. Evaluate your supplier's handling of complaints:
    i. The number of days required to resolve complaints.
    ii. The details of complaints.
    iii. The procedure for resolving complaints.
    iv. Actions taken to prevent similar complaints being made again.

Recommended reading

1. Gallegos, F., Manson, D.P., Senft, S., & Gonzales, C. *Information Technology Control and Audit*, 2nd ed. Auerbach Publications, 2004.

2. Global Harmonization Task Force. GHTF/SG3/N17:2008 Quality Management System – Medical Devices – Guidance on the Control of Products and Services Obtained from Suppliers.

3. International Pharmaceutical Excipients Council (IPEC). Good Manufacturing Practices Audit Guideline for Bulk Pharmaceutical Excipients, 1998. http://ipec-europe.org/

4. ISO 10011-2:1991 Guidelines for auditing quality systems; part 2: Qualification criteria for quality systems auditors.

5. ISO 10011-3:1991 Guidelines for auditing quality systems; part 3: Management of audit programmes.

6. PQG: "Pharmaceutical Auditing," Monograph, 1992, Institute of Quality Assurance.

7. WHO Technical Report Series, No. 823: WHO Expert Committee on Specifications for Pharmaceutical Preparations, 1992.

Self testing multiple choice questions

1. A company's own quality depends on _____ of materials and supplies from external vendors.
   a. The quantity
   b. The quality
   c. The shipping

2. The key procedure for selecting the right vendor is to:
   a. Obtain written documents that describe their manufacturing operations.
   b. Obtain opinions regarding the vendor's quality from other customers.
   c. Conduct an in-depth audit of the vendor's company.

3. It is important to choose vendors that have an established performance record when:
   a. Purchasing items that can change in quantity over time.
   b. The priority is price rather than quality.
   c. A quality record must be maintained during the production process.

4. The audit of external vendors should focus on their ability to:
   a. Deliver goods of the highest possible quality at the time they are required.
   b. Deliver goods on time but not necessarily to the highest quality.
   c. Respond to any changes required with goods of the highest quality, on time.

5. An investigation of the vendor's production capability includes:
   a. The maintenance and operation of their equipment.
   b. The performance of their staff because of salary incentives.
   c. Any documents containing intellectual property processing methods.

6. An investigation of how the vendor handles abnormalities includes their capability to:
   a. Eliminate all abnormalities from occuring.
   b. Respond immediately to deviations so they do not need to be documented.
   c. Prevent abnormalities arising.

7. An investigation of the vendor's manufacturing response capability includes the ability to:
   a. Develop new materials, equipment and technology independent of requests.
   b. Analyse the causes of defects and take the necessary countermeasures.
   c. Change the original specifications of products without a request.

8. If a non-conformity is documented during an inspection audit of the vendor, then:
   a. The inspector waits until his/her report is completed before informing the vendor of it.
   b. The inspector arranges a meeting with the vendor's CEO to inform the vendor of it.
   c. The inspector informs the vendor's QA at the exit meeting of the non-conformity.

9. Vendors must inform their customer's QA immediately of any changes:
   a. In the allocation of responsibility for the ordered product.
   b. In their internal complaint resolution system.
   c. In currency exchange rates as applied to their products.

10. Before carrying out a vendor audit, the inspector should:
    a. Prepare in advance an audit schedule of those items and areas to be reviewed.
    b. Wait until arriving at the vendor's premises before deciding on what to inspect.
    c. Request that the vendor provides a list of items and areas that will be inspected.

11. Ordering only a few raw materials and reagents from external suppliers will:
    a. Still have a major impact on the quality and performance of a company.
    b. Not have any impact on the quality and performance of a company.
    c. Only affect quality standing of a company if the materials are inferior.

12. The selection process of a vendor must:
    a. Follow established procedures defined in an SOP.
    b. Be made from at least three possible suppliers for a proper decision.
    c. Be based primarily on having the best price.

13. If a vendor has GMP accreditation, then they are a _____ supplier.
    a. Likely
    b. Possible
    c. Preferred

14. For back up and emergency situations a company should always:
    a. Have its own capacity for manufacturing the needed supplies.
    b. Have a secondary supplier.
    c. Have a contingency plan where product can be delayed.

15. In order to assess a vendor's quality capability a company should use:
    a. Objective methods.
    b. Quantitative methods.
    c. Subjective methods.

16. When negotiating a contract with the vendor, it is important that the price is:
    a. Sustainable for the long term.
    b. Fluctuating so a company is not tied into a long term arrangement.
    c. Negotiated constantly so a company can get the best deal.

17. The vendor must be able to meet:
    a. Any domestic or international regulatory standards.
    b. Any domestic standards as they are not responsible for international sales.
    c. Minimal standards as they are not the manufacturer of the final product.

18. It is important that the vendor's Quality Assurance department:
    a. Reports directly to management so it can make any changes to the process.
    b. Is in control of making the changes in the production process.
    c. Is in charge of handling changes in the production process.

19. Incoming supplies must always be examined:
    a. Just prior to being used.
    b. Immediately upon receipt.
    c. Only when they are used for final product.

20. A company SOP on external suppliers should be formulated through the cooperation and input from:
    a. QC, QA and Marketing.
    b. QC, Marketing and Purchasing.
    c. QC, Purchasing and QA.

21. When evaluating a vendor's ability to handle complaints there are _____ criteria.
    a. Three
    b. Four
    c. Five

22. Following the initial audit, subsequent audits of the vendor should be performed:
    a. Every year afterwards.
    b. Every two years.
    c. Every five years.

23. It is important that the initial audit of a vendor be performed:
    a. Only at a cursory level.
    b. Only regarding issues related to the product ordered.
    c. At an in-depth level.

24. One of the criteria for selection of a vendor is that they have programmes in place for:
    a. Continually increasing the technological base of their manufacturing.
    b. Social functions to improve the morale of a company.
    c. Cost reduction in their manufacturing process.

25. The primary reason of selection of vendors is not made solely from submitted responses is because:
    a. It is a requirement of the process that they can prove what they claim.
    b. Most vendors will lie about their capabilities.
    c. It provides an excuse for company auditors to have free trips.

26. The reason that the external supplier's organisational chart is checked on the audit is:
    a. To see whom is in charge so that the auditor does not waste time.
    b. To see that they have enough people in the right places for quality purposes.
    c. To see that they have the right departments listed on the chart.

27. It is important that external suppliers can show on the audit documentation:
    a.   Why it is not necessary for them to audit their suppliers.
    b.   The price they have paid for their materials.
    c.   The traceability of their raw materials.

28. When inspecting incoming supplies, the form should have a space for listing:
    a.   The weather conditions that day.
    b.   The level of inspection.
    c.   The name of the delivery service.

29. If analysis of incoming supplies shows any abnormalities or defects then a company:
    a.   Should inform the vendor of the results immediately.
    b.   Need not inform the vendor but only switches suppliers.
    c.   Should keep the information secret until the time of the next contract.

30. Quality verification points are:
    a.   Those points made by an auditor to help the vendor improve its quality rating.
    b.   Awarded points that reduce the audit requirements.
    c.   Predetermined points for an audit where quality can be verified.

Chapter 19. Post-treatment follow-up

ADRs and pharmacovigilance

The company's responsibility for its products does not end once the treatment is performed and the customer/patient leaves the hospital. The success of a company depends, above all, on whether the customers/patients are satisfied with the therapeutic procedure. No matter how good a company's quality and inspection systems are, some faulty products will likely get through to the customers/patients. The regulations stipulate that therapeutic companies must follow their customers/patients post-treatment in order to assess any adverse drug reactions (ADRs), efficacy of treatment, and any secondary problems.

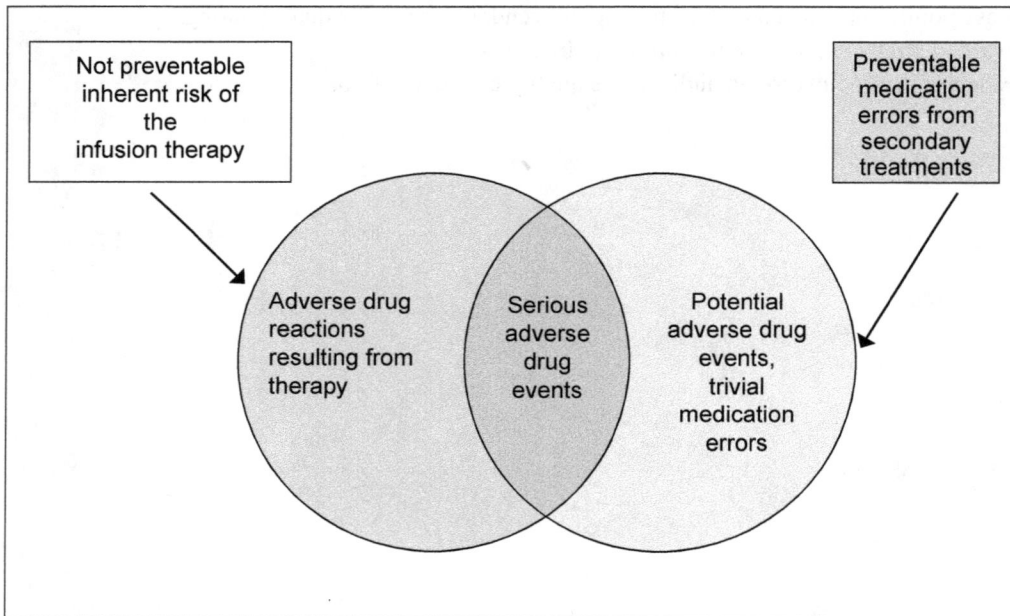

Relationship between immunotherapy and ADRs

19.1 No guarantees

There are no product guarantees when the product is immunotherapeutic. Since every patient can react differently, what is successful in one may only have moderate remission in another and absolutely no effect in a third case scenario. Worst case scenario is an adverse immune response similar to Graft vs. Host and the patient unexpectedly is placed into a life threatening situation. This is why it is important that a company follows up on its patients in order to determine the success and failure ratios for particular cell products, therapy types and treatment regimens. Overall, the success rate is determined by the ratio of remissions, partial recoveries and non responders against the specific illnesses for which the patient was being treated. Because of the current unknowns in immunotherapy, there is no way of knowing when adverse reactions may occur. Because of this, the investigation into the patient's health post-treatment becomes a life-long responsibility.

19.2 Customer claims

Quality Assurance needs to establish an organisational structure within the company to which patients can present claims about failed treatment. Quality Assurance must deal quickly and efficiently with any such claims: to receive the claim; to analyse the failed treatment; to take corrective measures and recurrence prevention measures; and to keep a record of all essential data. These are all requirements under the GTPs and must be strictly adhered to.

Take the following actions:

a. Set up an organisational structure:
   i. Assign a person to be in charge of claims and include a written job description to describe their role.
   ii. Decide on the procedures to be used for quick corrective actions and the process for distributing the recorded claim information to all the relevant people.
   iii. Clarify who is responsible for replying to the customer about the processing of the claim.
   iv. Simplify liaison between departments for effective communication.
   v. Prepare a claim handling process flow – a document which clearly describes the procedures for investigating and processing claims, and for repairing, replacing and disposing of any failed cell therapy products.
b. Standardise the flowchart showing the procedures for processing claims. These procedures should include:
   i. Record the receipt of a claim.
   ii. Maintain up-to-date information on the status of reported claims.
   iii. Record the analyses of failed therapies.
   iv. Record the corrective and recurrence prevention measures taken in each process and the names of those responsible. (Corrective measures are taken to deal with the specific product and customer; recurrence prevention measures are taken to prevent similar problems arising with other products.)
   v. Record the history of the manufacturing and storage of the product.
c. Prepare a form for reporting the processing of claims:
   i. Standardise the format of the report with spaces for the contents of claims, the analysis of results, and the actions to be taken.
   ii. Include columns that distinguish between corrective measures and recurrence prevention measures.
   iii. Include in the report the names of the responsible department and the person responsible for giving approval.

    iv. Ensure that the final written report of the claim covers causes, events and countermeasures.

d. Maintain a ledger (hardbound or computerised tamper-proof log) of reported claims covering:

    i.   Receiving claims.

    ii.   Checking the progress of claims.

    iii. Reporting each process.

Present these procedures clearly:

a. Ensure the standardised flowchart is easy to understand.

b. Make it clear who is in charge of each process and the time-limit for responding to the claim, and who is responsible.

c. Specify clearly the activities and data that are to be entered in the record forms.

d. Establish a communication method that will allow everyone to receive the final results of the processing of a claim.

Establish the following:

a. Standardise a system that will feed the claim information back to the original production department and affiliated hospital.

b. Establish a communication route between the Quality Assurance department and the production department.

c. Decide the department and person responsible for, and the person in charge of, conducting the failure analysis, and for reviewing and drafting recurrence prevention measures. Decide also who is to be responsible for making the final decision, and what points in this decision are to be recorded.

d. Set up a system for recording the failure analysis.

e. Record and control the effect of the recurrence-prevention measures.

f. Communicate the final conclusion of the recurrence prevention measures to all relevant departments; then decide who is to be responsible for the follow-up.

g. Standardise the registration of all records and their distribution to all those concerned.

h. Maintain these records in the standard format, and archive them properly.

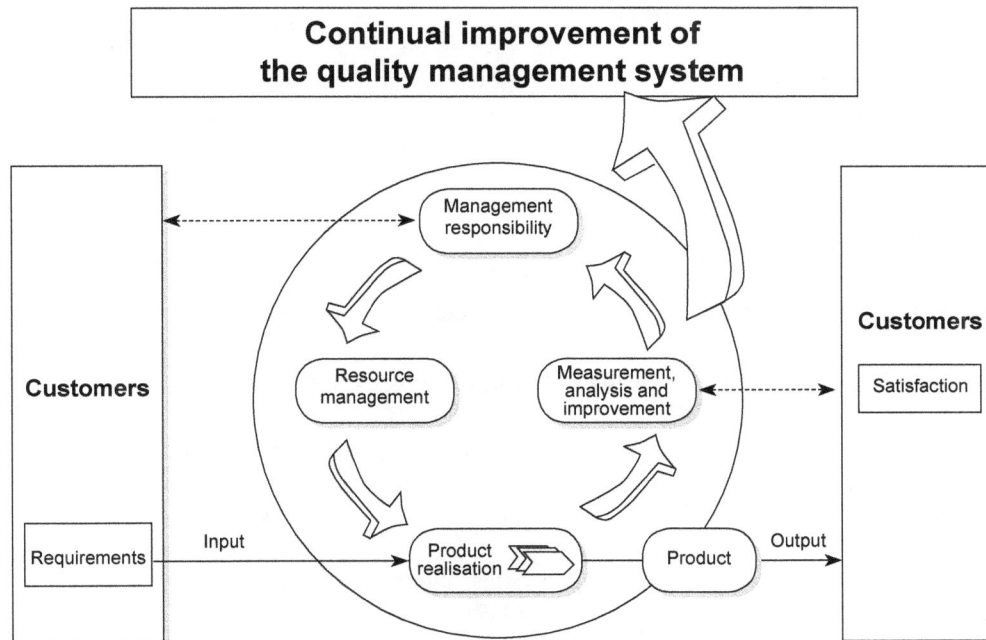

19.3 CEO responsibilities

The CEO should take an active interest in the after-treatment service, since he/she has an important quality assurance-like function that determines follow-up treatments, reputation in the international market, and ultimately affects future sales.

The CEO should:

a. Check the after-treatment feedback on therapeutic products that have been infused, and periodically give instructions to the production department based on this feedback.
b. Determine which treatments are most effective, moderately effective and least effective and base on this information develop the annual business plan as to which direction the company will move in the future and where it will concentrate its production capabilities.
c. Follow up on the progress of improvements and countermeasures implemented post complaints.
d. Whenever there are claims because of product failures, ask the department responsible to provide an explanation of the technical problem, and assess the feasibility of any corrective action needed to eliminate any future claims.
e. Facilitate the establishment and implementation of improvement plans for current problems.

19.4 Reasons for pharmacovigilance

An ongoing ADR-monitoring and reporting programme can provide benefits to the organisation, the affiliated hospitals, other health care professionals, and patients. Establishment of a quality programme of this nature is what is commonly referred to as pharmacovigilance.

The benefits include the following:

a. Providing an indirect measure of the quality of pharmaceutical care through identification of preventable ADRs and anticipatory surveillance for high-risk therapies or patients.
b. Complementing organisational risk-management activities and efforts to minimise liability.
c. Assessing the safety of drug therapies, especially those treatments that are clinically trialled approved drugs.
d. Measuring ADR incidence following treatments with specific immunotherapeutic types.
e. Educating health care professionals and patients about effects post infusion and increasing their level of awareness regarding ADRs.
f. Providing quality assurance screening findings for future use in cellular therapy-use evaluation programmes.

The pharmacovigilance system allows an organisation to fulfil its legal obligations and responsibilities in relation to patient safety. The system, is carefully structured, with established processes and defined outcomes.

19.5 Quality in pharmacovigilance

The quality of a pharmacovigilance system can be defined as all the characteristics of the system which are considered to produce, according to estimated likelihoods, outcomes relevant to the objectives of pharmacovigilance. In more general terms, quality is a concept that can be understood as a degree subject to measurement. By achieving that degree, then it can be assumed that quality has also been achieved.

The overall quality objectives of the pharmacovigilance system are:

a. Complying with the legal requirements for pharmacovigilance tasks and responsibilities.
b. Preventing harm from adverse reactions arising from the use of an authorised therapeutic product.
c. Promoting the safe and effective use of therapeutic products.
d. Contributing to the protection of patients' health.

Training of personnel for pharmacovigilance:

a. All personnel involved in performing pharmacovigilance activities must be provided with appropriate training on critical processes.

b. The training must ensure that employees have the appropriate qualifications and understanding of relevant pharmacovigilance requirements.

c. All staff members should receive and be able to seek information about what to do if they become aware of a safety concern.

d. An assessment of the employee's understanding and ability to conduct pharmacovigilance activities for the assigned tasks and responsibilities should be a requirement.

e. Adequate training should be considered by a company for those staff members to whom no specific pharmacovigilance tasks and responsibilities have been assigned but whose activities may have an impact on the pharmacovigilance system.

Facilities and equipment for pharmacovigilance:

a. Achieving the required quality for the conduct of pharmacovigilance processes is intrinsically linked to the control of facilities and equipment.

b. Facilities and equipment should be located, designed, constructed, adapted and maintained to suit their intended purpose in line with the quality objectives for pharmacovigilance.

c. Facilities and equipment which are critical for the conduct of pharmacovigilance should be subject to appropriate checks, qualification and/or validation activities to prove their suitability for the intended purpose.

d. Documented risk assessment should be used to determine the scope and extent of checks, qualification or validation activities.

e. A risk management approach should be applied throughout the lifecycle of the facilities and equipment taking into account such factors as impact on patient safety and data quality as well as the complexity of the concerned facilities and equipment.

19.6 Monitoring pharmacovigilance

There is a requirement for a company to monitor the performance and effectiveness of its pharmacovigilance system. In order to perform this task, then standard procedures will have to be written on how this is to be achieved and evaluated.

The list of SOPs for pharmacovigilance should include:

a. Reviews of the system only by responsible personnel designated by management.

b. SOP on performing an audit of the pharmacovigilance system.

c. SOP on evaluating the actions taken to minimise risks of an ADR.

d. A company should define possible indicators or clinical signs that might suggest an impending ADR.

e. Procedures for the review of the documentation produced by the inspectors of the pharmacovigilance system, as well as any deviations, abnormalities or non-conformances reported.

    i.   Corrective actions to the pharmacovigilance system.

    ii.   Introduction of preventive measures for patient safety.

    iii. Conducting the follow up audit.

f. A procedure defining how the report on ADRs should be written and to whom copies must be sent.

Recommended reading

1. European Medicines Agency and Heads of Medicines Agencies. EMA/541760/2011 Guideline on good pharmacovigilance practices (GVP): Module I – Pharmacovigilance systems and their quality systems, 2012.

2. Flowers, P., Dzierba, S., & Baker, O. A continuous quality improvement team approach to adverse drug reaction reporting. *Top Hosp Pharm Manage*, July 1992; 12(2):60-67.

3. Hartwig, S.C., Siegel, J., & Schneider, P.J. Preventability and severity assessment in reporting adverse drug reactions. *Am J Health Syst Pharm*, 1 September 1992; 49:2229-32.

4. Karch, F.E., & Lasagna, L. Adverse drug reactions: A critical review. *JAMA*, 22 December 1975; 234(12):1236-41.

5. Koch, K.E. Adverse drug reactions. In: Brown, T.R., ed. *Handbook of Institutional Pharmacy Practice*, 3rd ed. Bethesda, MD: American Society of Hospital Pharmacists, 1992.

6. Naranjo, C.A., Busto, U., Sellers E.M., et al. A method for estimating the probability of adverse drug reactions. *Clin Pharmacol Ther*, Aug. 1981; 30(2):239-45.

7. Nelson, R.W., & Shane, R. Developing an adverse drug reaction reporting program. *Am J Health Syst Pharm*, 1 March 1983; 40(3):445-46.

8. Prosser, T.R., & Kamysz, P.L. Multidisciplinary adverse drug reaction surveillance program. *Am J Hosp Pharm* 1990; 47(6):1334-39.

9. Requirements for adverse reaction reporting. Geneva: World Health Organization, 1975.

Self testing multiple choice questions

1. ADRs is an abbreviation for:
   a. Automatic Document Recognition.
   b. Adverse Drug Reactions.
   c. Anticipated Disease Responses.

2. The CEO should not participate actively in the:
   a. Improvement of the after-treatment service.
   b. Handling of complaints and implementing countermeasures.
   c. Establishment of an annual improvement scheme.

3. Immunotherapies cannot be guaranteed because:
   a. It is a known fact that immunotherapies have design and production faults.
   b. The individualistic nature of cellular therapies and host immune systems.
   c. Failures are most often caused by customer abuse.

4. To set up an effective organisational structure to deal with ADRs you should:
   a. Decide in advance which department is in charge but usually QA.
   b. Permit the complaint to be received by any department.
   c. Direct all complaints to the office of the CEO.

5. To set up ADR complaint handling procedures:
   a. Standardise and simplify the documentation for handling all adverse reactions.
   b. Standardise phased procedures for complaint processing.
   c. Control all temporary and permanent adverse effects separately.

6. The procedures for processing complaints should include:
   a. Identifying the technical staff responsible for the product so they can deal with it.
   b. Deferring any claims to the legal department until any actions are taken.
   c. Examining all previous records for similar complaints for similar treatments.

7. To handle the complaint procedures properly:
   a. Standardise an overall flowchart that permanently fixes who is involved.
   b. QA must make it clear which departments are involved and for what purpose.
   c. Establish communications that will allow everyone in a company to participate.

8. In a preventive programme for ADRs, a company should provide special attention to:
   a. The training of new employees.
   b. Training for new products or procedures.
   c. Training customer services staff to deal better with complaints.

9. To carry out periodic customer satisfaction surveys:
   a. Use a questionnaire on the quality of service as a customer follow-up sheet.

    b.   Prepare a sales index form that focuses on how many are repeat customers.

    c.   Establish customer service guidelines.

10. The regulations stipulate that biomedical therapeutic companies:
    a.   Do not need to do any post infusion monitoring because it is not necessary.
    b.   Must follow up on their customers/patients post-treatment.
    c.   Must only provide treatments that have a high percentage of success.

11. The department that usually deals with any patient complaints is:
    a.   The office of the CEO.
    b.   Quality Control.
    c.   Quality Assurance.

12. It is a requirement of the GTPs that Quality Assurance must:
    a.   Deal quickly and efficiently with any claims of failed treatment.
    b.   Identify who was responsible for the treatment failure and reprimand them.
    c.   Be responsible for any care or treatment resulting from the adverse reaction.

13. It is the responsibility of a company to maintain:
    a.   A ledger or log of all reported claims and ADRs.
    b.   A legal department that can deal with all reported ADRs.
    c.   An exit strategy in case too many ADRs are being reported.

14. The CEO should take an active interest in:
    a.   The social activities of patients after treatment.
    b.   After-treatment services as provided by their company.
    c.   Which patients have had returning treatments and focus on them.

15. Determining which treatments have been most effective allows the CEO to:
    a.   Make decisions on who should receive bonuses.
    b.   Incorporate them into the annual business plan.
    c.   Slow down research and development and focus only on what has worked.

16. Identification of preventable ADRs and anticipatory surveillance of high risk therapies:
    a.   Provides an indirect measure of the quality of pharmacovigilance.
    b.   Provides a direct measure of the quality of pharmacovigilance.
    c.   Provides a suitable replacement for pharmacovigilance.

17. Pharmacovigilance is all part of a company's:
    a.   Quality assurance policy.
    b.   Quality control system.
    c.   Risk management strategy.

18. The identification of ADRs is of value in providing:
    a.   An effective preventive measure to all future cell therapy developments.

    b.   Quality assurance screening for future cell therapy developments.

    c.   Only minimal information of little use for future cell therapy developments.

19. Quality in pharmacovigilance is defined as having those characteristics which:
    a.   Produce a steady output of ADRs.
    b.   Produce desirable estimated likelihoods.
    c.   Produce product with guaranteed likelihoods.

20. Promoting the safe and effective use of therapeutic products is part of:
    a.   Pharmaceutical science.
    b.   Biotechnology.
    c.   Pharmacovigilance.

21. A primary purpose of any pharmacovigilance system is to:
    a.   Contribute to the protection of patients' health.
    b.   Contribute to the overall earnings of a company.
    c.   Contribute to the enhanced reputation of a company.

22. It is imperative that personnel involved in pharmacovigilance are:
    a.   Able to deal with the emotion trauma caused by adverse reactions.
    b.   Trained in the proper social communicative techniques to deal with patients.
    c.   Trained in the appropriate critical processes.

23. The quality of the pharmacovigilance process is directly tied to:
    a.   The control of the organisational structure.
    b.   The control of facilities and equipment.
    c.   The control of personnel training.

24. Monitoring of the performance and effectiveness of pharmacovigilance requires:
    a.   Support from the CEO and senior management.
    b.   Sufficient personnel numbers to form teams to handle complaints.
    c.   The writing and implementation of appropriate SOPs.

25. In order to have foresight on an impending ADR, a company should have a policy of:
    a.   Investigating and defining possible indicators.
    b.   Recognising which days most likely to result in production problems.
    c.   Discovering the common denominator behind all ADRs.

26. The ledger or log should be a hardbound or computerised record that is:
    a.   Easily accessed by everyone.
    b.   Tamper proof.
    c.   Non-permanent.

27. At what point is it no longer the responsibility of a company to follow-up on its patients:
    a.   After six months.

    b.   After two years.

    c.   Never.

28. The expression of an ADR resulting from immunotherapy is in many ways no different:

    a.   From bacterial diseases.

    b.   From Graft Versus Host disease.

    c.   From viral diseases.

29. For the purposes of establishing success ratios, there are _____ possible outcomes.

    a.   Two

    b.   Three

    c.   Four

30. The contents of the final written report for a post-treatment investigation contain:

    a.   Causes, events and countermeasures.

    b.   Countermeasures, preventive measures and discussion notes only.

    c.   Events, treatments, and regulator.

Chapter 20. Product development

New therapies

Product design and development is the process of creating a new therapeutic product to be promoted as treatments for customers/patients. It involves identifying a market need, creating a specific therapy to meet this need, and testing and improving the therapeutic product until it is ready for production. It consists of a series of activities: research, analysis, and then testing for safety and efficacy, modifying, and re-testing until the product is ready to be infused into the patient. Analysis and development is usually carried out by a project team, with specialists and medical experts from both outside and within the company. In some countries, there is a specific requirement for the conduct of clinical trials, but when a treatment is autologous (developed from the cells of an individual and infuse back into that individual) then the requirement for such clinical trials is neither logical nor feasible. For allogeneic treatments, where the cells from one individual are being used to treat a large number of patients, then the conduct of limited feasible trials is viable but must keep in mind the fragile nature of the cells and their limited activation periods and/or lifespan. This text presents several procedures for managing the process of product design and development in a Total Quality framework that should be undertaken by biotechnologically based companies.

20.1 TQM in research and development (R&D)

In the past, research was viewed as an "anything goes" enterprise, with very little restrictions or controls placed upon a company in its pursuit of new products. Within a TQM system, research and development is seen as being no different from any other processes performed within a company and therefore must follow a quality approach. This challenge of finding and developing new therapies, yet not exceeding the requirements for efficacy and safety within a quality environment creates demands and challenges within R&D that are not encountered by personnel working in other departments.

In today's fast paced world where new discoveries become yesterday's news in a matter of months, not years, companies are under tremendous pressure to be at the forefront of the latest technologies and treatments. As competitive pressures grow and the search for new products and processes accelerates, the need for continuous improvement becomes imperative and therefore the systematic approach of Total Quality Management becomes even more critical. The business pressures placed on the acquisition of new therapies means that the R&D departments must perform at a level that provides little tolerance for failure. Only through a TQM approach, which promotes collaboration, interdepartmental communication, and constant feedback can an environment be created which is conducive to achieving the "no failure tolerated" goal. TQM encourages all departments to work together while focusing on the customer/patient the primary recipient of the benefits of the research. TQM assumes that even in research, the policies, practices, and intermediate products, can be improved incrementally on a continuous basis. To do this, the R&D cycle requires constant assessment, measurement and evaluation against a determined set of standards.

20.2 R&D requirements

Since TQM was initially designed to be applied to manufacturing process, then its application to R&D means that the philosophy of TQM will need some modification so that it can be adapted properly for use in R&D. In order to implement TQM in R&D, researchers and investigators must appreciate the social, technical, and management structures that exist within the organisation and not think of themselves as a stand-alone group. This requires a change in mindset since most researchers view their activities in terms of individual creativity but in an effective, modern corporate structure with time constraints, this must be discouraged and replaced with a collective mindset. The methods, practices, and reporting of the research must follow the strictly prescribed norms and expectations of Total Quality Management.

In the corporate world, the research paradigm practised and encouraged at most university research institutions, with freedom of expression, unwritten rules, no-constraints mentality, and not-for-profit (the pure science for science sake) orientation must be replaced with a patient needs it now focuses. TQM in R&D begins with strict attention to detail and applies quality outputs as its guiding light. The goal of the manager of the R&D department is to insure that appropriate methodology and scientific correctness is applied to a project, all of which has been validated. Accuracy, precision, efficacy and safety are the buzzwords of this paradigm shift. Ethical research produces accurate and reliable results. These ethical principles and practices constitute standards by which to judge the research management process. With constant testing and validation of the procedures, research becomes cyclic in nature within a continuous improvement process that is always examining the test results and then taking corrective measures in the form of adjustments to improve the development process.

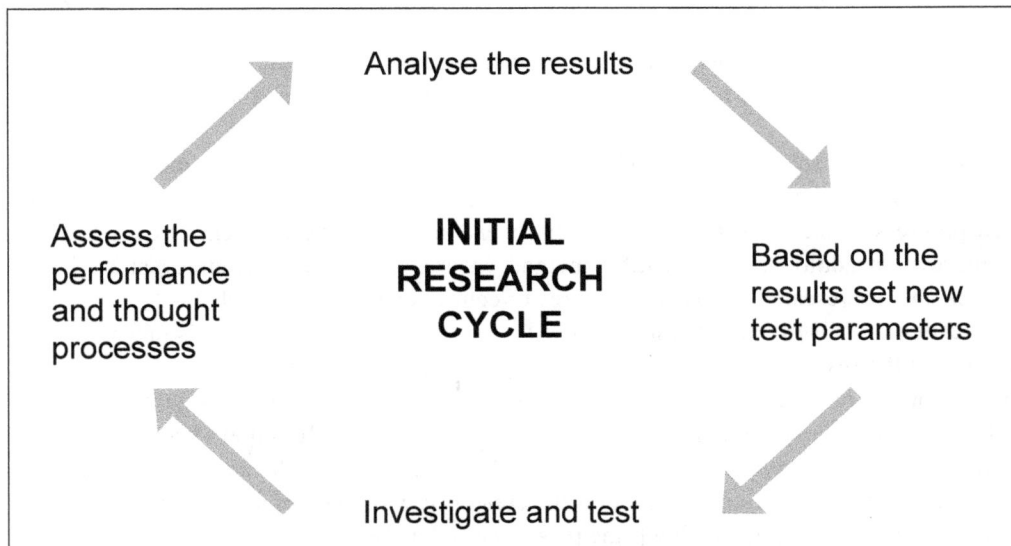

Initial research cycle

The emphasis is placed on the processes, ensuring that they are constantly improved and are the most current techniques available. The research cycle stresses learning, exploration, and further enhancements on a never-ending basis. Though the business model stresses bringing the new product to market as soon as possible, the TQM principles in R&D act as a balance so that at time of release, the product was the best it could possibly be.

Since in normal production most quality control testing occurs at the end of the manufacturing cycle, specific product-under-investigation tests have to be developed in R&D in order that TQM can be applied to this early stage development process. Accordingly, a lack of specific quality procedures may exist and these will have to be designed from scratch and incorporated into the research cycle. Researchers and investigators must develop measures that gauge efficiency and effectiveness during the creative process. At the same time, they must put into practice full documentation procedures to provide complete traceability of the development process which in turn will ultimately become the basis for developing operating procedures once the product achieves its final formulation. Therefore, R&D needs a system of indicators to track progress, prevent errors and mistakes, and reduce the cycle time needed to complete a project.

20.3 Creating a TQM environment in R&D

Innovation and creativity only come to fruition when there is an incentive-based, secure, and interactive environment, where researchers are motivated towards success by a highly developed quality framework. Support for this quality framework requires scientific input from colleagues within a company and from without.

The TQM environment when applied to research has these merits:
a. A reduction in fear of failure because of the PDCA cycle.
b. A system of core values based on quality standards.
c. Brings R&D in line with the organisational goals.
d. Failure becomes a motivator through CAPA thinking.
e. By having an in-the-end patient focus, any unwarranted risks are avoided.
f. Success results in profits which in turn provides rewards for achievement.

A combination of the scientific method along with TQM practices, provides a method by which researchers can assess, measure, and evaluate their own activities in a systematic manner. Each progressive step towards scientific discovery provides a new benchmark for measuring progress. In addition, the scientific method ensures a degree of accuracy and reliability, which enables researchers to exert a degree of control over their research.

The application of TQM is possible in any R&D environment as long as time is taken to train and educate the researchers and investigators. As the subsequent text in this chapter will highlight, as long as researchers can adopt some of the managerial and quality production systems used in TQM then the research facilities can benefit exponentially. To achieve success, the R&D manager/director must modify the traditional TQM practices to fit the unique needs of a research department. By facilitating innovation and creativity, with a combined managerial and technical structure that supports quality principles, success can be readily achieved.

20.4 Primary requirements for new products

A company will need to:

a. Have a good understanding on the part of the CEO of what therapeutic research, design and development involves.
b. Prepare written rules and procedures for managing research, design and development.
c. Prepare a new-product development project sheet.
d. Establish a standard for testing new products.
e. Calculate the costs of development.

It is important that the CEO understands what is involved in product design and development of an immunotherapy product, especially what it requires in manpower, capital, equipment and time. Under GMP, the CEO has full responsibility for providing the minimum resources that are needed. This is a regulatory requirement, mandated by international law and if a company is aware that it does not have the necessary resources to comply then it should not undertake the new project. It would be of benefit if the CEO also participates actively in planning right from the beginning, and has an understanding of the basic research.

The CEO must be aware that successful product development requires:

a. Sufficient time and expenditure on research.
b. Well-qualified research and development personnel.
c. A substantial programme of education and training in new cell therapies.
d. An annual review of the level of investment in, and the cost-effectiveness of, the new product development project.
e. That development personnel participate in cancer/disease market research to get a good understanding of what the market actually needs.
f. A recognition that basic research does not bring immediate benefits nor profits – it must be seen as part of a long-term plan.

Different departments and staff must be able to work together in the research and development of new products. This will require written rules and procedures. These should include the work procedures, the conceptual criteria governing the design, and a checklist of anticipated results.

The project outline should include:

a. The purpose and intent of the development project.
b. The name of the new therapeutic product, its selling points, its working title, and the related safety and environmental issues.
c. The state of the market: a forecast of demand, and information about what the competitors are doing.
d. A complete description of the new product: its specifications, design, and expected performance.

e. The evaluation items that have to be checked at each stage of development.
f. The target quality level, the target cost and the development schedule.
g. Approval by management.

The new product will have to be tested to ensure that it complies with the product development project sheet. It is therefore necessary to have a new-product testing standard to specify the minimum testing procedures required by the statues and regulations.

Calculate the time and cost of a new project:

a. The economic aspects:
  i. Preparing the original plan.
  ii. Training participants.
  iii. Changing over from the previous product type.
  iv. Research.
  v. Labour.
b. The technical aspects:
  i. Improving a company's own level of technology.
  ii. Improving the process capability.
  iii. Responding to external conditions.
  iv. Anticipating predictable technological progress.
c. The administration aspects:
  i. Organising a system for product development.
  ii. Establishing standards.
  iii. Organising education and training for engineers.
  iv. Clarifying responsibility and authority.
  v. Improving the administrative capability of the secretariat.

It will be necessary to implement cost control and budget allocation, and be prepared to make immediate adjustments to the cost if there is any change in the overall situation.

20.5 Plan for the design and development of a new product

To plan for the design and development of a new product:

a. Set up a unit to carry out the design work.
b. Establish design standards that will cover all the regulatory requirements.
c. Make clear the responsibility and authority of the developers.
d. Plan how corrective action should be taken when problems arise.
e. Establish procedures for changing the design standards when the need arises.

Set up a unit to ensure that the design work is carried out as planned:

a. Set up the unit.
b. Decide who will hold the key positions in this unit.
c. Decide how roles should be assigned and work shared: assignments should be made

by the person in charge, and should be based on ability.

d. Require the person in charge of the unit to check the state of progress periodically.

e. Train development personnel to raise their skills to the necessary level.

Always carry out complete quality assurance checks after any changes to specifications and procedure. Even though a research project, any changes should be done through the formal change control mechanism in order to ensure that there is full documentation for the history of development of the therapeutic product. Change control also ensures that any old documents are removed permanently from the locations where they have been distributed so that no confusion exists in the laboratory or the contract service providers.

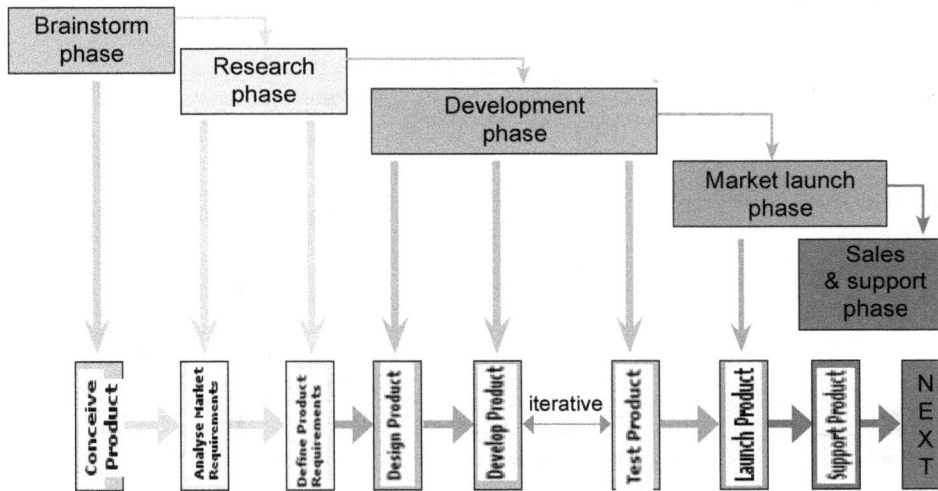

Quality phases in research and development

20.6 Implement the new design

To implement the design of a new product the design unit will need to:

a. Establish procedures for starting production of the new product.

b. Prepare a flowchart with the quality characteristics of the manufacturing processes.

c. Control the design and development budget.

d. Regulate the design and development process.

Prior to starting production of a new product, do the following first:

a. Check that the design and development phase was completed properly.

b. Carry out a very small scale run, remove any problems and improve the design of the production site if necessary.

c. Decide on the level and standard of quality required for production.

d. Decide on the quality characteristics required for each manufacturing process and specify these in an SOP.
e. Enter the required quality characteristics in the records.
f. Specify the roles of the related departments in the SOP.
g. Be sure to explain everything fully to the related departments and operators before starting production.

Prepare a flowchart of these processes. This flowchart should:

a. Specify the quality characteristics required in these processes.
b. Specify how the quality characteristics are to be checked.
c. Keep records of the quality characteristics.

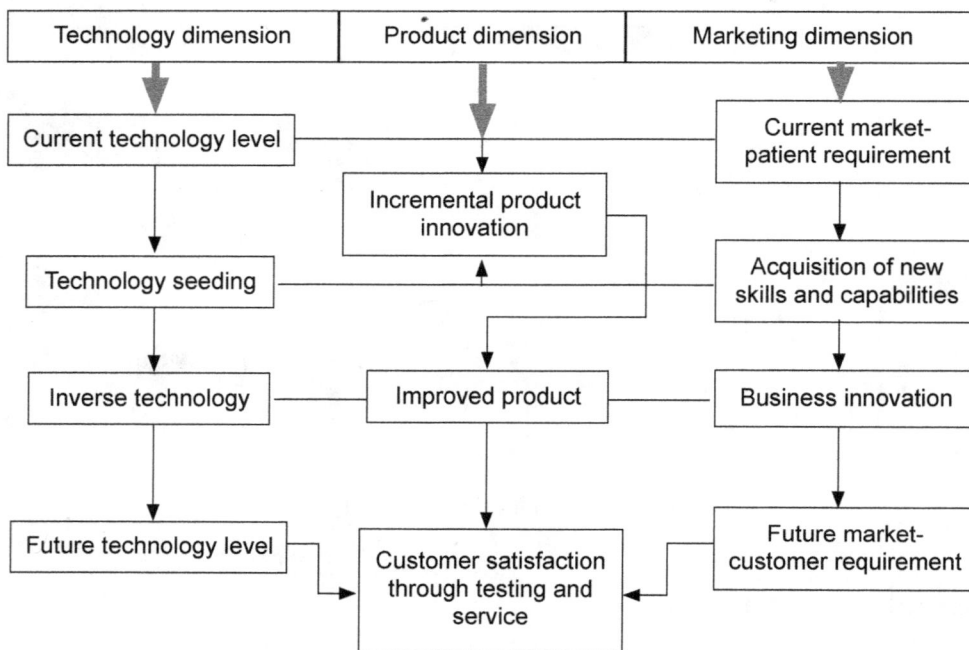

Flowchart of product development

Take the following steps:

a. Review the system for checking and recording quality characteristics.
b. Review the statistical quality records used to determine the acceptability of product.
c. Clarify any procedures for identifying quality characteristics.
d. If possible have an automatic system for checking the quality characteristics.

To regulate the design and development processes:

a. Have a good understanding of what the market needs.

b. Review these needs from both medium and long-term perspectives.
c. Understand the science and technology required.
d. Systemise the development processes.
e. Clarify the responsibilities and authorities.

20.7 Development of the new product

To prepare for the development of the new product:

a. Set up a new-product development system and establish standards.
b. Establish evaluation standards and criteria for reviewing the design and testing the product.
c. Check the process capability: Ensure that the production process has the capability to produce products with the required quality characteristics.
d. Carry out design reviews based on customer/patient complaints.

20.8 Complaints regarding the new product

a. Collect all information on customer complaints even if their claims appear to be mild or non related to infusion of the product.
b. Research the possible causes of these complaints from the catalogue of known adverse events.
c. Where the adverse event appears to be unique, then undertake an investigation to see if the cause can be identified and related to the product.
d. Take both corrective and preventive actions for each cause identified.
e. Check that the corrective actions decided on in the design review is adequate; carry out a follow-up check for each product.
f. Carry out a design review from product planning through product design as required.
g. Train specialists to ensure that the adverse event review is carried out correctly.

20.9 Long-term development

The design and development of a new product is normally at a significant cost. The criteria to be considered as having a good return on investment (ROI) is that the new product will hold a significant market share well into the future. For this to be true, then a company must build a long-term perspective into the planning of the product at its inception.

A company requires a system of evaluation to:

a. Ensure that it has a long-term development system.
b. Control the progress, quality and cost of design and development planning.
c. Make a continuous effort to reduce quality problems.
d. Base each phase of product planning on proper market research.

Actions for establishing a long-term development system:

a. Gain a good understanding of market needs and trends.
b. Check what high-level technology the company needs as compared to what is available.
c. Ensure that there are adequate resources of employees, materials, and money.
d. Determine a long-term development policy. (CEO function.)
e. Involve team members from outside a company that are specialists in those areas where the company is deficient in knowledge or experience.
f. Reserve the necessary funds in a long-term budget.

Recommended reading

1. Lock, D. *The Essentials of Project Management*. Gower, 1996.

2. McNair, C.J., & Leibfried, K.H.J. *Benchmarking: A Tool for Continuous Improvement.* New York: John Wiley & Sons, 1992.

3. Mesley, R.J., Pocklington, W.D., & Walker R.F. Analytical quality assurance: A Review, *Analyst* 1991; 116(10):975-90.

Self testing multiple choice questions

1. The CEO according to GMP is required to provide _____ needed for product development.
   a. None of the resources as that is production's responsibility
   b. The minimum resources
   c. The optimum resources

2. The procedures for managing design and development should be documented in order to:
   a. Reduce the costs of materials.
   b. Ensure that the different departments have traceability and work effectively.
   c. Improve the drawing methods.

3. The new product testing standard should:
   a. Ensure that the design parameters fully comply with the product-testing standard.
   b. Be adapted from pre-existing standards even if from an unrelated product.
   c. Make clear the design characteristics which ensure that the product is safe.

4. The assignment of roles for the development work should be based on the _____ of each person.
   a. Seniority
   b. Qualifications
   c. Ability

5. The follow-up development work is the responsibility of:
   a. The design staff.
   b. The production staff.
   c. The customer services staff.

6. The development standards should be approved by:
   a. The CEO.
   b. The senior design manager.
   c. The design unit.

7. Specification of the responsibility and authority of the development personnel should be approved by:
   a. The CEO.
   b. The senior design manager.
   c. The appropriate people in the development department.

8. To ensure that corrective action is taken as soon as problems arise:
   a. Inform the marketing department immediately, so they can prevent recurrence.
   b. Implement and confirm development changes.
   c. Establish the cause of the problem.

9. The procedures for changing the new project design standards, drawings and specifications include:
   a. Have the changes approved by marketing personnel before they are implemented.
   b. Be sure to collect old standards, drawings and specifications as soon as possible after the new ones are distributed.
   c. No need for a design change control notice since it is a new product in research.

10. The procedures for conducting research in cooperation with related departments should:
    a. Specify the related departments and their roles in an SOP.
    b. Decide on the quality characteristics each department is responsible for.
    c. Explain fully to the related departments why they are not involved.

11. The development processes should be documented fully so that:
    a. The quality characteristics of the processes can be determined later.
    b. The record of all the processes used are available for traceability.
    c. The success of the processes is guaranteed.

12. Prior to finalising the design and development processes:
    a. Get a good understanding of what the patient needs.
    b. Review the design standards for new products.
    c. Get a good idea of what technology will be required and if it is available.

13. To set up a new-product development system:
    a. Draw up a plan to identify what will be the company's market share.
    b. Draw up a development plan for the technology and equipment required.
    c. Carry out trial manufacture and development before any commitment.

14. To establish evaluation standards and criteria for initial design and product testing:
    a. Determine the final review procedures.
    b. Determine the prototype-testing standard.
    c. Determine the market-testing procedures.

15. Research projects should be designed that meet:
    a. Company standards.
    b. Business standards.
    c. Regulatory standards.

16. To ensure a company can produce products with the required quality characteristics:
    a. Select units of measurement that guarantee a pass.
    b. Select the appropriate measuring tests and equipment.
    c. Select only suitable processes that are achievable.

17. To carry out proper design reviews:
    a. Collect information on customer complaints after the new product is trialled.
    b. Carry out a design review from product planning through final formulation.
    c. Have an external agency perform the review.

18. To organise a long-term new-product development system:
    a. Examine the market and see which therapies are the most profitable.
    b. Ensure that there are adequate resources of employees, materials and money.
    c. Invest in the latest and most advanced technologies.

19. To reduce quality problems:
    a. Transfer any personnel reporting problems out of the R&D department.
    b. Only make corrections when critical elements that affect quality are identified.
    c. Design quality tests throughout the R&D phases and adjust via countermeasures.

20. Keep employees informed of quality problems so that:
    a. Employees know they are not receiving any incentives as a result.
    b. Employees will become sensitive to them and be afraid of reporting errors.
    c. Employees can be part of the PDCA cycle in R&D.

21. The development of a new product must be based first of all on:
    a. The patient needs.
    b. The business strategy.
    c. The product specifications.

22. Autologous treatments generally:
    a. Require clinical trials.
    b. Do not require clinical trials.
    c. Have no safety requirements.

23. Allogeneic treatments are designed for the treatment of:
    a. The individual from whom the cells were removed only.
    b. Diseases that are of a non-life threatening nature.
    c. Numerous people sharing the same illness as the donor of the cells.

24. Prior to the use of allogeneic therapies:
    a. A limited clinical trial is required.
    b. Payment has to be made to the donor of the cells.
    c. Irradiation of the cells must occur in order to render them harmless.

25. When considering clinical trials for immunocell therapies, the downside concerns:
    a. The nature of the placebo.
    b. The number of people to be entered into the trial.
    c. The fragility, length of the activation period and the actual cell lifespan.

26. In the past, the approach to research was considered in:
    a. An "anything goes" mentality.
    b. A safety first attitude.
    c. A research must be concluded rapidly approach.

27. Total Quality Management is essential in research in order to:
    a. Maximise profits.
    b. Achieve the "no failure tolerated" goal.
    c. Create the illusion that research is performed on behalf of patient requirements.

28. Ethical research provides:
    a. Guaranteed success in treatment.
    b. Accurate and reliable results.
    c. A good marketing edge.

29. The Research Cycle diagram in this chapter is based on:
    a. The Krebb's cycle.
    b. The PDCA cycle.
    c. The SPCA cycle.

30. In order to apply TQM to R&D effectively, researchers need to:
    a. Overcome the fear of failure.
    b. Overcome resistance to success.
    c. Overcome the inherent fallibility of testing.

31. In order to track progress, prevent errors and reduce cycle time, R&D needs:
    a. A system of work/leisure ratio.
    b. A system of post error detection.
    c. A system of early detection.

32. If the CEO cannot provide the minimum resources required for a project, then the project:
    a. Has to be conducted in phases.
    b. Should not be undertaken.
    c. Is performed until the resources are exhausted.

33. In a TQM environment applied to R&D, after any changes to any processes, carry out a complete _____ check.
    a. Quality assurance
    b. Quality control
    c. Engineering

34. The brainstorming phase of R&D refers specifically to:
    a. Conceiving the product concept.
    b. Documenting the product procedures.
    c. Identifying the key product personnel.

35. Prior to starting production of a new product, a company should:
    a. Do one final last check of the product design.
    b. Conduct a full scale manufacturing run.
    c. Conduct a very small scale run.

36. The three dimensions of any new project are:
    a. Research, product and quality.
    b. Technology, product and marketing.
    c. Development, product and testing.

37. Patient satisfaction with the treatment is impacted by improvements to the product based on:
    a. Lower cost of the technology.
    b. Future technology and patient needs.
    c. Regulatory requirements.

38. Customer/patient complaints should result in immediate:
    a. Design reviews.
    b. Removal of the product from the market.
    c. Compensation for any presumed harm done.

39. The long term development policy for research is a function of:
    a. The CEO.
    b. The research department.
    c. The quality departments.

40. Total Quality Management in R&D promotes:
    a. Individualism, limited communication and minimal feedback.
    b. Individualism, interdepartmental cooperation and continual feedback.
    c. Collaboration, interdepartmental communication and constant feedback.

41. The goal of TQM is to improve all aspects of research:
    a. Rapidly.
    b. Incrementally.
    c. Instantly.

42. The greatest difference between university research and corporate research is:
    a. In the corporate world there is freedom of expression without constraints.
    b. In the university world there is freedom of expression without constraints.
    c. There is no difference between the university and corporate worlds.

43. The previous "science for science's sake" attitude of research needs to be replaced in a TQM environment with:
    a. Science for the money's sake.
    b. Science for the patient's sake.
    c. Science for a company's sake.

44. The research cycle stresses:
    a. Products are never ready for market as they are always incomplete.
    b. Testing, investigation and restarting from zero on a never ending basis.
    c. Learning, exploration and enhancements on a never ending basis.

45. Total Quality Management in research acts as a counterbalance to:
    a.   The research model.
    b.   The business model.
    c.   The consumer model.

46. In normal production practices, most of the QC testing occurs at:
    a.   The end of the manufacturing cycle.
    b.   The end of the research cycle.
    c.   The end of the product life cycle.

47. Research and development must ensure it is in line with:
    a.   The organisational goals.
    b.   The organisational time-table.
    c.   The organisational profit structure.

48. When research is not patient focused:
    a.   Then it has better research practices.
    b.   Then it will not accept unwarranted risks.
    c.   Then it will accept unwarranted risks.

49. One of the main reasons research should always be viewed as part of a long-term plan is because:
    a.   Researchers need incredibly long time to think.
    b.   It does not result in immediate profits.
    c.   There is a necessary lag phase between conceptual design and proof of concept.

50. According to the Quality Phase Diagram for R&D, there are _____ phases in total.
    a.   Three
    b.   Five
    c.   Ten

In conclusion

In the preceding pages of this book, we have had the opportunity to examine how Total Quality Management can be applied in the biotechnology industries. Most importantly, the key purpose of introducing TQM is that a company can meet the quality expectations as defined by its customers or patients. TQM looks at quality very differently from the early days of quality processes where the entire focus was on the product in order to generate profit, which often meant reductions in time, components, and workforce in order to achieve that goal, while all the quality inspections were performed after the product was finished to determine that none of the "savings" resulted in a loss of quality. In companies where manufacturing items were always identical, never changing and inanimate, then risks to the customer could be minimised when implementing a reduction mentality, but the same is not true when dealing with biological products where variability, contamination, unknown risks, and individuality of the manufacturing processes are the norm rather than the exception. In fact, TQM in biotechnology is a dynamically changing, continually improving, varying of sources, and constant problem solving enterprise. A strong focus on anticipating problems and trying to prevent them from occurring through continuous inspection during the entire life cycle of the product is the core and mainstay of the industry.

The Plan-Do-Check-Act (PDCA) cycle is constantly in motion in biotechnological TQM, rotating in a never ending problem solving process. In order to achieve constant improvement there is a heavy reliance on the Quality Control Tools as described in the book and the use of diagrams, charts and check sheets. The appropriate use and understanding of cause and effect diagrams, flowcharts, Pareto Diagrams, histograms and scatter diagrams, makes the task of quality implementation much easier for even novices to comprehend and undertake.

But key to any TQM System is the involvement of senior management. If they are unwilling to commit, using excuses such as, "It will take too much time", or if they fail to show up to meetings, and they do not see quality failures as their problem, then a company will never be successful in implementing TQM. Numerous companies will talk the talk, but when it comes down to the reality, they are only "sleep-walking", going through the motions because they lack the support and commitment of senior management. If this is the prevailing attitude in your company, then I encourage you to get this book into the hands of your CEO and have him/her read it cover to cover until he/she appreciate that everyone within a company has a role to play in TQM.

How quality is incorporated into a company has changed dramatically from the early days of industrialisation, and even more rapidly since the onset of the 1980s when companies began to realise that the success of their companies was no longer guaranteed by simply bringing a product to market. Success was suddenly consumer driven, as more and more choices were available, permitting customers to become selective as to which products they would utilise. Selection became quality based with customers wanting products that met their needs for safety, performance, and longevity. These demands became a matter of law as soon as the regulatory bodies within the various manufacturing countries realised that consumer protection was the right of every citizen and instituted regulations that guaranteed these rights.

Since the 1980s we have witnessed the growth of manufacturing and production standards for the various industries at an unprecedented rate. Those companies that could not produce items to this level soon found themselves out of business.

TIME	Early 1900s	1940s	1960s	1980s and beyond
FOCUS	Inspection	Statistical sampling	Organisational quality focus	Customer driven quality
	←————————————————————————→			←————————————→
	Old concept of quality: Inspect for quality after production.			New concept of quality: Build quality into the process. Identify and correct causes of quality problems.

History of quality concepts

In today's environment, it is mandatory that top management reviews the organisation's quality management system, at planned intervals, to ensure its continuing suitability, adequacy and effectiveness. These reviews shall include assessing opportunities for improvement and the need for changes to the quality management system, including the quality policy and quality objectives. Management will identify and correct the causes of quality problems based on the inputs and information they receive from a company's internal auditing systems, feedback from customers and patients, the corrective and preventive action reports generated, as well as the follow-up from their previous management reviews. Therefore, in TQM a company is reliant on recommendations for improvement coming from the CEO and senior management levels.

This book was designed not only for those working at the coalface of biotechnology but also for the top levels of management to appreciate their role in making TQM work successfully and how they are responsible for making the decisions and taking the actions that will ultimately result in the improvement and effectiveness of the quality management system and its processes. Every decision made by senior management must be based on customer/ patient requirements and resource needs in order to fulfil those customer/patient requests. This shift of focus in the industry towards Customer Oriented Processes (COPs) means constant change in not only how the industry performs but also in how management must think. Once this focus has become the mindset of senior management, then TQM has the opportunity to realise its true potential in creating a quality environment.

If I was to summarise everything that this book attempted to convey in regards to TQM in biotechnology into a single flowchart diagram, specifically with a focus on cellular immunotherapy, then it would probably look like this:

Total Quality Management in immunotherapy process flowchart

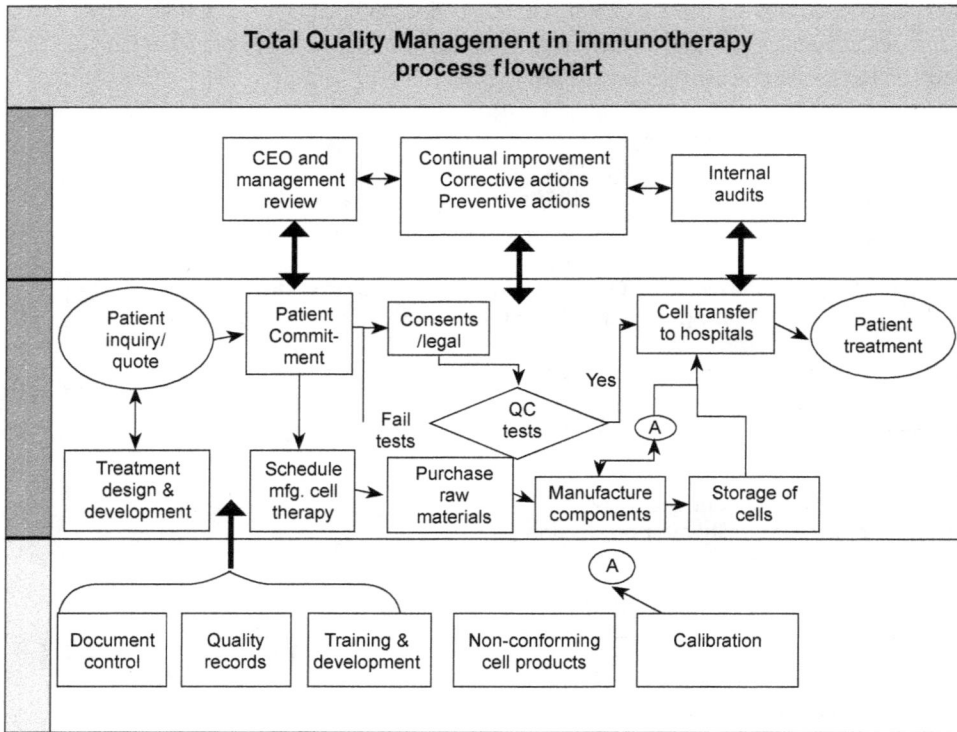

As can be seen in the opening chapters of Section 1 regarding the role of the CEO at the top, disseminating the TQM programme throughout a company is performed in a top-down management style. But as we saw in Section 2, the quality systems rely heavily on the internal auditing system, which continually feeds information back to the CEO and senior management so that decisions can be made rapidly and with the intent of continually improving the system and taking immediate corrective actions where and when necessary. Pumping into the overall system from the bottom-up is the support network established by the quality departments, ensuring that all the necessary working documents, control documents, forms and checklists are available, while continuous training and development takes place so that the entire workforce is integrated and part of the TQM System. The quality mechanical inputs (calibration, qualification, validation) and outputs (abnormalities, deviations, non-conformances) are also dealt with at this support level, ensuring that the flow remains constant, with a high level of safety and efficacy.

Meanwhile, the heart of the system as outlined in Sections 3 and 4, the Laboratories and the Production Areas, receive the benefits from the top layer inputs (Management) and the bottom layer inputs (Support) to ensure that the customer's/patient's needs are being met. After all, TQM is primarily about meeting the customer's needs, with a focus entirely on the patient in the biomedical industries and providing him/her with the best possible therapeutic products and services.

It is not surprising that in my many years of service within the biopharmaceutical and biotechnology industries that I have encountered very few companies that have successfully instituted TQM. Senior management in most of these companies was still profit oriented and technicians were more concerned with dealing with their day to day stresses and the work burden which was increasing disproportionately as management continued to pare down the workforce in order to maximise profits and make the bottom line look better at their executive meetings. Meanwhile the laminated posters outside all their offices stressed that quality begins with the customer/patient and elimination of product complaints was their number one performance goal. The reality is that as long as CEOs of companies have golden handshakes even when they fail and are rewarded even more handsomely for increasing sales of drugs, vaccines, and therapies, then the profit-driven focus will always be in conflict with the primary goal of TQM. In order to succeed in this direction, a paradigm shift is required and that hopefully is where this book can take root, in the next generation of enthusiastic, eager and quality minded graduates and fresh employees that truly do what to make a difference in the biomedical and biotechnological industries. Essentially, TQM, though requiring it to be adopted at the top levels of the company in order to become established, is only a force to be reckoned with when it is driven upwards from the grassroots level of the employees. Hopefully, I have conveyed that concept throughout this book, TQM involves everyone in a company, but it is this employee empowerment, the provision of knowledge and quality tools in their training package, which achieves the quality benchmarks that a company strives for in the four dimensions of Product/Services Design, Efficacy, Safety and Patient Aftercare. When these can be achieved at this penultimate level, when it is insisted upon and practised by the technicians and employees, then senior management must also comply and then and only then a company can truly say that it has, "Total Quality Management."

Suggested answers to self testing multiple choice questions

Before checking your answers the following information will help you understand why some of your answers may not agree with those that have been marked as being correct.

1. More than one correct answer for a question:

    There are questions that have more than one correct answer but Quality Management is about developing the ability to prioritise and select the best possible answer or solution. The reality is that in many situations you will be able to find multiple responses or solutions to a problem but you must be able to select which is the most important or most critical in achieving a resolution.

2. The question related to information from earlier chapters and was not directly related to the current chapter:

    Total Quality Management is about putting the totality of all that is being explained to you in the text together in a comprehensive manner as there will be many aspects that will impact one on another. You must develop the skill to apply information and solutions from one aspect to new or related areas. Therefore some questions are designed to compel you to join concepts in order to find solutions.

3. The information cannot be found in the textual content:

    One of the key characteristics for anyone working in the quality sphere is to be logical. Not everything you encounter will have an easily defined, readily available, already seen solution. That is where you have to develop your logical thinking skills in order to find a solution that meets the requirements of Total Quality Management.

Chapter 1

1.c	2.b	3.b	4.c	5.a	6.c	7.a	8.a	9.c	10.c
11.a	12.b	13.c	14.b	15.b	16.b	17.a	18.a	19.c	20.c
21.a	22.b	23.c	24.a	25.a	26.a	27.a	28.a	29.c	30.a
31.b	32.a	33.a	34.a	35.b	36.c	37.a	38.a	39.c	40.b
41.b	42c	43.c	44.c	45.a	46.c	47.b	48.a	49.c	50.c

Chapter 2

1.b	2.a	3.b	4.a	5.b	6.c	7.b	8.c	9.c	10.a
11.c	12.a	13.a	14.b	15.a	16.b	17.a	18.a	19.b	20.a
21.c	22.a	23.a	24.b	25.b	26.c	27.a	28.b	29.c	30.a
31.b	32.c	33.b	34.b	35.a	36.b	37.a	38.c	39.b	40.c
41.b	42.a	43.c	44.b	45.c	46.c	47.b	48.b	49.a	50.c

Chapter 3

1.b	2.b	3.b	4.b	5.c	6.c	7.b	8.b	9.a	10.a
11.b	12.a	13.c	14.b	15.c	16.c	17.a	18.a	19.c	20.a
21.a	22.a	23.c	24.a	25.a	26.a	27.b	28.c	29.b	30.b
31.b	32.a	33.a	34.b	35.c	36.b	37.a	38.b	39.a	40.a
41.c	42.b	43.b	44.b	45.b	46.c	47.a	48.c	49.b	50.b

Chapter 4

1.a	2.a	3.b	4.a	5.b	6.b	7.b	8.a	9.b	10.a
11.a	12.b	13.a	14.b	15.b	16.a	17.b	18.b	19.c	20.c
21.a	22.a	23.b	24.c	25.b	26.a	27.b	28.c	29.a	30.b
31.b	32.c	33.a	34.c	35.c	36.b	37.b	38.b	39.b	40.a
41.c	42.b	43.c	44.b	45.c	46.b	47.a	48.a	49.b	50.a

Chapter 5

1.b	2.b	3.a	4.c	5.a	6.b	7.a	8.b	9.c	10.a
11.c	12.b	13.c	14.b	15.c	16.b	17.c	18.a	19.c	20.a
21.a	22.c	23.b	24.b	25.c	26.b	27.a	28.b	29.a	30.b
31.c	32.c	33.b	34.b	35.a	36.b	37.a	38.c	39.b	40.c
41.b	42.a	43.b	44.a	45.c	46.a	47.a	48.c	49.b	50.a

Chapter 6

1.c	2.c	3.b	4.a	5.b	6.b	7.c	8.c	9.c	10.c
11.c	12.c	13.a	14.b	15.b	16.c	17.c	18.b	19.b	20.b
21.b	22.a	23.c	24.b	25.b	26.a	27.a	28.a	29.b	30.c
31.a	32.b	33.c	34.a	35.c	36.b	37.c	38.b	39.a	40.b
41.a	42.b	43.b	44.c	45.b	46.a	47.c	48.a	49.b	50.a

Chapter 7

1.c	2.c	3.a	4.c	5.c	6.b	7.a	8.a	9.a	10.a
11.c	12.b	13.a	14.b	15.c	16.b	17.b	18.b	19.b	20.a
21.b	22.b	23.b	24.b	25.b	26.b	27.c	28.a	29.c	30.a
31.a	32.c	33.a	34.a	35.c	36.a	37.a	38.c	39.b	40.a
41.a	42.c	43.b	44.a	45.b	46.a	47.c	48.b	49.c	50.a

Chapter 8

1.c	2.c	3.b	4.b	5.a	6.a	7.c	8.b	9.b	10.b
11.c	12.b	13.c	14.a	15.b	16.c	17.c	18.a	19.c	20.c
21.b	22.a	23.c	24.b	25.c	26.b	27.b	28.c	29.c	30.a
31.a	32.b	33.b	34.b	35.a	36.b	37.c	38.c	39.b	40.b
41.b	42.a	43.b	44.a	45.b	46.b	47.b	48.c	49.a	50.c

Chapter 9

1.b	2.a	3.c	4.b	5.a	6.b	7.c	8.b	9.b	10.a
11.b	12.c	13.c	14.c	15.b	16.a	17.b	18.b	19.c	20.b

21.c	22.b	23.b	24.a	25.b	26.c	27.b	28.a	29.c	30.a
31.b	32.a	33.c	34.a	35.c	36.c	37.a	38.a	39.c	40.b
41.b	42.c	43.a	44.b	45.b	46.a	47.c	48.c	49.c	50.b

Chapter 10

1.c	2.c	3.c	4.c	5.b	6.a	7.b	8.a	9.a	10.a
11.b	12.b	13.b	14.c	15.a	16.a	17.a	18.c	19.b	20.a
21.a	22.a	23.c	24.b	25.c	26.c	27.a	28.b	29.a	30.b
31.c	32.b	33.a	34.b	35.c	36.a	37.c	38.c	39.c	40.a
41.b	42.a	43.c	44.c	45.b	46.a	47.a	48.a	49.b	50.c

Chapter 11

1.b	2.a	3.a	4.c	5.a	6.b	7.b	8.a	9.b	10.c
11.c	12.a	13.c	14.c	15.b	16.a	17.a	18.c	19.a	20.a
21.b	22.c	23.c	24.b	25.b	26.a	27.c	28.a	29.a	30.b
31.b	32.a	33.b	34.c	35.a	36.c	37.a	38.b	39.a	40.b
41.b	42.b	43.c	44.a	45.b	46.a	47.a	48.c	49.a	50.b

Chapter 12

1.c	2.c	3.a	4.b	5.c	6.a	7.b	8.a	9.a	10.a
11.b	12.a	13.c	14.b	15.a	16.c	17.b	18.a	19.b	20.c
21.b	22.a	23.a	24.b	25.b	26.c	27.b	28.a	29.b	30.b
31.c	32.c	33.c	34.a	35.b	36.a	37.c	38.a	39.c	40.b
41.a	42.a	43.a	44.b	45.b	46.a	47.c	48.b	49.b	50.b

Chapter 13

1.a	2.b	3.c	4.b	5.b	6.b	7.c	8.c	9.b	10.a
11.b	12.a	13.b	14.c	15.b	16.a	17.b	18.a	19.a	20.c
21.c	22.a	23.b	24.c	25.b	26.b	27.a	28.a	29.c	30.a
31.b	32.b	33.a	34.c	35.b	36.c	37.a	38.b	39.b	40.c
41.a	42.b	43.c	44.a	45.a	46.b	47.a	48.c	49.a	50.b

Chapter 14

1.c	2.b	3.a	4.c	5.c	6.a	7.b	8.c	9.b	10.a
11.b	12.b	13.a	14.c	15.c	16.b	17.b	18.c	19.a	20.c
21.c	22.c	23.a	24.a	25.b	26.a	27.b	28.a	29.a	30.b
31.a	32.a	33.c	34.c	35.a	36.c	37.a	38.b	39.b	40.a
41.b	42.b	43.b	44.c	45.a	46.a	47.b	48.b	49.b	50.c

Chapter 15

1.b	2.a	3.b	4.a	5.b	6.a	7.c	8.c	9.b	10.a
11.b	12.a	13.a	14.a	15.a	16.c	17.b	18.c	19.a	20.c
21.c	22.a	23.c	24.b	25.a	26.b	27.b	28.b	29.a	30.b
31.a	32.a	33.a	34.c	35.b	36.b	37.a	38.b	39.c	40.a
41.b	42.a	43.b	44.b	45.b	46.a	47.a	48.b	49.a	50.b

Chapter 16

1.a	2.b	3.c	4.b	5.c	6.b	7.c	8.b	9.b	10.b
11.a	12.b	13.b	14.c	15.a	16.c	17.c	18.b	19.a	20.b
21.a	22.c	23.b	24.b	25.c	26.a	27.c	28.a	29.c	30.a
31.c	32.b	33.a	34.c	35.b	36.c	37.a	38.a	39.c	40.b
41.a	42.c	43.b	44.a	45.a	46.c	47.a	48.b	49.b	50.c

Chapter 17

1.b	2.b	3.c	4.a	5.c	6.b	7.b	8.a	9.a	10.c
11.a	12.b	13.b	14.b	15.a	16.b	17.c	18.b	19.b	20.a
21.b	22.a	23.b	24.b	25.b	26.a	27.c	28.a	29.b	30.c
31.b	32.c	33.a	34.c	35.a	36.b	37.b	38.a	39.c	40.a
41.c	42.b	43.b	44.c	45.b	46.a	47.c	48.a	49.c	50.b

Chapter 18

1.b	2.c	3.c	4.c	5.a	6.c	7.b	8.c	9.a	10.a
11.a	12.a	13.c	14.b	15.a	16.a	17.a	18.c	19.b	20.c
21.b	22.b	23.c	24.a	25.a	26.b	27.c	28.b	29.a	30.c

Chapter 19

1.b	2.b	3.b	4.a	5.b	6.c	7.b	8.b	9.a	10.b
11.c	12.a	13.a	14.b	15.b	16.a	17.a	18.b	19.b	20.c
21.a	22.c	23.b	24.c	25.a	26.b	27.c	28.b	29.b	30.a

Chapter 20

1.b	2.b	3.a	4.c	5.b	6.c	7.a	8.c	9.b	10.a
11.b	12.a	13.b	14.b	15.c	16.b	17.b	18.b	19.c	20.c
21.a	22.b	23.c	24.a	25.c	26.a	27.b	28.b	29.b	30.a
31.c	32.b	33.a	34.a	35.c	36.b	37.b	38.a	39.a	40.c
41.b	42.b	43.b	44.c	45.b	46.a	47.a	48.c	49.b	50.b